T0176960

GREEN CARBON DIOXIDE

GREEN CARBON DIOXIDE

ADVANCES IN CO_2 UTILIZATION

Edited by

Gabriele Centi
Siglinda Perathoner

University of Messina, ERIC aisbl and CASPE/INSTM, Italy

Published by John Wiley & Sons, Inc., Hoboken, New Jersey
Published simultaneously in Canada

For general information on our other products and services or for technical support, please contact our
Customer Care Department within the United States at (800) 762-2974, outside the United States at
(317) 572-3993 or fax (317) 572-4002.

Wiley also publishes its books in a variety of electronic formats. Some content that appears in print
may not be available in electronic formats. For more information about Wiley products, visit our web
site at www.wiley.com.

Library of Congress Cataloging-in-Publication Data:

Green carbon dioxide : advances in CO2 utilization / edited by Gabriele Centi, Siglinda Perathoner.
    pages cm
 Includes bibliographical references and index.
 ISBN 978-1-118-59088-1 (cloth)
 1. Carbon dioxide–Industrial applications. 2. Carbon dioxide mitigation. I. Centi, G. (Gabriele),
1955- editor of compilation. II. Perathoner, Siglinda, 1958- editor of compilation.
 TP244.C1G735 2014
 665.8′9–dc23

                                                                                    2013034528

Printed in the United States of America

10 9 8 7 6 5 4 3 2 1

■ CONTENTS

Mitigating climate change, preserving the environment, using renewable energy, and replacing fossil fuels are among the grand challenges facing our society that need new breakthrough solutions to be successfully addressed. The (re)use of carbon dioxide (CO_2) to produce fuels and chemicals is the common factor in these grand challenges as an effective solution to contribute to their realization. Reusing CO_2 not only addresses the balance of CO_2 in the Earth's atmosphere with the related negative effects on the quality of life and the environment, but represents a valuable C-source to substitute for fossil fuels. By using renewable energy sources for the conversion of CO_2, it is possible to introduce renewable energy into the production chain in a more efficient approach with respect to alternative possibilities. The products derived from the conversion of CO_2 effectively integrate into the current energy and material infrastructure, thus allowing a smooth and sustainable transition to a new economy without the very large investments required to change infrastructure. As a longer-term visionary idea, it is possible to create a CO_2-economy in which it will be possible to achieve full-circle recycling of CO_2 using renewable energy sources, analogous to how plants convert CO_2 to sugar and O_2, using sunlight as a source of energy through photosynthesis. Capture and conversion of CO_2 to chemical feedstocks could thus provide a new route to a circular economy.

There is thus a new vision of CO_2 at the industrial, societal, and scientific levels. Carbon dioxide is no longer considered a problem and even a waste to be reused, but a key element and driving factor for the sustainable future of the chemical industry. There are different routes by which CO_2 can be converted to feedstocks for the chemical industry by the use of renewable energy sources, which also can be differentiated in terms of the timescale of their implementation. CO_2 is a raw material for the production of base chemicals (such as light olefins), advanced materials (such as CO_2-based polymers), and fuels (often called solar fuels).

There are many opportunities and needs for fundamental R&D to realize this new CO_2 economy, but it is necessary to have clear indications of the key problems to be addressed, the different possible alternative routes with their related pro/cons, and their impact on industry and society. The scope of this book is to provide to managers, engineers, and chemists, working at both R&D and decision-making levels, an overview of the status and perspectives of advanced routes for the utilization of CO_2. The book is also well-suited to prepare advanced teaching courses at the

Masters or Ph.D. level, even though it is not a tutorial book. Over a thousand references provide the reader with a solid basis for deeper understanding of the topics discussed.

It is worthwhile to mention that this book reports perspectives from different countries around the world, from Europe to the US and Asia. CO_2 is becoming, in fact, a primary topic of interest in all the countries of the world, although with different priorities, which are reflected here.

Chapter 1 introduces the topic with a perspective on producing solar fuels and chemicals from CO_2 after having introduced the role of CO_2 (re)use as an enabling element for a low-carbon economy and the efficient introduction of renewable energy into the production chain. Two examples are discussed in a more detail: (i) the production of light olefins from CO_2 and (ii) the conversion of CO_2 to fuels using sunlight. The final part discusses outlook for the development of artificial leaf-type solar cells, with an example of a first attempt at a photoelectrocatalytic (PEC) solar cell to go in this direction.

Chapter 2, after introducing some background aspects of CO_2 characteristics and the photocatalytic chemistry on titania, focuses the discussion on the analysis of photo- and electrochemical pathways for CO_2 conversion, discussing in detail the role of free radical-induced reactions related especially to the mechanism of methane (and other products) formation from CO_2 during both photo- and electro-induced processes.

Chapter 3 also provides a critical analysis of the possible reduction pathways for synthesis of useful compounds from CO_2, with a focus especially on photo- and electrocatalytic routes. This chapter not only offers the readers a general overview of recent progress in the synthesis of useful compounds from CO_2 but provides new insights in understanding the structure-component-activity relationships. It highlights how new nanostructured functional materials play an important role in photo- and electrocatalytic conversion of CO_2, with a series of examples showing how rather interesting results could be obtained by tuning the catalysts' characteristics.

Chapter 4 focuses the discussion on the analysis of the reaction mechanisms of heterogeneous catalytic hydrogenation of CO_2 to produce products such as methane, methanol, and higher hydrocarbons. In CO_2 methanation, CO_{ads} is the key intermediate for methanation. In methanol synthesis, two possible pathways are discussed in detail: (i) direct hydrogenation of CO_2 via formate and (ii) the reduction of CO_2 to CO with subsequent hydrogenation to methanol. Depending upon the partial pressure of CO and CO_2, either the hydrogenation of CH_3O species or the formation of CH_3O can be rate-limiting for methanol formation. The mechanism of formation of higher alcohols may proceed through the reaction of CO insertion with hydrocarbon intermediates (RCH_n-) or through a direct nondissociative hydrogenation of CO_2. In the hydrogenation of CO_2 through a modified Fischer–Tropsch synthesis (FTS) process, the different effects of carbon dioxide on Co- and Fe-based catalysts are analyzed, showing also how the nature of the catalyst itself changes, switching from CO to CO_2 feed. This chapter thus gives valuable insights on how to design new catalysts for these reactions.

Chapter 5 analyzes in detail the recent developments in the metal oxide catalysts for the direct synthesis of organic carbonates such as dimethyl carbonate (DMC) from alcohol and CO_2. Ceria, zirconia, and related materials can catalyze the reaction with high selectivity under the conditions of the reaction without additives. Surface monodentate monoalkyl carbonate species are important intermediates. The yield is generally very low because of the equilibrium limitation. Combination of the reaction with organic dehydrating agents such as nitriles has been applied in order to overcome the equilibrium control. About 50% maximum methanol-based yield of DMC can be obtained when benzonitrile is used as a dehydrating agent. This chapter also analyzes future challenges for the design of catalysts and for the use of dehydrating agents to suppress the catalyst deactivation and the side reactions involving the dehydrating agents and the hydrated products.

Chapter 6 discusses in detail the theory and application of the STEP (solar thermal electrochemical production) process for the utilization of CO_2 via electrosynthesis of energetic molecules at solar energy efficiency greater than any photovoltaic conversion efficiency. In STEP the efficient formation of metals, fuels, and chlorine and carbon capture is driven by solar thermal-heated endothermic electrolyses of concentrated reactants occurring at a voltage below that of the room temperature energy stored in the products. As one example, CO_2 is reduced to either fuels or storable carbon at solar efficiency over 50% due to a synergy of efficient solar thermal absorption and electrochemical conversion at high temperature and reactant concentration. Other examples include STEP iron production, which prevents the emission of CO_2 occurring in conventional iron production, STEP hydrogen via efficient solar water splitting, and STEP production of chlorine, sodium, and magnesium.

Chapter 7 analyzes the electrochemical reduction of CO_2 in organic solvents used as the electrolyte medium, with a focus on understanding the effects of various parameters on electrolytic conversion of CO_2: Electrode materials, current density, potential, and temperature are examined, with methanol as electrolyte. A methanol-based electrolyte shows many advantages in the electrocatalytic reduction of CO_2 over other aqueous and nonaqueous solvents. CO_2 is completely miscible with methanol, and its solubility in methanol is five times higher than in water. The concentration of CO_2 can be increased as liquid CO_2 is made in a methanol electrolyte by increasing the electrolytic pressure. The faradaic efficiency of reduction products mainly depends on nature of the electrolyte. The strategy for achieving selective formation of hydrocarbons is also discussed.

Chapter 8 analyzes the conversion of CO_2 to synthetic fuels via a thermochemical process, particularly the reforming of CO_2 with hydrocarbons to form syngas. Aspects discussed include catalyst selection, possible operation, and potential application. In addition, research approaches for the conversion of syngas to methanol, DME, and alkane fuel (which is commonly known as gas-to-liquid or GTL) are also analyzed.

Chapter 9 discusses in detail the photocatalytic reduction of CO_2 with water on TiO_2-based nanocomposite photocatalysts. In particular, it is shown how the rate of CO_2 conversion can be improved by several means: (i) incorporation of metal or metal ion species such as copper to enhance electron trapping and transfer to the catalyst surface; (ii) application of a large-surface-area support, such as mesoporous silica, to enhance better dispersion of TiO_2 nanoparticles and increase reactive surface sites; (iii) doping with nonmetal ions such as iodine in the lattice of TiO_2 to improve the visible light response and charge carrier separation; and (iv) pretreatment of the TiO_2 catalyst in a reducing environment like helium to create surface defects to enhance CO_2 adsorption and activation. Combinations of these different strategies may result in synergistic effects and much higher CO_2 conversion efficiency. The final section also provides recommendations for future studies.

Recent updates on the photocatalytic mechanism of CO_2 reduction, with focus on novel carbon-based AgBr nanocomposites, are discussed in Chapter 10. Aspects analyzed include the efficiency of photocatalytic reduction of CO_2 and stability under visible light ($\lambda > 420$ nm). Carbon-based AgBr nanocomposites were successfully prepared by a deposition-precipitation method in the presence of cetyltrimethylammonium bromide (CTAB). The photocatalytic reduction of CO_2 on carbon-based AgBr nanocomposites irradiated by visible light gives as main products methane, methanol, ethanol, and CO. The photocatalytic efficiency for CO_2 reduction is compared with that of AgBr supported on different materials such as carbon materials, TiO_2, and zeolites.

While Chapters 1–10 look mainly in a medium-long term R&D perspective, it is necessary to have practical solutions also for the short term, because the climate changes associated with the increase in greenhouse gas (GHG) emissions have already started to become an issue in several countries, with an intensification of extreme weather events. Chapter 11 thus is focused on a topic different from those discussed in the other chapters. It provides an analysis of the state of the art in enhanced oil recovery (EOR) and carbon capture and sequestration (CCS) and their role in providing a stable energy supply and reduction in CO_2 emissions. EOR increases oil production by using CO_2, thus achieving both a stable energy supply and CO_2 reduction simultaneously. In contrast, CCS reduces CO_2 emissions even for non-oil producers. This chapter provides the background, fundamental mechanisms, and challenges associated with EOR and CCS, and shows that there are still several issues that need to be resolved, including recovery or storage efficiency, the cost of CO_2 capture, transport, and injection, and the CO_2 leakage risk. More research is required on fundamental mechanisms of the dynamics of EOR and CCS to allow significant improvements in the efficiency and safety of these techniques.

This book thus provides an overview on the topics of CO_2 (re)use from different perspectives, with strong focus on aspects related to industrial perspectives, catalyst design, and reaction mechanisms. Most of the contributions are related to photo- and electrocatalytic conversion of CO_2, because these are considered the new directions for achieving a sustainable use of CO_2, and the basis for realizing over the long term artificial leaf-type (artificial photosynthesis) devices.

The editors are very grateful to all the authors for their authoritative participation in this book. A special thanks goes to Dr. Maria D. Salazar-Villalpando, formerly of the National Energy Technology Laboratory (NETL-DoE, US), who originally initiated this book, inviting all authors to contribute the different chapters.

The Editors
G. CENTI AND S. PERATHONER
May, 2013

ACKNOWLEDGMENTS

The Editors and all the authors contributing to the book wish to express their sincere thanks to Maria D. Salazar-Villalpando, who started this editorial project.

■■■■■■ CONTRIBUTORS

Mudar Abou Asi School of Environmental Science and Engineering, Sun Yat-sen University, Guangzhou, China

S. Assabumrungrat Department of Chemical Engineering, Faculty of Engineering, Chulalongkorn University, Thailand

Gabriele Centi Department of Electronic Engineering, Industrial Chemistry and Engineering, CASPE/INSTM and ERIC, University of Messina, Messina, Italy

Burtron H. Davis University of Kentucky, Center for Applied Energy Research, Lexington, Kentucky, USA

G.R. Dey Radiation and Photochemistry Division, Bhabha Atomic Research Centre, Trombay, Mumbai, India

K. Faungnawakij National Nanotechnology Center (NANOTEC), Pathumthani, Thailand

Muthu K. Gnanamani University of Kentucky, Center for Applied Energy Research, Lexington, Kentucky, USA

Chun He School of Environmental Science and Engineering, Sun Yat-sen University, Guangzhou, China; Guangdong Provincial Key Laboratory of Environmental Pollution Control and Remediation Technology, Guangzhou, China

Shuichiro Hirai Department of Mechanical and Control Engineering (Research Center for Carbon Recycling and Energy), Tokyo Institute of Technology, Tokyo, Japan

Masayoshi Honda Department of Applied Chemistry, School of Engineering, Tohoku University, Sendai, Miyagi, Japan

Boxun Hu Department of Chemistry, University of Connecticut, Storrs, CT

Yanling Huang School of Environmental Science and Engineering, Sun Yat-sen University, Guangzhou, China

Gary Jacobs University of Kentucky, Center for Applied Energy Research, Lexington, Kentucky, USA

Satoshi Kaneco Department of Chemistry for Materials, Graduate School of Engineering, Mie University, Tsu, Mie, Japan

Hideyuki Katsumata Department of Chemistry for Materials, Graduate School of Engineering, Mie University, Tsu, Mie, Japan

M. Kumaravel Department of Chemistry and Applied Chemistry, PSG College of Technology, Peelamedu, Coimbatore, Tamilnadu, India

N. Laosiripojana The Joint Graduate School of Energy and Environment, King Mongkut's University of Technology Thonburi, Thailand

Ying Li Mechanical Engineering Department, University of Wisconsin-Milwaukee, Milwaukee, Wisconsin

Stuart Licht Department of Chemistry, George Washington University, Washington, DC, USA

Wenping Ma University of Kentucky, Center for Applied Energy Research, Lexington, Kentucky, USA

M. Murugananthan Department of Chemistry and Applied Chemistry, PSG College of Technology, Peelamedu, Coimbatore, Tamilnadu, India

Yoshinao Nakagawa Department of Applied Chemistry, School of Engineering, Tohoku University, Sendai, Miyagi, Japan

V. R. Rao Pendyala University of Kentucky, Center for Applied Energy Research, Lexington, Kentucky, USA

Siglinda Perathoner Department of Electronic Engineering, Industrial Chemistry and Engineering, CASPE/INSTM and ERIC, University of Messina, Italy

Dong Shu Base of Production, Education & Research on Energy Storage and Power Battery of Guangdong Higher Education Institutes, School of Chemistry and Environment, South China Normal University, Guangzhou, China

Steven L. Suib Department of Chemistry, University of Connecticut, Connecticut, USA

Tohru Suzuki Environmental Preservation Center, Mie University, Tsu, Mie, Japan

Keiichi Tomishige Department of Applied Chemistry, School of Engineering, Tohoku University, Sendai, Miyagi, Japan

Shohji Tsushima Department of Mechanical and Control Engineering (Research Center for Carbon Recycling and Energy) Tokyo Institute of Technology, Tokyo, Japan

Suguru Uemura Department of Mechanical and Control Engineering (Research Center for Carbon Recycling and Energy) Tokyo Institute of Technology, Tokyo, Japan

Ya Xiong School of Environmental Science and Engineering, Sun Yat-sen University, Guangzhou, China; Guangdong Provincial Key Laboratory of Environmental Pollution Control and Remediation Technology, Guangzhou, China

Zuocheng Xu School of Environmental Science and Engineering, Sun Yat-sen University, Guangzhou, China

Jingling Yang School of Environmental Science and Engineering, Sun Yat-sen University, Guangzhou, China

Qiong Zhang School of Environmental Science and Engineering, Sun Yat-sen University, Guangzhou, China

Linfei Zhu School of Environmental Science and Engineering, Sun Yat-sen University, Guangzhou, China

Perspectives and State of the Art in Producing Solar Fuels and Chemicals from CO_2

GABRIELE CENTI and SIGLINDA PERATHONER

1.1   INTRODUCTION

The last United Nations Climate Change Conference (COP17/CMP7, Durban, Dec. 2011) and the Intergovernmental Panel on Climate Change (IPCC) Fifth Assessment Report under preparation [1], two of the actual reference points regarding the strategies for the reduction of greenhouse gas (GHG) emissions, are still dedicating minor attention to the question of reusing CO_2. We discuss here how reusing CO_2 is a key element in strategies for a sustainable development as well as a nonnegligible mitigation option for addressing the issue of climate change. There is a somewhat

Green Carbon Dioxide: Advances in CO_2 Utilization, First Edition.
Edited by Gabriele Centi and Siglinda Perathoner.
© 2014 John Wiley & Sons, Inc. Published 2014 by John Wiley & Sons, Inc.

rigid separation between the discussion on the reduction of the emissions of GHG, based mainly on the introduction of renewable or alternative sources of energy and on the increase of efficiency in the use/production of energy, and the strategies for cutting current GHG emissions, based essentially only on the carbon capture and sequestration (CCS) option. The use of carbon dioxide as a valuable raw material is considered a minor/negligible contribution for the issue of climate change and thus not a priority to address.

The World Energy Outlook 2010 [2] report prepared by the International Energy Agency (IEA) has discussed different options and scenarios for GHG emissions, proposing a reduction of CO_2 emissions in the 2.3–4.0 $Gt \cdot y^{-1}$ range within one decade (by the year 2021) and in the 10.8–15.4 $Gt \cdot y^{-1}$ range in two decades (by the year 2031) with respect to the business-as-usual scenario. About 20% of this reduction would derive from CCS. According to this estimation, about 400–800 $Mt \cdot y^{-1}$ of CO_2 in a decade and about 2100–3000 $Mt \cdot y^{-1}$ of CO_2 in two decades will be captured. The McKinsey report [3] estimated the global potential of CCS at 3.6 $Gt \cdot y^{-1}$ and the potential in Europe at 0.4 $Gt \cdot y^{-1}$ —around 20% of the total European abatement potential in 2030.

With these large volumes of CO_2 as raw material at zero or even negative cost (the reuse of CO_2 avoids the costs of sequestration and transport, up to about 40–50% of the CCS cost, depending on the distance of the sequestration site from the place of emission of CO_2) soon becoming available, there are clear opportunities for the utilization of CO_2. In addition to direct use, many possibilities exist for its conversion to other chemicals, in addition to already-existing industrial processes.

A number of recent articles, reviews, and books have addressed the different options for converting CO_2 [4–15]. Scientific and industrial initiatives toward the chemical utilization of CO_2 have increased substantially over the last few years [6a], and there is increasing attention to the use of CO_2 to produce

- Advanced materials (for example, a pilot plant opened at Bayers Chempark in Leverkusen, Germany in February 2011 to produce high-quality plastics—polyurethanes— based on CO_2 [16]);
- Fine chemicals [5,10,11,14] (for example, DNV is developing the large-scale electrochemical reduction of carbon dioxide to formate salts and formic acid [17]);
- Fuels [6,9,15] (for example, Carbon Recycling International started in September 2011 in Svartsengi, Iceland, a plant for producing 5$Mt \cdot y^{-1}$ of methanol from CO_2 using H_2 produced electrolytically from renewable energy sources—geothermal, wind, etc. [18]).

The chemical transformation of CO_2 is a dynamic field of research, in which many industrial initiatives also thrive, even if it is not always straightforward to grasp the real opportunities and limitations of each option. CO_2 utilization as a raw material is expected in a short- to medium-term perspective to continue its progression, with several new products coming onto the market (e.g., polycarbonates). In the long term, CO_2 recycling can become a key element of sustainable carbon-resource management in chemical and energy companies, combined with curbing consumption. CO_2 can also become a strategic molecule for the progressive introduction of renewable energy resources into the chemical and energy chain, thus helping to slowly lessen our consumption of fossil fuels. Thus the prospects for large-scale utilization [6a] indicate that CO_2 recycling can become an important component of the strategy portfolio necessary for curbing CO_2 emissions (with an estimated potential impact of hundred millions of tons of CO_2 recycling, similar to the impact of CCS) and at the heart of strategies for sustainable chemical, energy, and process industries, for a resource and energy efficiency development, for example, as a key enabling technology and backbone for the resource-efficient Europe flagship initiative of the Europe 2020 Strategy [19].

The concept of "Green Carbon Dioxide" [20], considering CO_2 not a "devil" molecule (a problem or even a possible reuse of a waste) but a key element for sustainable strategies of energy and chemical companies, is thus the new emerging vision that we emphasize in this chapter, because the new strategies toward resource and energy efficiency development need to be advanced in both the industrial and scientific communities. The concept of solar fuels is a key part of this vision [21–26], but we will not limit the discussion on the state of the art and perspectives to this area, because CO_2 recycling is an enabling element for a low-carbon economy and the efficient introduction of renewable energy in the production chain. This chapter thus analyzes these aspects and describes the opportunities offered by CO_2 recycling and solar fuels in this more general vision and context.

1.1.1 GHG Impact Values of Pathways of CO_2 Chemical Recycling

For a correct evaluation of the real impact of recycling CO_2 via chemical conversion with respect to alternative options such as storage (CCS) or even direct use in applications such as enhanced oil recovery (EOR), which, however, can be applied only in specific locations, food use, and intensive agriculture (to enrich the atmosphere in greenhouses), etc. it is necessary to discuss the impact value of the chemical recycling of CO_2.

The GHG impact value (GIV) indicates the effective amount of CO_2 eliminated from contribution to the GHG effect (over a given time frame, for example, 20 years) on a life-cycle assessment (LCA) basis. For example, for CO_2 storage (CCS) the energy necessary for the recovery, transport, and storage of CO_2 must be calculated for each ton of stored CO_2. GIV for CCS clearly depends on a number of factors, from the type and composition of the emissions, to the distance of the capture site from the storage site, the modalities of transport, etc. Detailed studies are not available, but on average, it is a realistic estimate that around 0.5–0.6 tons of CO_2-equivalent energy is necessary for the capture, transport, and storage of 1 ton of CO_2 sequestrated [21,27]. In fact, capture with the amine-absorption technologies (the most used today) accounts for about 0.2 tons of CO_2 and transport/storage accounts for an additional 0.3–0.4 tons of CO_2 (per ton of CO_2 sequestrated). These are average values, because in many places, such as in various areas of Europe, it is necessary to transport CO_2 for over 150–200 km and pipelines are not available.

The storage will be long term, and thus over a time frame of 20 years the average GIV for CCS will be around 0.4–0.5.

There are different options for the reuse of CO_2. We may roughly distinguish two main routes for reusing CO_2 to produce commercially valuable products, apart from the routes involving bacteria and microorganisms:

1. Those reactions incorporating the whole CO_2 moiety in organic or inorganic backbones;
2. Those involving the rupture of one or more of the C–O bonds.

This classification is important in term of energy balance and applications. The first class of reactions (both organic and inorganic) is not energy intensive and sometimes may also occur spontaneously, although with low kinetics, as in the production of inorganic carbonates. The second class, reactions involving the cleavage of the C–O bond, is energy intensive and requires the use of reducing agents, typically H sources such as H_2. For a CO_2 resource- and energy-efficient management, the energy necessary for these reactions should derive from renewable (solar, wind, geothermal energy, etc.), or at least from non-carbon-based (nuclear energy) sources or, eventually, waste-energy sources.

There are two typical potentially large-scale examples for the first class of reactions [6a], the production of saleable precipitated carbonate and bicarbonates or carbonates from minerals and the production of polymers incorporating CO_2 units. An example of the first case is the mineralization via aqueous precipitation (MAP) process developed by Calera [28] in a 10MW demonstration unit in Moss Landing, California (US) and followed by an Australian demonstration project in Latrobe Valley, Victoria (Australia). The flue gas from fossil fuel combustion is reacted with alkaline solutions heavy in calcium and/or magnesium, such as certain minable brines, to form a stable carbonate solid with a by-product of relatively fresh water that would be suitable for desalination. When suitable brines are not readily available, an alkaline solution of sodium hydroxide must be manufactured via, for example, chemical electrolysis. Once the CO_2 has been absorbed into a bicarbonate solution, it can be stored underground or transformed into a carbonate (building material). A full LCA does not exist, and also in this case the exact value depends on the specific process characteristics (the alkalinity sources, for example) and use of the final product [29]. The production of the alkaline solution is the energy-intensive step, and energy estimation indicates a GIV value of about 0.6–0.8 [30] for the production of building materials (with thus a lifetime over 20 years). This average value is similar for the other CO_2 mineralization technologies, where, for example, the critical energy-intensive step is the mining and crushing of the minerals (for example, olivine) used as the raw material.

For the production of CO_2-based polymers, correct LCA assessments also do not exist, and the GIV value depends on specific process characteristics that have not yet been developed on a commercial scale, apart from Asahi Kasei's phosgene-free process to produce aromatic polycarbonate starting from ethylene epoxide, bisphenol-A, and CO_2 [31]. Polypropylene and polyethylene carbonate as well as polyhydroxyalkanoate (PHA) from CO_2 and linear epoxides are currently developed by Novomer on a pilot-plant scale [32], while Bayer is developing on a pilot scale a process for producing polyurethane from polyether polycarbonate polyols, as already cited [16]. These are the more relevant examples of CO_2-based polymers, but additional examples also exist [6a].

The difference with respect to inorganic carbonates is that these polymers will substitute for polymers derived from fossil fuel sources, although it must be noted that the weight content of CO_2 ranges from about 17%wt. for aromatic polycarbonate to about 30%wt. for polyurethane and about 50%wt. for polypropylene carbonate. In addition, the use of these polymers will bring further benefits. For example, polyurethane foams are one of the most efficient insulation materials on the market today for roof and wall insulation, insulated windows and doors, and air barrier sealants. Spray polyurethane foams can cut yearly energy costs upwards of 35% with respect to alternative insulation material, but their use is still limited by their cost. The expansion of their market through larger availability at low cost from production using CO_2 as raw material will thus have direct benefits in terms of use of CO_2 as raw material and reduction of the use of fossil fuels and indirect benefits in terms of energy saving. Polycarbonate multiwall structures also offer a real advantage in thermal insulation. As clear as glass, polycarbonate has superior characteristics as to energy savings, safety, and practicality for civil and industrial buildings, with the use of panels in structures that need reduced weight, abundant light, and high resistance to atmospheric agents. The indirect impact on energy saving is difficult to estimate, because it will depend on market development, incentives for energy saving, etc. It may be thus given only a very approximate estimation. We consider realistic a conservative average GIV value (the lifetime of these CO_2-based polymers is over 20 years) of about 2–4.

Production of fuels from CO_2, which involves the rupture of one or more of the C–O bonds and thus the need to supply renewable energy to make the conversion sustainable, is a different case. In fact, CO_2 recycling via incorporating renewable energy introduces a shorter path (in terms of time) to close the carbon cycle compared to natural cycles and an effective way to introduce renewable energy sources in the chemical/energy chain. In addition, it will reduce the use of fossil fuels for these chemical/energy uses. Let us consider the simple case of methane production from CO_2, although as discussed below this is not the ideal energy vector into which carbon dioxide can be transformed. However, it is the simplest example for considering the fuel life cycle and the related GIV. If we capture CO_2 from the emissions deriving from the combustion of methane, we must spend energy in the capture (similar to the CCS case), but if we use renewable energy for the conversion of CO_2 to methane (as discussed below), the net effect is that we introduce renewable energy into the energy chain. Considering that (i) 0.2 tons of CO_2 are necessary for capturing each ton of CO_2 (in terms of CO_2-equivalent energy), (ii) 0.2 tons of CO_2 are associated with the loss of energy in the conversion process, and (iii) 0.1–0.2 tons of CO_2 are necessary to produce the renewable energy necessary for the conversion of CO_2 to fuels, we have still a positive value of saving about 0.4–0.5 tons of CO_2. We must note that the carbon footprints for the different renewable resources are different and on the average not negligible, but to simplify the discussion we consider only an average value. When an excess of energy is used in the transformation, for example, to store the excess of energy in the form of chemical fuel, the impact value could be even better. Each time that a cycle is completed (capture of CO_2, conversion of CO_2 to CH_4, for example, by using H_2 produced from a renewable energy sources—photovoltaic, wind, etc., storage/transport of methane, use of methane to produce energy and CO_2), there is a saving of at least 0.4–0.5 tons of CO_2 per amount of CO_2 sequestrated. However, the cycle could be repeated several times. In a 20-year time frame, a single molecule of CO_2 is recycled virtually several thousand times, with thus a continuous mechanism of reintroduction of renewable energy. From a practical aspect, the number of cycles will depend on the cost differential with respect to use of fossil fuels, incentives in limiting GHG emissions, carbon taxes, technology development, etc. It is thus quite difficult to estimate a correct GIV, but we consider a conservative average a GIV value of about 10–12 over 20 years. There is thus a large amplification of the impact value of chemical utilization of CO_2 to produce polymers or fuels, with respect to the CCS or mineralization cases, even within the limits resulting from the absence of more specific studies.

Figure 1.1 summarizes the discussed average impact value on GHG for the different routes of chemical CO_2 recycling with respect to CCS and CO_2 mineralization. It may be noted that this concept is the opposite of that used in the IPCC report [33] which indicated that "the lifetime of the chemicals produced is too short with respect to the scale of interest in CO_2 storage. Therefore, the contribution of industrial uses of captured CO_2 to the mitigation of climate change is expected to be small." This statement is not correct, because the chemicals/fuels produced from

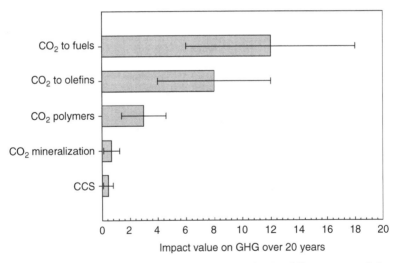

Figure 1.1 Average indicative impact values estimated for the different routes of chemical CO_2 recycling with respect to CCS and CO_2 mineralization.

the conversion of CO_2 and incorporation of renewable energy have an effective impact value for the reduction of GHG emissions at least one order of magnitude higher than that for CCS. Thus the effective potential of carbon capture and recycle (CCR) technologies in GHG control is at least similar to that of CCS technologies and estimated to be around 250–350 Mt·y^{-1} in the short to medium term [6a]. This amount represents about 10% of the total reduction required globally and is comparable to the expected impact of CCS technologies, but with additional benefit in terms of (i) fossil fuel savings, (ii) additional energy savings (e.g., the cited insulating effect of polyurethane foams), and (iii) acceleration of the introduction of renewable energy into the chemical production and energy chains.

1.1.2 CO_2 Recycling and Energy Vectors

Solar energy is abundant, accounting for over three orders of magnitude the current global consumption of primary energy, but it must be converted to electrical, thermo-mechanical, or chemical energy to be used. Energy must be also stored/transported to be used when and where it is necessary. Storage of energy in chemical form, that is, fuels, is still the most efficient way for storage and transport of energy, in terms of energy density and cost-effectiveness [21]. Energy density in a typical fuel is about two orders of magnitude larger than in batteries, and even with the possible future developments in batteries, it will be not possible to fill this gap. This is the main reason that (chemical) energy vectors will be still dominating the future energy scenario. Even in the blue-sky scenarios of the IEA [2], chemical energy will still have a predominant role with respect to other forms (electrical, etc.) in the year 2050. Thus our society is and still will be in the future largely based on the use of liquid hydrocarbons as the energy vector. Even today, with a very limited fraction of energy produced from renewable energy sources (solar, wind, etc.), it is not possible to fully use the renewable energy produced outside peak hours. When this fraction of renewable energy will exceed about 5–10%, it will become imperative to find efficient ways to convert electrical to chemical energy [34,35].

Suitable energy vectors must fulfill a number of requirements: (i) have a high energy density both by volume and by weight; (ii) be easy to store without need of high pressure at room temperature; (iii) be of low toxicity and safe in handling and show limited risks in their distributed (nontechnical) use; (iv) show a good integration in the actual energy infrastructure without need of new dedicated equipment; and (v) have a low impact on the environment in both their production and use.

H_2 has been often presented as the ideal energy vector [36,37], but still has two main drawbacks: (i) it has too low energy density for practical large-scale use, and the expected targets in H_2 storage materials to overcome this issue are very difficult to be met, and (ii) too large investments are necessary to change the energy infrastructure required for its use. Solar H_2 [38] may be a better and more sustainable alternative, when combined with the possibility of forming liquid fuels that are easily transportable and with high energy density [9,39]. This possibility is offered by the use of solar H_2 to produce fuels from CO_2 or, even better, to integrate directly a solar cell able to produce protons/electrons (the equivalent of H_2) from water oxidation using sunlight with an electrocatalyst able to use the protons/electrons to convert CO_2 efficiently to fuels. This is the photoelectrocatalytic (PEC) approach, in which the two reactions of water oxidation using sunlight and CO_2 reduction using the electrons/protons generated in the light-illuminated side occur in two different cell compartments separated by a proton-conducting membrane [21c,40]. Producing solar fuels by recycling CO_2 is a carbon-neutral approach to store and transport solar energy that can be well integrated into the current energy infrastructure. It is an effective path to introduce renewable energy into the energy chain and, as has been discussed recently, also in the chemical production chain [41].

1.2 SOLAR FUELS AND CHEMICALS FROM CO₂

The actual average global energy consumption is about 16 TW, and it is estimated to increase to about 25 TW by the year 2050 [2]. A conservative estimation of the potential for solar energy is at least 5–10 times higher than this estimated consumption, while significantly lower for other renewable sources: 2–4 TW for wind, 2–3 TW for tides, 5–7 TW for biomass, and 3–6 TW for geothermal energy [42]. Of these different renewable energy sources only biomass can be converted to liquid fuels, while almost all the others produce electrical energy. The biomass-to-fuel approach, however, is complex and costly, and there are many concerns regarding its effective contribution to limiting GHG emissions. It is thus a transitional, not long-term, solution for the GHG question. As pointed out in the previous section, the issue is thus how to convert electrical to chemical energy in a resource- and energy-efficient approach.

The main example of the use solar light of to convert CO_2 to chemicals is photosynthesis. Plants or algae use solar energy and CO_2 to generate the several molecules necessary for life (cellulose, hemicellulose, lignin, starch, lipids, oils, etc.) through quite complex machinery. It is then necessary to convert these molecules to fuels according to different possible routes, all characterized by many steps and a large energy consumption [43–46]. The efficiency in using solar light for the chemical transformation is low, around 1% or less for green plants. Some algae are more efficient in the process, up to about 10%. When the LCA efficiency is considered, including the energy necessary to grow and harvest the biomass and its transformation to useful chemicals, the efficiency is drastically reduced (below 0.1%) because of the many steps necessary, reflecting also in the effective impact on the reduction of the emissions of CO_2. For example, LCA shows that production of biodiesel from algae may be even negative compared with production of diesel from fossil fuels in terms of global warming potential (GWP) or climate change power (CCP), if the emissions of N_2O associated with the intensified use of fertilizers, the effect of substitution in land use, and other factors are taken into consideration in the LCA [47].

Bio-solar fuels may be directly produced from bio-organisms, and not through the transformation of the primary bio-products (lipids, cellulose, starch, etc.), but current developments are essentially limited to the production of H$_2$ using cyanobacteria or some green algae [48,49]. Interesting recent results showed that genetically modified cyanobacteria (*Synechococcus elongatus* PCC7942) consume CO$_2$ in a set of steps to produce directly a mixture of isobutaraldehyde (primarily) and isobutanol [50]. However, productivities are still quite low. Carbon Sciences [51] has announced that it has developed a breakthrough enzyme-based technology to transform CO$_2$ into low-level fuels, such as methanol, but more precise information and data are lacking. Other biotech companies, such as Joule Biotechnologies, Gevo, and Global Bioenergies, have also announced that they have developed genetically modified microorganism pools able to use CO$_2$ to produce directly fuels or chemicals, but data are not available to estimate productivities and pro/cons of the proposed technologies. Thus biotech routes will probably have a relevant role in converting CO$_2$ to solar fuels and chemicals in the future, but currently available data are not sufficient to really prove the potential. Algae and plants are already used in producing some very high-value chemicals industrially, but this has not had a significant impact on CO$_2$ emissions, because of the very low value of these products. We discuss here the case of solutions using CO$_2$ for potentially large-volume products (fuels) or chemicals having a large impact value.

1.2.1 Routes for Converting CO$_2$ to Fuels

There are different routes to convert CO$_2$ back to fuels. The most-studied area is the hydrogenation of CO$_2$ to form oxygenates and/or hydrocarbons. Methanol (CH$_3$OH) synthesis from CO$_2$ and H$_2$ has been investigated up to pilot-plant stage, with promising results [6a,b]. The CAMERE process [carbon dioxide hydrogenation to methanol via a reverse water-gas shift (RWGS) reaction] is based on a first stage in a RWGS reactor where part of the CO$_2$ feed is converted to CO (>60%). After elimination of the water produced, the resulting H$_2$/CO$_2$/CO mixture is then fed to a methanol synthesis (MS) reactor, where it is converted to CH$_3$OH, according to:

$$CO_2 + 3H_2 \rightleftharpoons CH_3OH + H_2O \qquad (1.1a)$$

$$CO + H_2O \rightleftharpoons CO_2 + H_2 \qquad (1.1b)$$

The two-step process allows 25% reduction of volume of the MS reactor with respect to a single stage, higher process efficiency, and lower operational costs. Zinc aluminate- and Cu/Zn/Al$_2$O$_3$-based catalysts are used in the two steps. The CAMERE process was developed on a bench scale (50–100 kg MeOH/day) [52]. The Japanese company Mitsui Chemicals Inc. completed in 2009 a pilot (100 t·y^{-1}) methanol plant in Osaka [53]. Mitsui Chemicals is also planning a first commercial-scale (600,000 t·y^{-1}) methanol plant based on Mitsui's Green House Gases-to-Chemical Resources (GTR) technology. Methanol is intended as feedstock for olefins and aromatics. The previously cited company Carbon

Recycling International [18] also started in September 2011 a plant in Svartsengi, Iceland for producing 5 Mt·y^{-1} of methanol from CO_2.

The alternative possibility is the production of dimethyl ether (DME), a clean-burning fuel that is a potential diesel pool additive. Kansai Electric Power Co. and Mitsubishi Heavy Industries (Japan) have developed a bench-scale unit (100 cm^3 catalyst loading) for DME synthesis via the reforming of methane by CO_2 and steam [54]. Ni/MgO-Al$_2$O$_3$ and Ru/MgO-Al$_2$O$_3$ catalysts were used for the methane reforming with a CH_4 conversion almost complete over 1073 K in 800-h durability tests; carbon deposition on the catalyst was very low.

Ethanol formation, either directly or via methanol homologation, and conversion of CO_2 to formic acid are also potentially interesting routes. Methanol, ethanol, and formic acid may be used as feedstock in fuel cells, providing a way to store energy from CO_2 and then produce electricity. Alcohols are in principle preferable over hydrocarbons because their synthesis requires less hydrogen per unit of product. In fact, the key problem in this route is the availability of H_2. If the latter is produced from hydrocarbons (the main current route is by steam reforming of methane) there are no real advantages in converting CO_2. H_2 must thus derive from renewable sources. The possible options are the following:

- Water electrolysis, coupled with a renewable source of electrical energy (photovoltaic cells, wind, or waves, etc.). This technology is already available, but the need for multiple steps, the overpotential in the electrolyzer, and other issues limit the overall efficiency. The technology is mature, with a limited degree of further possible improvements.
- Biomass conversion, preferably using waste materials and in conditions that require low energy consumption. An example is the catalytic production of H_2 directly in liquid phase from aqueous solutions (ethanol waste streams, for example). This option could be a way for valorization of side waste streams in biorefinery, but it is not an efficient way if considered alone. In fact, if we consider the whole life cycle including growing the plant, harvesting, fermentation, etc., and finally H_2 production (from bioethanol, for example), the overall energy consumption (and thus amount of CO_2 produced) is higher than the advantage in hydrogenating CO_2 back to fuels.
- Production of H_2 via biogas produced from anaerobic fermentation of biomass. Also in this case, it could be a valuable option using waste biomass, but it is a quite complex process considering the whole production chain. There are also problems of purification of biogas.
- Production of H_2 using cyanobacteria or green algae. This is an interesting option, but still with low productivity and under development.
- Direct H_2 production by water photoelectrolysis, which still suffers from low productivity and in some cases the need for further separation/recovery of hydrogen but has great potentiality for development to reach industrial feasibility.

The following section further discusses the production of H_2 using renewable energy, abbreviated as "renewable H_2" hereinafter. Hydrogen, after eventual compression and heating to the requested reaction temperature, may then be used for the hydrogenation of CO$_2$ to produce CO via the RWGS reaction. Carbon monoxide and hydrogen, the so-called syngas, may then be converted to methanol and/or DME, or Fischer–Tropsch (FT) products (hydrocarbons, mainly) by known catalytic processes [6]. These processes may be also combined in a single process with the RWGS reaction, but the formation of water in the latter is an issue in syngas transformation. The cited transformations are technologies essentially available, although some further improvement is necessary in terms of both catalysts and reactor technologies. The key aspect in all these routes is to produce renewable H$_2$ through economic and eco-/energy-efficient processes. Dry reforming of methane with CO$_2$ is an alternative possibility to produce syngas:

$$CH_4 + CO_2 \rightleftharpoons 2CO + 2H_2 \qquad (1.2)$$

However, this is a strong endothermic reaction suffering from fast deactivation due to carbon formation. Coupling with the reaction of steam reforming of methane and of partial methane combustion (the so-called tri-reforming process) reduces the issue of deactivation and allows autothermic operations. The process is interesting and is developed up to pilot-scale operations [12]. However, the CO$_2$ recycling effectiveness is low, even if specific LCA studies are not present in the literature.

There are other routes and options in converting CO$_2$ to fuels, as discussed in detail elsewhere [5a,6a,7,9,12,21]. When sources of renewable H$_2$ below a cost of about \$2–3/kg are available, either because it is used as excess energy or for other reasons, these processes of CO$_2$ conversion to fuels could be already close to commercialization, with minor technological aspects that still need to be developed further [6a]. However, different process steps and large chemical plants are necessary, and in general terms it is not an energy-efficient process technology. A longer-term vision would be a technology that will couple and integrate directly the stage of hydrogen generation (from water using solar energy) to the stage of CO$_2$ reduction and conversion to fuels in mild reaction conditions. This is the concept of *artificial leaf* discussed in the following section.

1.2.2 H$_2$ Production Using Renewable Energy

Converting CO$_2$ to fuels or chemicals via breaking of the C–O bond requires H$_2$, and the latter should be produced with the use of renewable energy sources to make the process effective in terms of GHG impact, as discussed in the previous sections. The current method of producing H$_2$ is mainly based on methane (or other fossil fuels) steam reforming:

$$CH_4 + 2H_2O \rightleftharpoons CO_2 + 4H_2 \qquad (1.3)$$

Four moles of H$_2$ are produced per mole of CO$_2$, that is, 5.5 kg CO$_2$ per kg H$_2$, but including the energy required in this endothermic reaction this value rises to about 9 kg CO$_2$/kg H$_2$ [55]. The impact on CO$_2$ emissions of producing H$_2$ from biomass needs to include many factors, from the growing of biomass to biomass harvesting/transport, conversion etc. On average 5–6 kg CO_2 per kg H_2 are produced [56].

For the direct production of H_2 using renewable energy, LCA data are not very reliable, because all technologies are still at the development stage. For wind/electrolysis, a value below 1 kg CO_2/kg H_2 was estimated [57]. Utgikar and Thiesen [58] reported life-cycle CO_2 emissions for various hydrogen production methods and indicated values for hydroelectric/electrolysis or solar thermal production of H_2 around 2 kg CO_2/kg H_2 and values for photovoltaic/electrolysis production around 6 CO_2/kg H_2. However, the latter value appears overestimated and does not include the latest developments in the field. We thus assume an average of 1–2 kg CO_2/kg H_2 for hydrogen produced via electrolysis using excess electrical energy.

The production of H_2 by electrolysis is already well established, with the electrical energy deriving from photovoltaic (PV) cells, wind turbines, etc. Today, the efficiency of the PV-electrolysis system is at best about 12% [59]. H_2 could be produced under pressure in modern electrolyzers, while other routes produce H_2 at atmospheric pressure. Polymer electrolyte membrane (PEM) water electrolysis technology is today a safe and efficient way to produce renewable hydrogen, with stack efficiencies close to 80% at high (1 $A·cm^{-2}$) current densities [60], but overpotential is still a major problem limiting the cost-effectiveness. PEM electrolyzers show a number of advantages over the alternative and well-established alkaline technology: the absence of corrosive electrolytes and better integration with solar and wind power. A target cost of about \$2–4/kg H_2 necessary to make the technology competitive has still not been reached, but the current trend in cost reduction looks promising, especially when the problem of overpotential could be solved by a better understanding of the processes at the electrode surface.

1.2.3   Converting CO_2 to Base Chemicals

Light olefins (ethylene and propylene), currently produced in an amount of about 200 $Mt·y^{-1}$, are the building blocks of current chemical production. They are synthetized today from fossil fuels, but this process is the single most energy-consuming production in the chemical industry, and the specific emission of CO_2 per ton of light olefin ranges between 1.2 and 1.8 [61]. It is thus interesting to explore the possibility of using CO_2 as the carbon source to produce olefins, using H_2 produced from renewable energy sources [61]. Because of their high energy of formation, C2-C3 olefins represent an excellent opportunity to store solar energy and incorporate it into the value chain for chemical production instead of that for energy. The high value of the energy of formation of olefins also explains why their actual process of formation is the most energy-consuming process in the chemical industry, with a large impact on CO_2 emissions.

The largest part of ethylene and propylene production is currently used to produce polymers, directly (polyethylene and polypropylene; polypropylene production, for example, accounts for >60% of the total world propylene consumption) or indirectly (for example, main products of propylene are acrylonitrile, propylene oxide, acrylic acid, and cumene, which are intermediates in the production of polymers).

The synthesis of light olefins from CO_2 requires the availability of H_2. Ethylene and propylene have a positive standard energy of formation with respect to H_2, but water forms in the reaction and thus the process essentially does not need extra energy with respect to that required to produce H_2. From the energetic point of view, the energy efficiency of the process is thus related to the energy efficiency of the production of H_2. The process for olefin synthesis from CO_2 may be described as the combination of a stage of RWGS (Eq. 1.1b) and a consecutive stage of FT synthesis:

$$CO + H_2 \rightarrow C_nH_{2n} + C_nH_{2n+2} + H_2O + CO_2 (n = 1, 2, \dots) \qquad (1.4)$$

The FT catalyst should be modified in order to minimize the formation of alkanes (especially CH_4) and increase selectivity to C2-C3 olefins. The above two stages may be combined together, but water should be preferably removed in situ to shift the equilibrium and avoid FT catalyst reversible inhibition. Current catalysts, derived mainly from the doping with alkaline metals of conventional FT catalysts or catalysts combining a zeolite with a methanol or FT catalyst, give at best selectivities around 70–80% and around 40% conversion [61]. Further improvements in catalysts, reactor design, and process operations are thus necessary, but it is reasonable to consider possible a further optimization with a target in selective synthesis of light olefins over 80% at higher conversion (>70%).

The process flow sheet in light olefin production from CO_2 [61] (Fig. 1.2) is based on a first step of production of H_2 by electrolysis and electrical energy derived from renewable sources. The core of the process is the combination of RWGS and modified FT reactions. The two steps are in separate stages for the optimization

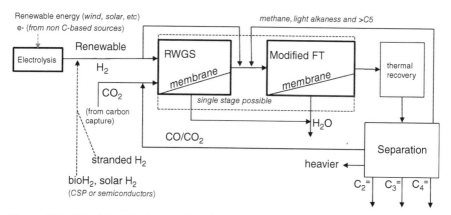

Figure 1.2   Simplified block flow sheet for the CO_2 to olefin process. Adapted from ref. 61.

of the relative catalysts and reaction conditions. Inorganic membranes permeoselective to water are integrated in the reactors to improve the performance of the process. H_2 is added in part between the two stages, and the methane, light alkanes, and >C5 hydrocarbons produced in the process are recycled. The overall process, because of the formation of water, results in a slightly exothermic process. After thermal recovery, the light olefins are separated in a sequence of columns similar to the steam cracking process, while CO and CO_2 are recycled to the RWGS unit.

A techno-economic analysis of this possibility [61] indicates that for a predicted renewable H_2 cost as target for the year 2020 (US$2–3/kg H_2), the process is economically valuable. The current renewable H_2 cost is still higher, but not so high not to consider in more detail the possibility of producing light olefins from CO_2.

1.2.4   Routes to Solar Fuels

Although the concept of solar fuels is broad and typically also includes H_2 production using solar energy, we do not here consider hydrogen strictly a solar fuel but an intermediate to produce the true solar fuels in the form of liquid (easily storable and with high energy density) chemicals, that is, energy vectors that can be more easily stored and transported (see Section 1.2). As briefly discussed in Section 2.1, there are different possible energy vectors that derive from the hydrogenation of CO_2, either directly or through the intermediate stage of the RWGS reaction (Eq. 1.1b), to produce syngas (mixture of CO/H_2), which can be then converted through already-established and commercially applied routes, although the syngas is produced from hydrocarbons instead of from CO_2 and H_2:

- Formic acid, which may be used in formic acid fuel cells or as a vector to store and transport H_2 (the reaction of synthesis is reversible, and formic acid can be catalytically decomposed under mild conditions to form back H_2 and CO_2 [62])
- Methanol and dimethyl ether (DME)
- Methane (substituted natural gas, SNG)
- >C1 alcohols or hydrocarbons

These different routes have been discussed in detail elsewhere [6a]. The main routes are chemical (catalytic), but electrochemical or solar thermal routes are also possible, even if the latter two routes are still not sufficiently developed. Syngas may also be produced by reaction with hydrocarbons (particularly methane) through so-called dry reforming. The main potential advantage of this route is that it can be applied directly to flue gases (even if technical problems exist), while all the other routes require a first step of separation of CO_2 from the flue gases. However, dry reforming is an endothermic reaction occurring at high temperature (about 900–1000°C) and with formation of carbon (which deactivates the catalyst) as a side reaction. An overview of the different routes is given in Figure 1.3.

The catalytic chemistry of the RWGS reaction, the following transformation to methanol/DME, or hydrocarbons via Fischer–Tropsch synthesis, and the

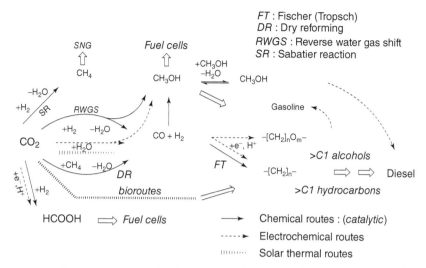

Figure 1.3 Conversion routes of CO$_2$ to solar fuels. The renewable energy is used either directly (in the solar thermal production of syngas) or indirectly, according to two main routes: (i) production of renewable H$_2$ or (ii) production of electrons or electron/protons (by water photo-oxidation), used in the electrochemical routes. Adapted from ref. 6d.

subsequent production of gasoline (methanol to gasoline, MTG) or of diesel via hydrocracking of the alkanes produced in the FT process (using Co-based catalysts) is well established, even if there is still need for development because of the change of feed composition starting from CO$_2$ rather than from syngas. Also, in terms of process development, most of the necessary knowledge is available. Minor technological barriers to development of these routes are thus present. Only for the synthesis of formic acid, either catalytically or electrocatalytically, is there still need for development in terms of productivity and stability.

The main gap in the catalytic routes of CO$_2$ conversion to fuels is economic, with the cost of production of renewable H$_2$ as the key factor. However, opportunities already exist in terms of available (low cost) sources of renewable H$_2$ that make the production of fuels from CO$_2$ interesting. Mitsui Chemicals and Carbon Recycling International are two companies that are running pilot plant projects to exploit the conversion of CO$_2$ to methanol, as discussed above, while Mantra Venture Group and DNV are exploring at pilot-plant scale the electroreduction of CO$_2$ to formic acid. RCO2 AS has instead developed at pilot scale a process based on recovery of CO$_2$ from flue gas and its conversion to methane using renewable H$_2$. Details on these processes and the related chemistry and catalysis are provided elsewhere [6]. A book discussing the various possible routes for CO$_2$ conversion was recently published by Aresta [5a]. Other authors have also recently published reviews on this topic [6–15].

The catalytic synthesis of higher alcohols from CO$_2$ is an interesting route but is still not competitive. New interesting catalysts, however, have been developed.

Conversion of CO_2 to higher alcohols and hydrocarbons ($\geq$C2) by biocatalysis or electrocatalysis methodologies is also an interesting route, but is still at a preliminary stage [21]. Genetically modified cyanobacteria have been recently reported to consume CO_2 in a set of steps to produce a mixture of isobutyraldehyde (primarily) and isobutanol. With a gas-phase electrocatalysis approach CO_2 may be reduced to a mixture of $\geq$C2 hydrocarbon and alcohols, mainly isopropanol [63]. Artificial metabolic pathways involving enzymes or cyanobacteria have been proposed to use NADPH and ATP from photosynthesis for the synthesis of *n*-butanol (UCLA) or isobutene (Global Bioenergies) directly from CO_2 and water.

1.3 TOWARD ARTIFICIAL LEAVES

The leaf is a highly complex machinery that utilizes solar light to oxidize water and produce electrons/protons used in a different part of the cell to reduce CO_2 to carbohydrate, lipid, and other components necessary for plant life and growth. In an artificial leaf it is necessary to mimic the various steps in this hierarchical process (capture of sunlight photons, electron–hole separation with long lifetimes, energy transduction, etc.) while developing a new functional and robust design that realizes two goals:

- Intensify the process, thus allowing higher productivity and efficiency in converting sunlight (in plants the quantum yield is typically below 1%);
- Use solid components that keep functionalities but are more robust, scalable, and cost-effective.

It is also necessary to separate the two reactions, water oxidation using sunlight and CO_2 reduction using the electrons/protons generated in the light-illuminated side. They should occur in two different cell compartments separated by a proton-conducting membrane [21c], in order to reduce back reactions, achieve high efficiency, and, importantly from the practical perspective, have separate production of O_2 and of the products of reduction of CO_2. This is a critical issue both for safety and to avoid the costs of separation. The same is valid in the simpler case of production of O_2 and H_2 in two compartments. This is the photoelectrocatalytic (PEC) approach.

An artificial leaf should thus be composed of the following main elements:

- An *anode* exposed to sunlight carrying a photocatalyst able to oxidize water and supported on a conductive substrate that allows fast collection of the electrons and is permeable to protons, in order to transport the electrons and protons to the cathode side;
- A *membrane* enabling fast transport of protons with a minimum transfer resistance, good contact with the anode and cathode sites, and an effective barrier action to O_2 diffusion;
- A *cathode*, which is formed in the simplest approach by a conductive substrate (in contact with the membrane and permeable to protons) containing active

centers for proton and electron recombination to H_2. In a more challenging approach, the cathode contains centers able to chemisorb CO_2 and convert it catalytically (in the presence of electrons and protons) to fuels or valuable chemicals.

An alternative for the anode side is to have an electrocatalyst able to perform water electrolysis, with the electrons supplied by semiconductor, preferably active in the visible region of the sunlight and in direct contact with the electrocatalyst. This alternative solution is preferable to the first case, because the issues related to charge separation are in principle lower. Sensitizers for the semiconductor, if robust enough to anodic oxidation conditions and not quenching the water oxidation processes, allow extension of the range of wavelengths of activities (*antenna* effect).

1.3.1   PEC Cells for CO_2 Conversion

A number of proposals have been made recently for artificial leaf solar cells, although often they are still at a conceptual level and their feasibility of realization and behavior have not been proved [64]. Most of them are designed only for producing H_2 from water, and often also without separate production of O_2 and H_2. In principle, an artificial leaf-type PEC solar cell for converting CO_2 using sunlight requires a different design [40]. One of the few examples reported specifically for this objective has a design close to that of commercial PEM fuel cells, to take advantage of the large knowledge portfolio on their engineering and mass/charge transfer optimization. The scheme of the solar fuel cell is shown in Figure 1.4, which also shows the practical realization of the cell that is currently under testing [63,65].

The cathodic part operates in gas phase, because this (i) simplifies the recovery of the reaction products (they can be collected by cooling the gas outlet from the flowing cell), (ii) allows continuous operations and the use of large concentrations of CO_2, (iii) greatly reduces mass transfer limitations, and especially (iv) changes the type of products that are formed.

The electrochemical reduction of CO_2, in both aqueous or organic electrolytes, gives mainly C1 products of reduction and is limited in productivity by the solubility of CO_2 in the electrolyte and the mass transfer in solution [66]. Conversely, gas-phase operations allow formation of >C2 products [21,63,65] (never detected in liquid-phase operations) and eliminate the problem of solubility of CO_2 and mass transfer in liquid phase, as well as avoiding the formation of a double layer and related effects. The electrocatalyst, however, should be different. Instead of using conventional electrodes, the solar PEC cell shown in Figure 1.4 uses an electrode based on metal nanoparticles dispersed over conductive doped carbon nanotubes (CNT) and then deposited over a carbon-cloth (CC) conductive material acting as electron transport net [40,67]. With a Fe/N-CNT-based gas diffusion electrode, it was shown that isopropanol is formed, being the main reaction product of the CO_2 electrocatalytic reduction [21,63].

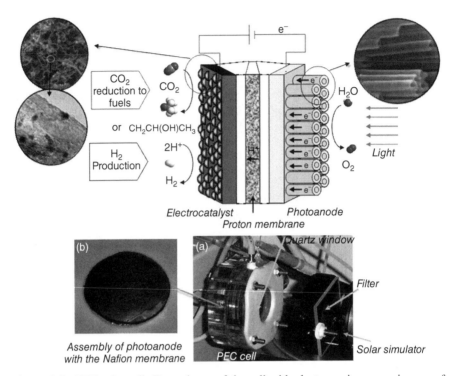

Figure 1.4 PEC solar cell. *Top*: scheme of the cell with electron microscopy images of an example of the TiO_2-nanotube array electrode and of the Fe-nanoparticles on N-doped carbon nanotubes, used as photocatalyst for water oxidation and electrocatalyst for CO_2 reduction, respectively. It is also shown that it may be possible to use this cell for the production of H_2/O_2 in separate compartments by water photoelectrolysis. *Bottom*: photo of the experimental cell and the assembly of a photoanode with a Nafion membrane. Adapted from refs. 40,63,65.

The photoanode is instead based on an array of vertically-aligned doped titania nanotubes, produced by anodic oxidation of thin Ti layers [68], in order to meet the following demanding requirements for a porous photoanode for PEC solar cells:

- Cost-effective and easily scaled to large operations;
- Robust for stable operations;
- Having an optimal nanostructure allowing (i) enhanced light harvesting (possible over the entire sunlight spectrum, with effective use of the radiation for creating a photocurrent), (ii) low rate of charge recombination and reduced interfaces/grain boundaries, which favors the charge recombination, (iii) negligible defects and centers that favor thermal or radiative pathways (which reduce quantum efficiency), and (iv) fast transport of the electrons to a conductive substrate;
- Having a porous nanostructure that allows fast transport of protons (generated from water oxidation) to the underlying proton-conductive membrane,

avoiding surface recombination between protons and electrons (which should have different paths of transport), and an optimal interface with the membrane.

Because of self-doping during preparation and/or the creation of surface phononic heterostructures (by deposition of very small gold nanoparticles), these titania nanotubes ordered thin films are active in the visible light region [69], although still with not enough performance. Commercial Nafion is used as the proton-conductive membrane. There are still a number of problems at the interface between this membrane and the titania photo anode, and the effectiveness in the transport of protons is limited. In addition, transient measurements [63b,70] indicate the presence of significant surface quenching processes (associated to the formation of surface peroxo-species) that limit the steady-state productivity in water oxidation. This cell design could be transferred to application as an artificial leaf, but productivity of both electrodes must be improved in terms of (i) response to visible light, (ii) reducing surface self-quenching during reaction, (iii) presence of several interfaces that limit the mass/charge transfer and cell efficiency, and (iv) rate of CO_2 reduction. A new advanced cell design has been proposed and is under investigation to solve these issues [64].

1.4 CONCLUSIONS

This brief discussion of the perspectives and state of the art of production of solar fuels and chemicals from CO_2 has presented some of the trends and issues regarding the recycling of CO_2 as an enabling element for a low-carbon economy and the efficient introduction of renewable energy into the production chain. The aim was not to provide a systematic review of the state of the art but only to highlight the opportunities offered by this new vision of CO_2 recycling to solar fuels and chemicals.

After introducing the general context and the motivations, and general issues regarding the energy vectors and the problem of introducing renewable energy in the energy and chemical production chains, this chapter has analyzed two main aspects:

- The challenge of using CO_2 for the production of light olefins (ethylene, propylene) as an example of the possible reuse of CO_2 as a valuable carbon source and an effective way to introduce renewable energy into the chemical industry value chain, improve resource efficiency, and limit greenhouse gas emissions;
- The conversion of CO_2 to fuels using sunlight (solar fuels), with a concise presentation of the possible routes with some indication also of the industrial developments in the field, in order to highlight the possible routes for the storage of solar energy and discuss briefly the pro/cons and how they integrate into the existing energy infrastructure for a smooth, but fast, transition to a more sustainable energy future.

These examples show that it is possible to create CO_2-based resource-efficient chemical and energy production. Thus CO_2 must no longer be a waste, but rather an enabling factor for the effective introduction of renewable energy into the chemical production chain, and solar fuels are an effective driving force toward a more sustainable energy.

From a longer-term perspective, the objective should be the photoelectrochemical activation of CO_2 in artificial leaf-type PEC cells to fully enable the potential of solar radiation by collecting energy in the same way that natural leaves do, but in an intensified process directly producing chemicals/fuels. Because of the complexity of the problems, a fundamental understanding is the key for advancing, taking into consideration system engineering and integration. The fast advances in the development of nano-tailored materials will be key to progress in this field, but only when combined with the integration between catalysis and electrode concepts to achieve a real breakthrough in the understanding of the reaction mechanisms of these fast surface processes.

ACKNOWLEDGMENTS

This work was realized in the frame of the PRIN10-11 project "Mechanisms of activation of CO_2 for the design of new materials for energy and resource efficiency."

REFERENCES

1. Intergovernmental Panel on Climate Change web site. Accessed on Dec. 5, 2011: www.ipcc.ch.
2. International Energy Agency (IEA) (2011). *World Energy Outlook 2011*, IEA Pub: Paris (France).
3. Nauclér T, Campbell W, Ruijs T (2008). *Carbon Capture & Storage: Assessing the Economics*, McKinsey & Company, McKinsey Climate Change Initiative, Sept. 22.
4. (a) Graham-Rowe D (2008). Turning CO_2 back into hydrocarbons. *New Scientist*, March 03. (b) McKenna P (2010). Emission control: Turning carbon trash into treasure, *New Scientist*, Sept. 29. (c) Ritter SK (2007). What can we do with carbon dioxide?, *Chem. & Eng. News*, 85: 11–17.
5. (a) Aresta M (ed.) (2010). *Carbon Dioxide as Chemical Feedstock*, Wiley-VCH, Weinheim, Germany. (b) Aresta M, M Dibenedetto A (2010). *Industrial utilisation of carbon dioxide*, in *Developments and Innovation in Carbon Dioxide Capture and Storage Technology: Carbon Dioxide Storage and Utilisation*, Volume 2, Maroto - Valer M (ed.), Woodhead Publishing Limited, Abington Hall, Granta Park, Cambridge, UK, pp. 377–410. (c) Aresta M, M Dibenedetto A (2007). Utilisation of CO_2 as a chemical feedstock: opportunities and challenges, *Dalton Trans.*, 28: 2975–2992.
6. (a) Quadrelli EA, Centi G, Duplan J-L, Perathoner S (2011). Carbon dioxide recycling: emerging large-scale technologies with industrial potential, *ChemSusChem*, 4: 1194–1215. (b) Centi G, Perathoner S (2009). Opportunities and prospects in the chemical recycling of carbon dioxide to fuels, *Catal. Today*, 148: 191–205. (c) Centi G,

Perathoner S (2004). Heterogeneous catalytic reactions with CO_2: status and perspectives, *Studies in Surface Science and Catal.*, 153: 1–8. (d) Centi G, Quadrelli EA, Perathoner S (2013). Catalysis for CO_2 conversion: a key technology for rapid introduction of renewable energy in the value chain of chemical industries. *Energy Environ. Sci.*, 2013, DOI: 10.1039/C3EE00056G.

7. (a) Peters M, Köhler B, Kuckshinrichs W, Leitner W, Markewitz P, Müller TE (2011). Chemical technologies for exploiting and recycling carbon dioxide into the value chain, *ChemSusChem*, 4: 1216–1240. (b) Peters M, Müller T, Leitner W (2009). CO_2: from waste to value, *Tce*, 813: 46–47.

8. Mikkelsen M, Jorgensen M, Krebs FC (2010). The teraton challenge. A review of fixation and transformation of carbon dioxide, *Energy & Env. Science*, 3: 43–81.

9. (a) Olah GA, Goeppert A, Prakash GK Surya (2009). Chemical recycling of carbon dioxide to methanol and dimethyl ether: from greenhouse gas to renewable, environmentally carbon neutral fuels and synthetic hydrocarbons, *J. Org. Chem.*, 74: 487–498. (b) Olah GA, Goeppert A, Surya Prakash GK (2009). *Beyond Oil and Gas: The Methanol Economy*, 2nd Edition, Wiley-VCH, Weinheim, Germany.

10. Sakakura T, Choi J-C, Yasuda H (2007). Transformation of carbon dioxide, *Chem. Rev.*, 107: 2365–2387.

11. Ma J, Sun N, Zhang X, Zhao N, Xiao F, Wei W, Sun Y (2009). A short review of catalysis for CO_2 conversion, *Catal. Today*, 148: 221–231.

12. Song C (2006). Global challenges and strategies for control, conversion and utilization of CO_2 for sustainable development involving energy, catalysis, adsorption and chemical processing, *Catal. Today*, 115: 2–32.

13. Whipple DT, Kenis PJA (2010). Prospects of CO_2 utilization via direct heterogeneous electrochemical reduction, *J. Phys. Chem. Lett.*, 1: 3451–3458.

14. Dorner RW, Hardy DR, Williams FW, Willauer HD (2010). Heterogeneous catalytic CO_2 conversion to value-added hydrocarbons, *Energy & Env. Science*, 3: 884–890.

15. Jiang Z, Xiao T, Kuznetsov VL, Edwards PP (2010). Turning CO_2 into fuel, *Phil. Trans. of the Royal Society, A: Math., Phys. & Eng. Sciences*, 368: 3343–3364.

16. BayNews (The Bayer Press Server), Accessed on Dec. 5, 2011: http://www.press.bayer.com/baynews/baynews.nsf/id/F7D95A62E946F894c0125783A002F2CB5

17. Agarwal AS, Zhai Y, Hill D, Sridhar N (2011). The electrochemical reduction of carbon dioxide to formate/formic acid: engineering and economic feasibility, *ChemSusChem*, 4: 1301–1310.

18. Carbon Recycling International web site. Accessed on Dec., 5th, 2011: www.carbonrecycling.is.

19. Resource-efficient Europe—Flagship initiative of the Europe 2020 Strategy. Accessed on Dec. 5th, 2011: ec.europa.eu/resource-efficient-europe.

20. Quadrelli EA, Centi G (2011). Green Carbon Dioxide, *ChemSusChem*, 4, 9: 1179–1181.

21. (a) Centi G, Perathoner S (2011). CO_2-based energy vectors for the storage of solar energy, *Greenhouse Gases: Science and Techn.*, 1: 21–35. (b) Centi G, Perathoner S (2010). Towards solar fuels from water and CO_2, *ChemSusChem*, 3: 195–208. (c) Centi G, Perathoner S, Passalacqua R, Ampelli C (2012). *Solar production of fuels from water and CO_2*. In N. Z. Muradov & T. N. Veziroglu (Eds.), *Carbon-Neutral Fuels and Energy Carriers*, Boca Raton, FL (US): CRC Press (Taylor & Francis Group), pp. 291–323.

22. Scholes GD, Fleming GR, Olaya-Castro A, van Grondelle R (2011). Lessons from nature about solar light harvesting, *Nature Chem.*, 3: 763–774.

23. Haije W, Geerlings H (2011). Efficient production of solar fuel using existing large scale production technologies. *Env. Science & Techn.*, 45: 8609–8610.

24. Moore GF, Brudvig GW (2011). Energy conversion in photosynthesis: a paradigm for solar fuel production, *Ann. Review of Cond. Matter Phys.*, 2: 303–327.

25. Cogdell RJ, Brotosudarmo THP, Gardiner AT, Sanchez PM, Cronin L (201). Artificial photosynthesis—solar fuels: current status and future prospects, *Biofuels*, 1: 861–876.

26. Roy SC, Varghese OK, Paulose M, Grimes CA (201). Toward solar fuels: photocatalytic conversion of carbon dioxide to hydrocarbons, *ACS Nano*, 4: 1259–1278.

27. Parsons W (2009). *Strategic Analysis of the Global Status of Carbon Capture and Storage*. Global CCS Institute, Canberra, Australia.

28. Calera Co. web site. Accessed on Dec. 5th, 2011: www.calera.com.

29. (a) Kolstad C, Young D (2010). Cost Analysis of Carbon capture and storage for the Latrobe Valley, Donald Bren School of Environmental Science and Management, University of California, Santa Barbara. Accessed on Dec. 5th, 2011: fiesta.bren.ucsb .edu/~kolstad/HmPg/papers/CCS%20Costs%20Latrobe.pdf. (b) Zaelke D, Young O, Andersen SO (2011). Scientific Synthesis of Calera Carbon Sequestration and Carbonaceous By-Product Applications, Donald Bren School of Environmental Science and Management, University of California, Santa Barbara. Accessed on Dec. 5th, 2011: www.igsd.org/climate/documents/Synthesis_of_Calera_Technology_Jan2011.pdf

30. Thybaud N, Lebain D (2010). Panorama des voies de valorisation du CO_2, Report for l'Agence de l'Environnement et de la Maîtrise de l'Energie (Ministère de l'Ecologie, de l'Energie, du Développement Durable et de la Mer - ALCIMED, France.

31. Fukuoka S, Fukawa I, Tojo M, Oonishi K, Hachiya H, Aminaka M, Hasegawa K, Komiya K (2010). A Novel non-phosgene process for polycarbonate production from CO_2: green and sustainable chemistry in practice, *Catal. Surveys from Asia*, 14: 146–163.

32. Novomer web site. Accessed on Dec. 5th, 2011: www.novomer.com.

33. Rubin E, Meyer L, de Coninck Heleen (2005). IPCC Special Report, Carbon Dioxide Capture and Storage, Cambridge University Press, New York.

34. Nozik AJ, Miller J (2010). Introduction to solar photon conversion, *Chem. Reviews*, 110: 6443–6445.

35. Sanborn D (2011). Electrochemical technologies: at the summit of civilization's energy conversion and storage evolution—where they will stay, *Int. J. Hydrogen Energy*, 36: 9401–9404.

36. Züttel A, Borgschulte A, Schlapbach L (2008). *Hydrogen as a Future Energy Carrier*, Wiley-VCH: Weinheim (Germany).

37. Farrauto RJ (2009). Building the hydrogen economy, *Hydrocarbon Eng.*, 14: 25–30.

38. Nowotny J, Sheppard LR (2007). Solar-hydrogen, *Int. J. Hydrogen Energy*, 32: 2607–2608.

39. Bockris OM (2008). Hydrogen no longer a high cost solution to global warming: New ideas, *Int. J. Hydrogen Energy*, 33: 2129–2131.

40. Centi G, Perathoner S (2011). Nanostructured Electrodes and Devices for Converting Carbon Dioxide Back to Fuels: Advances and Perspectives. In L. Zang (ed.), *Energy*

Efficiency and Renewable Energy Through Nanotechnology, Springer-Verlag; London, UK, pp. 561–584.

41. Centi G, Iaquaniello G, Perathoner S (2011). Can We Afford to Waste Carbon Dioxide? Carbon Dioxide as a Valuable Source of Carbon for the Production of Light Olefins. *ChemSusChem*, 4: 1265–1273.

42. International Energy Agency (2010). *Energy Technology Perspectives 2010*, OECD/IEA, Paris (France).

43. Centi G, van Santen RA (2007). *Catalysis for Renewables*, Wiley-VCH: Weinheim, Germany.

44. Stöcker M (2008). Biofuels and biomass-to-liquid fuels in the biorefinery: Catalytic conversion of lignocellulosic biomass using porous materials, *Angew Chemie Int. Ed.*, 47: 9200–9211.

45. Gallezot P (2008). Catalytic conversion of biomass: challenges and issues, *ChemSusChem*, 1: 734–737.

46. Huber GW, Iborra S, Corma A (2006). Synthesis of transportation fuels from biomass: chemistry, catalysts, and engineering, *Chem. Rev.*, 106: 4044–4098.

47. Stephenson AL, Kazamia E, Dennis JS, Howe CJ, Scott SA, Smith AG (2010). Life-cycle assessment of potential algal biodiesel production in the united kingdom: a comparison of raceways and air-lift tubular bioreactors, *Energy Fuels*, 24: 4062–4077.

48. Eroglu E, Melis A (2011). Photobiological hydrogen production: recent advances and state of the art, *Bioresource Techn.*, 102: 8403–8413.

49. Ghirardi ML, Mohanty P (2010). Oxygenic hydrogen photoproduction-current status of the technology, *Current Science*, 98: 499–507.

50. Atsumi S, Higashide W, Liao JC (2009). Direct photosynthetic recycling of carbon dioxide to isobutaraldehyde, *Nature Biotechn*, 27: 1177–1180.

51. Carbon Science web site. Accessed on Dec. 5th, 2011: www.carbonsciences.com.

52. (a) Joo O-S, Jung K-D, Jung Y (2004). Camere Process for methanol synthesis from CO_2 hydrogenation, *Studies in Surface Science and Catal*, 153: 67–72. (b) Choi M-J, Cho D-H (2008). Research Activities on the Utilization of Carbon Dioxide in Korea, *CLEAN–Soil, Air, Water*, 36: 426–432.

53. Mitsui Chemicals web site. Accessed on Dec. 5th, 2011: www.mitsuichem.com/csr/report/pdf/14_15_mk09en.pdf.

54. Hirano M, Tatsumi M, Yasutake T, Kuroda K (2007). Dimethyl ether synthesis via reforming of steam/carbon dioxide and methane, *J. Japan Petroleum Inst.*, 50: 34–43.

55. Dufour J, Serrano DP, Gàlvez J. Moreno J, Gonzàlez A (2011). Hydrogen production from fossil fuels: life cycle assessment of technologies with low greenhouse gas emissions, *Energy Fuels*, 25: 2194–2202.

56. Djomo SN, Blumberga D (2011). Comparative life cycle assessment of three biohydrogen pathways, *Bioresource Techn.*, 102: 2684–2694.

57. Spath PL, Mann MK (2004). Life Cycle Assessment of Renewable Hydrogen Production via Wind/Electrolysis, NREL/MP-560-35404 Report. Accessed on Dec 5th, 2011: www.nrel.gov/ docs/fy04osti/35404.pdf.

58. Utgikar V, Thiesen T (2006). Life cycle assessment of high temperature electrolysis for hydrogen production via nuclear energy, *Int. J. Hydrogen Energy*, 31: 939–944.

59. Gibson TL, Kell NA (2008). Optimization of solar powered hydrogen production using photovoltaic electrolysis devices, *Int. J. Hydrogen Energy*, 33: 5931–5940.

60. Millet P, Ngameni R, Grigoriev SA, Mbemba N, Brisset F, Ranjbari A, Etievant C (2010). PEM water electrolyzers: from electrocatalysis to stack development, *Int. J. Hydrogen Energy*, 35: 5043–5052.

61. Centi G, Iaquaniello G, Perathoner S (2011). Can we afford to waste carbon dioxide? carbon dioxide as a valuable source of carbon for the production of light olefins, *ChemSusChem*, 4: 1265–1273.

62. (a) Ferenc J (2008). Breakthroughs in hydrogen storage—formic acid as a sustainable storage material for hydrogen, *ChemSusChem*, 1: 805–808. (b) Leitner W (1995). Carbon dioxide as a raw material: the synthesis of formic acid and its derivatives from CO_2, *Angew. Chemie Int. Ed.*, 34: 2207–2221.

63. (a) Gangeri M, Perathoner S, Caudo S, Centi G, Amadou J, Begin D, Pham-Huu C, Ledoux MJ, Tessonnier JP, Su DS, Schlögl R (2009). Fe and Pt carbon nanotubes for the electrocatalytic conversion of carbon dioxide to oxygenates. *Catal. Today*, 143: 57–63. (b) Ampelli C, Centi G, Passalacqua R, Perathoner S (2010). Synthesis of solar fuels by a novel photoelectrocatalytic approach. *Energy Environ. Sci.*, 3: 292–301.

64. Bensaid S, Centi G, Garrone E, Perathoner S, Saracco G (2012). Artificial leaves for solar fuels from CO_2, *ChemSusChem*, 5: 500–521.

65. Centi G, Perathoner S, Wine G, Gangeri M (2007). Electrocatalytic conversion of CO_2 to long carbon-chain hydrocarbons. *Green Chem.*, 9: 671–678.

66. (a) Lvov SN, Beck J., LaBarbera MS (2012). Electrochemical Reduction of CO_2 to Fuels. In T. N. V. N. Z. Muradov (Ed.), *Carbon-Neutral Fuels and Energy Carriers*, CRC Press (Taylor & Francis Group), pp. 363–400. (b) Gattrell M, Gupta N, Co A (2006). A review of the aqueous electrochemical reduction of CO_2 to hydrocarbons at copper. *J. Electroanal. Chem.*, 594: 1–19. (c) Bockris JOM, Wass J C. (1989). The photoelectrocatalytic reduction of carbon dioxide. *J. Electrochem. Soc.* 136: 2521–2528.

67. Centi G, Perathoner S (2009). The role of nanostructure in improving the performance of electrodes for energy storage and conversion. *Eur. J. Inorg. Chem.*, 26: 3851–78.

68. (a) Centi G, Passalacqua R, Perathoner S, Su DS, Weinberg G, Schlögl R (2007). Oxide thin films based on ordered arrays of one-dimensional nanostructure. A possible approach toward bridging material gap in catalysis. *Phys. Chem. Chem. Phys.*, 9: 4930–4938. (b) Centi G, Perathoner S (2009). Nano-architecture and reactivity of titania catalytic materials. Part 2. Bidimensional nanostructured films. *Catalysis*, 21: 82–130. (c) Ampelli C, Passalacqua R, Perathoner S, Centi G, Su DS, Weinberg G (2008). Synthesis of TiO_2 thin films: Relationship between preparation conditions and nanostructure. *Top. Catal.*, 50: 133–144. (d) Perathoner S, Passalacqua R, Centi G, Su DS, Weinberg G (2007). Photoactive titania nanostructured thin films. Synthesis and characteristics of ordered helical nanocoil array. *Catal. Today*, 122: 3–13.

69. Centi G, Perathoner S (2012). Nanostructured titania thin films for solar use in energy applications. In G. Rios, G. Centi & N. Kanellopoulos (Eds.), *Nanoporous Materials for Energy and the Environment*, Pan Stanford Pub.: Singapore, pp. 257–282.

70. (a) Genovese C, Ampelli C, Perathoner S, Centi G (2013). Electrocatalytic conversion of CO_2 to liquid fuels by using nanocarbon-based electrodes, *J. Energy Chem.*, 22:202–213. (b) Ampelli C, Passalacqua R, Genovese C, Perathoner S, Centi G (2011). A novel photo-electrochemical approach for the chemical recycling of carbon dioxide to fuels, *Chem. Eng. Trans.*, 25:683–688. (c) Ampelli C, Passalacqua R, Perathoner S, Centi G (2009). Nano-engineered materials for H_2 production by water photo-electrolysis, *Chem. Eng. Trans.*, 17:1011–1016.

CHAPTER 2

Transformation of Carbon Dioxide to Useable Products Through Free Radical-Induced Reactions

G. R. DEY

2.1 INTRODUCTION

The reduction of carbon dioxide (CO_2), as reversal of the oxidative degradation of organic materials, is a great challenge. Conversion of CO_2 into useful substances is necessary in developing alternative fuels and/or various raw materials for different industries. Additionally, this will assist in preventing the rise in tropospheric temperature due to the greenhouse effect, which at present is known as global warming.

Solar energy is the earth's ultimate power supply. All energy forms, except geothermal or nuclear, such as fossil fuels, biomaterials, hydropower, and wind, are just transformations of solar energy. The ultimate goal is to demonstrate that artificial photosynthesis may be replicated via photoreduction of CO_2 employing a suitable photocatalyst to produce hydrocarbons (mainly methane) and methanol. In this way, solar energy can be transformed and stored in the form of chemical energy. Moreover, methanol is the most promising photolytically reduced product

Green Carbon Dioxide: Advances in CO₂ Utilization, First Edition.
Edited by Gabriele Centi and Siglinda Perathoner.
© 2014 John Wiley & Sons, Inc. Published 2014 by John Wiley & Sons, Inc.

from CO_2 following free radical reactions. It can be transformed into other useful chemicals with conventional chemical technologies and can easily be transported and used as fuel, approximating renewable energy. In addition to the advantage of energy storage, this process can contribute to the fixation of CO_2 and prevent the increase in atmospheric concentration of CO_2.

Concurrently, electrochemical methods also have had significant application in the chemical reduction of CO_2 through free radical-induced reactions. Utilization of dissolved CO_2 and/or gaseous phase CO_2 for the generation of CO, HCHO, etc. is one of the successful applications of electrochemical technology for clean environmental programs.

In this chapter a short review of the recent developments in the field of CO_2 chemical reduction is presented. The main topics of discussion include (i) the photochemical methods (using photocatalyst, mainly TiO_2) and (ii) the electrochemical methods to produce different intermediates and products such as CO, HCHO, HCOOH, CH_3OH, CH_4, etc.

2.1.1 Background

It is worthwhile to recall some general information regarding the physical and chemical characteristics of CO_2. CO_2 is a colorless, odorless gaseous molecule consisting of two oxygen atoms and a carbon atom with a linear arrangement. It does not have significant absorption in the $-UV-vis$ regions. However, the absorption spectrum of CO_2 exhibits a gradual increase in absorption beyond 175 nm and reaches a maximum value around 135 nm. At λ_{max}, the absorption band is more diffused with a low absorption coefficient value (~ 20 cm^{-1}) [1,2]. Nevertheless, being a linear molecule, CO_2 is infrared (IR) sensitive because of some of its active vibrations, which produce an oscillating dipole.

The IR spectrum of CO_2 is shown in Figure 2.1. There are two different kinds of IR active vibrations in the molecule, stretching and bending (two each) modes, making a total of four types of vibrations (total number of vibration = $3n$-5, where n represents the number of atoms existing in the molecule). Symmetrical stretching vibration does not produce any change in the dipole moment of the molecule and hence is not shown by IR spectroscopy, whereas the asymmetrical stretching vibration causes activity in the molecule, resulting in infrared absorption in the 2349 cm^{-1} region. The deformation vibrations (two bending vibrations) of CO_2 having identical frequencies are said to degenerate and appear at the same region with doubly degenerate 667 cm^{-1} as shown in Figure 2.1 [3–8].

CO_2 is nontoxic, and it can easily be stored, transported, and handled without hazard. However, 1% (10,000 ppm) CO_2 will make some people feel drowsy. Moreover, at a high concentration of $\sim 5\%$ by volume, it is toxic to humans and animals; on inhalation, it causes dizziness and visual and hearing dysfunction because it binds with hemoglobin. Finally, at $\sim 10\%$ by volume it poses a danger of asphyxiation [9].

In natural atmosphere, CO_2 is present at around 0.03% and its solubility is about 90 cm^3 per 100 ml of water. In aqueous solution, it exists in many forms such

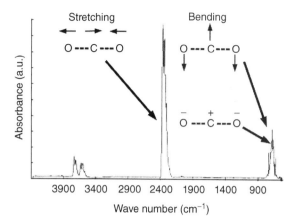

Figure 2.1   IR spectrum of CO_2. Adapted from ref. 5.

as $CO_2(g), CO_2(aq), H_2CO_3(aq), HCO_3^-(aq)$, and $CO_3^{2-}(aq)$ [10,11]. Solution in water changes the solution pH, reaching maximum acidity of up to pH ~3.76, the pK_{a1} of H_2CO_3. On dissolution of CO_2, the pH of the resulting solution decreases, depending on the extent of concentration of CO_2 dissolved. The solubility of CO_2 decreases with increase in temperature and decrease in CO_2 pressure [12,13].

Plants play the most significant role in the fixation of CO_2 on the Earth through a natural cyclic process. CO_2 is used to produce food through photosynthesis, in which chlorophyll present in green plants converts CO_2 to carbohydrate ($C_6H_{12}O_{11}$) in the presence of water and sunlight. Plants feed animals, and the decay of plants and/or animals leads to the release of CO_2 into the atmosphere. Another portion of CO_2 from oceans/seas is used by aquatic marine life. Understanding the dissolution behavior of CO_2 in ocean/seawater is complicated. Solubility of CO_2 is a strong inverse function of seawater temperature (i.e., solubility is greater in cooler water); since deep water (that is, seawater in the ocean's interior) is formed under the same surface conditions that promote CO_2 solubility, it contains a higher concentration of dissolved inorganic carbon $[CO_2(aq), H_2CO_3(aq), HCO_3^-(aq), CO_3^{2-}(aq)]$ than one might otherwise expect. Consequently, these two processes act together to pump carbon from the atmosphere into the ocean's interior [14,15]. After marine plants and animals decay, it is converted to fossil fuels, which on use/burning returns to the atmosphere as well as to marine water as soluble CO_2 [14,15]. A schematic diagram of overall processes in carbon fixation (CO_2 cycle) is presented in Figure 2.2.

In the Earth's history of CO_2 levels, the amount of CO_2 in the atmosphere has varied significantly. It is believed that the Earth's early atmosphere was probably composed mostly of CO_2 [16,17]. At the end of the last ice age (approximately 14,000 years ago), the level of CO_2 in the air increased to about 50%. A rise in global temperature indicated a significant role for CO_2 regarding the same. After this global climatic transition, the atmospheric CO_2 concentration remained almost constant at about 280 ppm until the end of the eighteenth century [9]. Since then,

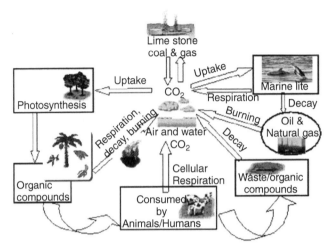

Figure 2.2 Schematic presentation of CO_2 cycle.

the anthropogenic (man-made) emission of CO_2 from various activities such as burning of fossil fuels (coal, gas, and oil used in transport and heating buildings, etc.), deforestation, waste incineration, and the manufacture of cement altogether have disturbed the equilibrium between natural sources and CO_2 sinks. All living things also give off CO_2, but, in general, plants capture as much CO_2 as animals and microorganisms generate (Fig. 2.2). CO_2 emission has been rising sharply in the atmosphere since 1900 through the aforementioned activities, which cannot be captured by green plants. Some other natural sources of CO_2 include decay of dead animals and plant matter, volcanic eruptions, evaporation from oceans, and respiration. Consequently, the concentration of CO_2 in the air has increased to about 380 ppm and continues to increase at the rate of about 1.2 ppm each year [18]. At present, CO_2 concentration stands at 390 ppm, which has grown with a rate of about 2 ppm per year since 2009 [19]. Because of CO_2 emissions since the beginning of the Industrial Revolution nearly 200 years ago, the strength of the Earth's natural greenhouse effect has been enhanced. CO_2 is the largest individual contributor to the enhanced greenhouse effect, accounting for about 60% of the increase in heat trapping [17]. Berner showed the large variation of CO_2 with peaks, over geological time, according to the estimation done with GEOCARB modeling [18].

A considerable rise in abnormal weather events such as flooding, summer droughts, etc., has been seen in recent years and is correlated with the global climatic change. Meteorologists and scientists believe that the most likely cause of the changes is the anthropogenic emission of the "greenhouse gases" that can trap heat in the Earth's atmosphere. Greenhouse gases such as H_2O and CO_2 in the Earth's atmosphere decrease the escape of terrestrial thermal infrared radiation. Increasing CO_2 in the environment therefore effectively increases radiative energy input to the earth. The hypothesis that the increase in atmospheric CO_2 is related to apparent changes in the climate has been tested with modern methods of time-series analysis. The average global temperature is increasing at the rate of $0.0055 \pm 0.00096°C$ per year, and the temperature rise and atmospheric CO$_2$ have been significantly correlated over the past few decades [20–24].

2.2 CHEMICAL REDUCTION OF CO$_2$

Interest in CO$_2$ chemical reduction has increased because of its potential use as a C1 source, the increase of its concentration in the atmosphere, and efforts to mimic photosynthetic carbon assimilation [25]. Chemical reduction of CO$_2$ with different energy inputs such as light energy through photochemical reactions using semiconductor photocatalysts [26–32], high-energy ionizing radiation (radiolysis) [33–37], electrochemical reactions (electrolysis) [38–42] (discussed below), sonolysis [43–45], and dielectric barrier discharge [46–50], etc. has been studied by researchers all over the world in the last decades. This chapter focuses on two different techniques: (i) photochemical and (ii) electrochemical methods for the chemical reduction of CO$_2$.

It is useful to consider the thermodynamic data, for example, the heat of formation of CO$_2$ = 336 kcal mol^{-1} [51] and its electron affinity = -0.6 ± 0.2 eV [52], in order to understand the stepwise reduction shown below:

$$CO_2(aq) + 2H^+ + 2e^- \rightarrow HCOOH \tag{2.1}$$

$$HCOOH + 2H^+ + 2e^- \rightarrow HCHO + H_2O \tag{2.2}$$

$$HCHO + 2H^+ + 2e^- \rightarrow CH_3OH \tag{2.3}$$

$$CH_3OH + 2H^+ + 2e^- \rightarrow CH_4 + H_2O \tag{2.4}$$

From a thermodynamic perspective, transforming one mole of CO$_2$ into a mole of HCOOH or HCHO or CH$_3$OH or CH$_4$ requires 2 or 4 or 6 or 8 electrons as shown in Eqs. (2.1)–(2.4) [53].

2.2.1 Photochemical Reduction of CO$_2$

Photochemical reduction of CO$_2$ may be considered the reverse reaction of the widely utilized photochemical degradation of organic species [54–60], which has been studied in a more extensive way. It is thus useful to recall some aspects of the photochemical processes occurring on photocatalysts to destroy organic pollutants through oxidation. Various photocatalysts such as TiO$_2$, ZnO, SnO$_2$, WO$_3$, Fe$_3$O$_4$, CdS, and ZnS have been employed in different forms such as loose particles, anchored particles, and suspended particles, etc. TiO$_2$ is the most common photocatalyst used in a range of applications for environment-related problems. The band gap energy (E$_{bg}$) between conduction and valence bands of TiO$_2$ is ~3 eV (E$_{bg}$ for anatase is 3.2 eV and that of rutile, another titania crystalline form, is 3 eV) [28,60]. On photo-irradiation (if E$_{hv}$ > E$_{bg}$; where E$_{hv}$ is the energy of the incident photo light), TiO$_2$ produces an electron–hole pair (charge carrier). The preliminary step in TiO_2 photocatalytic oxidation is the formation of both hydroxyl radicals ($^{\bullet}OH$) ($E^{\circ} = 2.72$ V, a strong oxidant) [61,62] and superoxide radical anions ($O_2^{\bullet-}$) ($E^{\circ} = -0.33$ V, a weak reductant) [63] through the reactions of the photo-generated electron–hole pair with solutes (water and air) present in the system (see Eqs. (2.5)–(2.7)) [55]. The generation of $O_2^{\bullet-}$ requires oxygen in the system during photolysis:

$$TiO_2 \text{ -hv} \rightarrow e^- + h^+(\text{electron-hole pair}) \tag{2.5}$$

$$h^+ + H_2O \rightarrow {}^{\bullet}OH + H^+ + TiO_2 \tag{2.6}$$

In the presence of air,

$$e^- + O_2 \rightarrow O_2^{\bullet-} \tag{2.7}$$

Rothenberger and coworkers [64] have studied the absorption spectrum of the trapped electron, an intermediate produced after picosecond laser light excitation of colloidal TiO_2. The recombination coefficient between the trapped electrons and the free holes has been reported to be $(3.2 \pm 1.4) \cdot 10^{-11}$ cm^{3}s^{-1}[64,65]. The generation of $^{\bullet}OH$ in photo-activated TiO_2 has been confirmed with the formation of hydroxylated intermediates of halogenated aromatic compounds during photocatalytic degradation [65] and, more recently, in benzene transformation to phenol using TiO_2 photocatalyst [27]. In the latter system, the phenol yields in benzene photolysis depend on various parameters such as oxygen concentration, amount of TiO_2 catalyst, amount of benzene, and ambient conditions. It is understood that $^{\bullet}OH$ radicals alone cannot convert benzene to phenol to a large extent. The overall reaction mechanism means that the initial reaction of hole with benzene is significant for the enhancement of phenol yields. Phenol has been separated from semiconductor suspended solutions through an ultrafiltration method, which also explains that the hole (h^+)-generated free $^{\bullet}OH$ radicals easily migrate to liquid phase from solid catalyst surface and react with benzene, yielding phenol after subsequent reaction with O_2. Alternatively, it is also possible that the surface $^{\bullet}OH$ radical migrates to aqueous phase before or after its reaction with benzene. These two processes are difficult to distinguish. Finally, it has been resolved that in TiO_2 suspended aqueous system on photo-irradiation $^{\bullet}OH$ radicals generated are available in aqueous phase as unbound/free reacting species. These radicals can be utilized to convert aromatic compounds to their respective hydroxylated products for specific end use. Moreover, EPR studies have also confirmed the existence of hydroxyl and hydro-peroxy radicals in aqueous solutions of light-illuminated TiO_2 [57,66,67].

In the absence of oxygen, $^{\bullet}OH$ radical and e^-, having divergent properties, are the reactive species (with very high recombination capacity completing within ~ 10 ns [57]) generated on photolysis of suspended TiO_2 in aqueous medium. All processes including photolytic generation of charge carriers within conduction and valence bands along with redox processes taking place therein on the TiO_2 semiconductor catalyst are presented in Figure 2.3. In the presence of an $^{\bullet}OH/h^+$ scavenger (mostly alcohols), this system is converted into a reducing condition, whereas, on

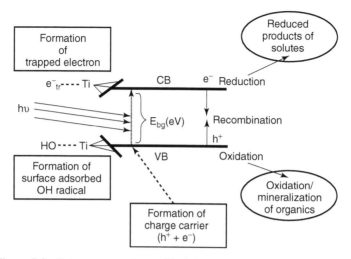

Figure 2.3   Primary steps along with their reactions in photolysis of TiO$_2$.

addition of an electron scavenger (generally N$_2$O or O$_2$), oxidation due to the reaction with the hole occurs. In this way, an absolute reduction or oxidation condition of the TiO$_2$ photocatalyst system can be achieved.

UV–vis absorption spectra of thin films containing nanocrystalline anatase TiO$_2$ particles of different sizes have been reported previously [68], in which the absorption band due to TiO$_2$ exhibited red shift with increase in particle size. The UV–vis absorption spectrum of TiO$_2$ as a suspension in water exhibits absorption ≤ 400 nm. To convey the clear spectral properties, the UV–vis absorption spectra of TiO$_2$ on surface-deposited films on quartz surfaces (curves a and b) and suspended in water (curve c) are shown in Figure 2.4, which exhibits an absorption peak in the UV region starting with low absorption around 380 nm.

Size of the particle plays a significant role in its photocatalytic activity [54]. Zang et al. [69] have suggested that the pure TiO$_2$ grains with diameter ranging from 11 to 21 nm have a maximum photocatalytic efficiency. Absorption spectra of the catalysts exhibit a blue shift for smaller TiO$_2$ particles (<10 nm), indicating a size quantization effect, which is operating in extremely small-size particles. However, if the particle size is about 5–10 nm, surface recombination of electron–hole pairs is more prominent, resulting in low photocatalytic efficiency.

Commercially available as well as laboratory-synthesized TiO$_2$ photocatalysts in different forms have been used for CO$_2$ reduction [26,28–31,70–75] to generate different products such as HCOOH, HCHO, methanol, hydrocarbons, CO, etc. Extensive research on photochemical reduction of CO$_2$ has been carried out in the recent past to produce the above-mentioned products, which depend on various factors. Some important results are summarized below.

2.2.1.1 Effects of Photocatalysts.
The photocatalytic reduction of CO$_2$ with H$_2$O on various TiO$_2$ forms such as highly dispersed, anchored, and finely

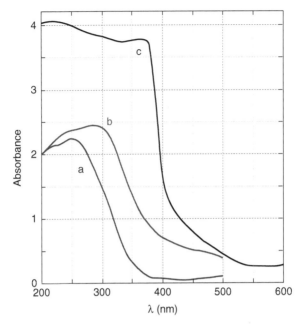

Figure 2.4 UV–vis absorption spectra for (a, b) different types of TiO$_2$ films on fused quartz surfaces and (c) suspended TiO$_2$ particles in water. Curves a and b are reprinted with permission from ref. 68.

powdered varieties leads to the formation of CH$_4$, CH$_3$OH, and CO [76]. Their formation efficiencies depend on various factors such as the ratio of H$_2$O and CO$_2$, the kind of catalyst, and the reaction temperature. An increase in temperature and H$_2$O-to-CO$_2$ ratio enhances the catalytic activity, while an excess of H$_2$O inhibits the reaction rate. The total yields of CH$_4$, CH$_3$OH, and CO (under UV irradiation) are larger at 323 K than at 275 K. In situ spectroscopic study of the system indicated that the photocatalytic reduction of CO$_2$ with H$_2$O is related to the formation of reduced TiO$_2$ surface sites generated by UV irradiation. The charge transfer excited state (Ti^{3+}-O$^-$)* of tetrahedral coordinated titanium oxide species shows the higher reactivity [70].

The nature of the photocatalytic products depends on the crystalline plane of the catalyst. For example, photo-irradiation of CO$_2$ in the presence of H$_2$O on TiO$_2$ (100) gives a high yield of CH$_4$ and CH$_3$OH, while a TiO$_2$ (110) surface gives only CH$_3$OH with low yield.

2.2.1.2 *Effects of CO$_2$ Concentration/Pressure.*

CO$_2$ concentration has a significant effect on the yield of the different photo-generated products. As discussed above, in both gas and liquid phases, its concentration varies because of increase in CO$_2$ pressure. The liquid phase includes both liquid and its solution; solvent is used in the latter case for CO$_2$ intake. The solubility of CO$_2$ depends on the nature of the solvents. The different products and yields reported in CO_2 reduction with variations of CO_2 concentration/pressure are discussed below.

Use of CO_2 at Ambient Pressure. Low yields of methane have been reported on photoreduction of CO_2 over TiO_2 suspension in water [77]. However, metal doping of titania leads to an increase in the methane yield. The highest yield has been reported with Pd-doped TiO_2 (Pd-TiO_2). Interestingly, acetic acid was produced in substantial amounts when photocatalysts such as Rh-TiO_2, Cu-TiO_2, and Ru-TiO_2 were used for CO_2 reduction. In another report, Dey et al. [26] have noted a low yield of methane during photolysis of ambient-pressure CO_2 in the presence of suspended TiO_2 in water. Notably, on addition of 2-propanol (a hole scavenger) into the system, the CH_4 yield became 2.5 times higher.

High-Pressure CO_2. At high-pressure CO_2 (~2.5 MPa pressure), acetic acid (CH_3COOH), methanol (CH_3OH), and formic acid ($HCOOH$) in the liquid phase and methane (CH_4) as a major product with ethane (C_2H_6) and ethylene (C_2H_4) (low yield) in the gas phase have been observed as photolytic products on TiO_2 suspended aqueous solutions. In the presence of $NaOH/Na_2SO_4$, the yield of total gaseous product was enhanced. At the same time, H_2 yield decreased with photolysis time, which is due to the consumption of H_2 either by oxidation with a hole or by the reaction with transient intermediates of CO_2 reduction leading to the production of hydrocarbons and alcohols [31]. In other photoreduction studies [30] at high pressure of CO_2, methane or other photolysis products were not produced in the absence of any electron donor. However, in the presence of 2-propanol (a hole scavenger), methane has been found as a photoreduction product.

Increase in CO_2 pressure accelerates the formation of an intermediate $^{\bullet}C$-radical (detected/identified through ESR measurements) [30,54]. This radical species is the intermediate during the photo-irradiation to form methane through the mechanism presented in Figure 2.5. Interestingly, the photocatalytic reduction of liquid CO_2 in the presence of TiO_2 powders gives rises to formic acid as the main photolysis product. In this system, formic acid is generated through protonation of the carbon dioxide anion radical (first intermediate, see Fig. 2.5) [29].

In Situ Generated CO_2. The presence of gases such as N_2 and O_2 is important for chemical reduction of CO_2 [26] in the presence of 2-propanol wherein CO_2 has been generated (except in a CO_2-purged system) in the system itself. The variation of the yields of CH_4, in N_2-purged, aerated, O_2, and CO_2-purged systems gives rise to a very interesting chemistry. Methane yields under the study have been reported as follows:

methane yield in CO_2-purged = aerated > oxygenated > N_2-purged systems

A higher yield of CH_4 in aerated and CO_2-saturated systems in contrast to O_2 and N_2-purged systems highlights the importance of the 2-propanol-containing systems. 2-Propanol (acting primarily as a hole scavenger) reacts with h^+,

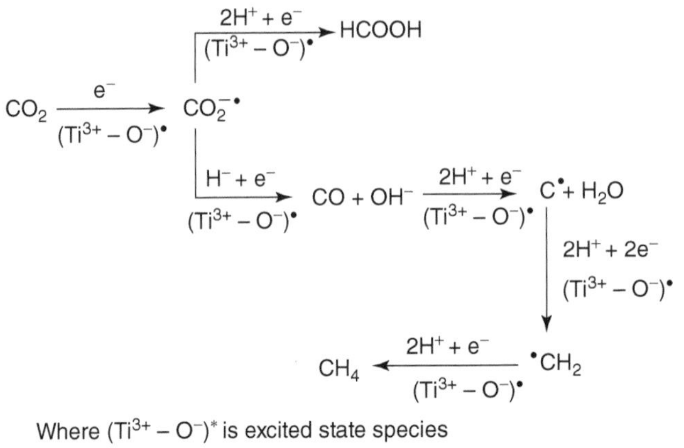
2H⁺ + e⁻
(Ti3+ – O⁻)•
HCOOH
CO2
e⁻
(Ti3+ – O⁻)•
CO2⁻•
H⁻ + e⁻
(Ti3+ – O⁻)•
CO + OH⁻
2H⁺ + e⁻
(Ti3+ – O⁻)•
C• + H2O
2H⁺ + 2e⁻
(Ti3+ – O⁻)•
CH4
2H⁺ + e⁻
(Ti3+ – O⁻)•
•CH2
Where (Ti3+ – O⁻)* is excited state species

Figure 2.5 Reaction mechanism showing the generation of methane through carbon ($^{\bullet}C$) radical.

generating an organic radical [$(CH_3)_2C^{\bullet}OH$] through H abstraction. In N_2-purged systems, the absence of oxygen prevents the mineralization of 2-propanol to CO_2 to a greater extent. In other words, under this condition a small amount of CO_2 has been produced during photolysis. The mineralization of organic compounds/alcohols is enhanced in the presence of a photocatalyst [78]. Thus the CO_2 generated in situ because of mineralization of alcohol on the surface of the photocatalyst performs an important role in the formation of CH_4 in systems other than those that are CO_2-purged. In the case of an oxygenated system (~100% oxygen), where the concentration of oxygen is ~5 times higher than in aerated system (20% oxygen), the reported CH_4 yield is relatively low [26]. This suggests the involvement of photo-generated e^- in CH_4 formation. Therefore, the yield of methane under such reaction conditions depends on both the oxygen and the hole scavenger concentrations.

In an oxygenated system, most e^- react with O_2 (due to high O_2 concentration), generating $O_2^{\bullet-}$ species rather than reducing CO_2, with consequent low CH_4 yields. Under this condition, O_2 helps in faster and enhanced mineralization of organic compounds. On the other hand, in the aerated system, although CO_2 yield is lower than that in the oxygenated system, relatively more photo-generated e^- are available for photoreduction of CO_2 because of lower competition of O_2 in free radical-induced reactions. The surface adsorption of CO_2 is higher than that of O_2, and the surface-adsorbed CO_2 has a better chance of getting reduced to CH_4. Nevertheless, in such systems, it is difficult to distinguish the surface-adsorbed CO_2 from the in situ generated CO_2 at the surface of the photocatalyst. In another set of experiments, the reduction of CH_3OH does not lead to the formation of CH_4, suggesting that the pathway for methane formation is not through the methanol route under such experimental conditions [71].

As discussed above, in the CO_2-purged TiO_2 photocatalyst suspended aqueous system, the environment is vital to control of the system for the production of different photolytic products. H$_2$ yields are different in the presence or absence of 2-propanol in the TiO$_2$ suspension in water under different ambient conditions. The yield of H$_2$ is usually higher in the N$_2$-purged (deoxygenated) system containing 2-propanol compared with the system without 2-propanol. 2-Propanol, acting as a hole scavenger, inhibits the h$^+$-e$^-$ recombination reaction, the reverse of reaction 2.5, which is normally very fast and completed within 10–100 ns [57].

Photo-generated holes (as discussed above) are also considered as $^\bullet$OH radicals or surface-attached $^\bullet$OH radicals generated on TiO$_2$ photolysis in an aqueous environment. The $^\bullet$OH radical reaction with 2-propanol is very fast in condensed media. The high concentration of 2-propanol (>0.1 M) makes the system more favorable for reducing conditions by rapid scavenging $^\bullet$OH radicals/holes and subsequently allows photo-generated e$^-$ to react with H$^+$ to yield more H$_2$. Because of the negligible reaction between e$^-$ and 2-propanol, the wastage of e$^-$ was nil because of 2-propanol. A similar mechanism was explained above for CH$_4$ formation during chemical reduction of CO$_2$. Low yield of H$_2$ in the system without 2-propanol has been explained as due to the recombination reaction (reverse of reaction 2.5), which is more favorable in the absence of any added hole scavenger. This emphasizes the necessity of a hole scavenger for the enhancement of the reduction reaction and product yields.

Also, in oxygen-containing systems such as aerated and O$_2$-purged systems, H$_2$ has not been observed in the absence of 2-propanol. In this case, O$_2$ present in both aerated and O$_2$-purged systems ranging from 0.2 to 1.2 mM is good enough to scavenge the photo-generated e$^-$, preventing H$_2$ formation. This way O$_2$ present in these systems acts as an e$^-$ scavenger following reaction 2.7 and thereafter produces H$_2$O$_2$.

N$_2$O is also a well-known electron scavenger [79], and the methane yield decreases considerably on changing the ambient air, O$_2$, N$_2$, or CO$_2$ to N$_2$O. H$_2$ yield is nil in the N$_2$O-purged system; moreover, in the presence of 2-propanol, H$_2$ yield has been identical to that in other O$_2$-containing systems [80]. This suggests that even in the presence of 2-propanol (having efficient h$^+$ scavenging properties), a fraction of photo-generated e$^-$ takes part in the reduction reaction, resulting in a low yield of H$_2$ other than its scavenging reaction due to O$_2$ or N$_2$O. As mentioned elsewhere [81], the H$_2$ formation in the presence of methanol is proceeding through photocatalytic reactions of methanol. A few important reactions involving various products during chemical reduction of CO$_2$ are listed in Table 2.1. Furthermore, the H$_2$ generation in such systems takes place through reduction of protons by photo-generated e$^-$, not through photochemical/photodecomposition reactions of solute molecules. H$_2$ generation has also been reported in photocatalytic reduction of CO$_2$ on TiO$_2$ (0.5 g/80 ml) suspension in aqueous solution containing 4 M methanol, and its yield is enhanced on metal powder addition[82]. The decrease in H$_2$ production in photocatalytic reduction of CO$_2$ (at high-pressure CO$_2$) in the presence of 0.2 N NaOH has been explained previously as due to the formation of hydrocarbons [31]. The explanations for the decreasing trends of H$_2$ in both the CO$_2$-purged systems with and without 2-propanol are given as follows:

TABLE 2.1   List of Significant Reactions with Their Rate Constant Values in Condensed Phase

Reactions	Rate constant ($M^{-1}s^{-1}$)	References
$e^- + N_2O \rightarrow N_2 + O^{\bullet-}$	9.1×10^9	87
$O^- + H_2O \rightarrow OH^- + {}^{\bullet}OH$	9.4×10^7	87
${}^{\bullet}OH + (CH_3)_2CHOH \rightarrow (CH_3)_2C^{\bullet}OH + H_2O$	1.9×10^9	87
${}^{\bullet}OH + H_2 \rightarrow H_2O + H$	3.4×10^7	87
$e^- + H^+ \rightarrow H$		87
$H + H \rightarrow H_2$	5×10^9	88
$e^- + O_2 \rightarrow O_2^{\bullet-}$	1.8×10^{10}	89
$O_2^{\bullet-} + H^+ \rightarrow HO_2$	5×10^{10}	90
$2HO_2 \rightarrow H_2O_2 + O_2$	1×10^6	91
$e^- + CO_2 \rightarrow CO_2^{\bullet-}$		92
$CO_2^{\bullet-} + O_2 \rightarrow CO_2 + O_2^{\bullet-}$	4.2×10^9	90
$CO_2^{\bullet-} + CO_2 + e^- \rightarrow CO + CO_3^{2-}$		
${}^{\bullet}CH_3 + (CH_3)_2CHOH \rightarrow products$	3.4×10^3	84
$CO + H^{\bullet} \rightarrow {}^{\bullet}CHO$	1.7×10^7	93
$CO_2^{\bullet-} + CO_2^{\bullet-} \rightarrow products$	6.5×10^8	94
$CO_2 + e^- \rightarrow CH_4 (via\ CO\ intermediate)$		
$2CO_2^{\bullet-} + 2H^+ \rightarrow (COOH)_2 \rightarrow 2CO_2 + H$		
$h^+ + (CH_3)_2CHOH \rightarrow (CH_3)_2C^{\bullet}OH(OR) + H^+$	Similar to ${}^{\bullet}OH$ reaction	
$2(CH_3)_2C^{\bullet}OH \rightarrow Products$		95
$(CH_3)_2C^{\bullet}OH + O_2 \rightarrow (CH_3)_2C(OH)O_2^{\bullet}$	4.5×10^9	96
$CO_2^- + e^- + H_2O \rightarrow HCO_2^- + OH^-$	$\sim 1 \times 10^9$	97
$CO_2^{\bullet-}{}_{(ads)} + H^+ + e^- \rightarrow CO + OH^-$		
$CH_3^{\bullet} + O_2 \rightarrow CH_3O_2^{\bullet}$	4.1×10^9	98
$CH_3^{\bullet} + CH_3^{\bullet} \rightarrow C_2H_6$		99

OR = Organic radical

1. In the absence of 2-propanol, H_2 is consumed mainly for the generation of hydrocarbons such as methane, etc. The yield of CH_4 increases with photolysis time [26].
2. In systems both with and without 2-propanol, H_2 yield is reduced through its reaction either with h^+ or ${}^{\bullet}OH$ radical (see Table 2.1), and the generated H atom must be consumed by either CO_2 or both CO_2 and 2-propanol and/or photodegradation products of 2-propanol in addition to minor H_2 regeneration through H-H recombination reaction.

The methane yield decreases when N_2, air, or O_2 is replaced with N_2O in 2-propanol-containing aqueous TiO_2 suspended systems, which shows the involvement of photo-generated e^- in product (CH_4) formation under photo catalytic reduction of in situ generated CO_2. The methane formation in the photo-irradiated aerated/O_2-purged system containing 2-propanol has been explained previously [26]. In brief, the generation of methane via methyl radical (${}^{\bullet}CH_3$) formation has been ruled out because:

1. Methane yield has been reported to be higher in the aerated system in which •CH$_3$ if formed should produce peroxymethyl radical (CH$_3$O$_2$•) after its reaction with oxygen and subsequently undergo mineralization to CO$_2$.

2. As discussed above, 2-propanol has negligible absorbance at ≥ 300-nm light; hence the experimental light, i.e., 350 ± 50 nm (or if it is >300-nm light) is not favoring •CH$_3$ generation through direct photodissociation.

3. In the N$_2$O-purged system the yield of CH$_4$ decreases

4. •CH$_3$, if generated, should react with either water or alcohol (H source) present in the system to generate CH$_4$ through the H-abstraction reaction.

The generation of •CH$_3$ radical has been studied thoroughly in aqueous solution; therefore, its reaction with water is either very slow or negligible [83,84]. On the other hand, H-abstraction reactions of •CH$_3$ with alcohols are too slow. The low reaction rate constant for •CH$_3$ with 2-propanol, 3.4×10^3 dm^{3}mol^{-1} s^{-1} [83,84], is not favoring H-atom abstraction under the experimental condition (especially in an aerated/oxygenated system) to generate methane. Again, the possibility of CH$_4$ formation caused by photo-Kolbe reaction (the Kolbe reaction is a decarboxylation reaction of carboxylic acids yielding hydrocarbons) has been ruled out, because in this case •CH$_3$ radical is produced and it undergoes dimerization to produce ethane or abstracts H to yield CH$_4$ depending on the reaction conditions. The most significant part of the photo-Kolbe reaction is the ambient condition; according to Kraeutler and Bard, the system should be either in an inert (helium) atmosphere or under a high-vacuum condition [85,86]. Nevertheless, in the chemical reduction of CO$_2$ in presence of 2-propanol, different environments (air and oxygenated systems) including N$_2$-purged environments (an inert atmospheric condition) have been employed. The interesting observation highlighted in the reported study is the methane yield, which is higher when the system contains air or oxygen than in the inert atmosphere (N$_2$ in their case). According to Kraeutler and Bard, the traces of oxygen suppress the hydrocarbon yield [85,86]. Therefore, methane generation does not follow through the decarboxylation of carboxylic acids produced as an intermediate during the mineralization of alcohol (2-propanol).

It is noteworthy to mention here that if •CH$_3$ generates, then a trace of ethane should also be reported as a product in gas phase analysis. As the observation is dissimilar, it has been confidently concluded that the methane has not been generated through •CH$_3$ radical intermediate under their experimental conditions [71]. A very low yield of CH$_4$ has been reported in the photocatalytic redox reaction of methanol in the presence of TiO$_2$ under diverse ambient conditions [26,71]. This pointed out that the methanol intermediate-generating pathway:

$$CO_2 \rightarrow HCOOH \rightarrow HCHO \rightarrow CH_3OH \rightarrow CH_4$$

is not the CO$_2$ reduction route under this condition.

The formation rates and the yields of CO and CH$_4$ in N$_2$O-purged 2-propanol aqueous solution containing TiO$_2$ suspensions have been reported to be different [72]. The higher formation rate and the yield of CO compared with CH$_4$ under identical conditions indicate that the all of the CO generated during CO_2 reduction does not convert to CH_4 instantly. Methane yield in N_2O-purged system is quite low compared with CO_2-purged, aerated, oxygenated, and N_2-purged systems. In the CO_2-purged system containing TiO_2 suspension in water in the presence of 2-propanol, the yields of CO and CH_4 and their formation rates are higher than those obtained in the system without 2-propanol. CO generation has been known previously in TiO_2 system but in different systems, and its yield depends mainly on the reaction conditions [71,72,100]. The generation of methane in these systems has been explained previously as due to the free radical-induced reduction reaction of CO_2. In the absence of 2-propanol, the yield of CH_4 is less because of the prominent recombination reaction (reverse of reaction 2.5), and part of the photo-generated e^- is only able to react with CO_2, yielding CO and CH_4. In this system the generation of CO reveals that it is an intermediate produced on CO_2 (only source of carbon in the system) photocatalytic reduction through either reduction or photodecomposition reaction. The former process is more likely since the latter does not exist because of negligible absorption of experimental light (≥ 350 nm) by CO_2. Therefore, the reaction route is through a CO intermediate:

$$CO_2 \rightarrow CO \rightarrow C^\bullet \rightarrow CH_2 \rightarrow CH_4$$

This is why CO yield remains higher than the end-product CH_4.

The above-mentioned reaction pathway has also been supported by the results reported especially for the low methane yields during the photolysis of formic acid (HCOOH) and formate ion ($HCOO^-$) in the presence of TiO_2 [101]. In aerated systems, CO yield is higher in contrast to the deoxygenated (Ar-purged) systems under identical conditions. It has been proposed that the formation of CO takes place during the chemical reduction of in situ generated CO_2, a photo-mineralized product of $HCOOH/HCOO^-$, but not through the direct photodecomposition or photo-dehydration ($CO + H_2O$) of solute molecules as the photo-light ≥ 350 nm used was not absorbed by the solute molecules. Under the study, the emission of H_2 was also observed and its yield was significantly higher in the Ar-purged system as compared to CO yields.

During photocatalytic studies on the system containing 2-propanol, acetone has also been observed as a photocatalytic product along with the above-mentioned products (CO_2, CH_4, CO, and H_2). Organic radical (OR) generated contributes in the reduction reaction because of its high reduction potential ($E°(CH_3)_2{}^\bullet COH/(CH_3)_2CO, H^+ = -1.7$ V) [102] to a lesser extent due to its other successive decay reactions. The formations of both CO and CH_4 support the previously elucidated mechanism for the process of CO_2 transformation to methane through CO and $C^\bullet$ radical [30,70].

2.2.2 Electrochemical Reduction of CO_2

Electrochemistry deals with the conversion of chemical energy to electrical energy and vice versa. Electrochemical oxidation or reduction of compounds, at well-controlled electrode potentials, by adding or withdrawing electrons (free radicals) offers many interesting possibilities in environmental engineering. A typical electrochemical cell [103] consists of three electrodes (one working electrode, one counterelectrode, and one reference electrode), a conducting medium (electrolyte), and cell boundaries with optional supporting facilities for heating/cooling of the entire cell and stirring of electrolyte to achieve the desired experimental temperature and homogeneous distribution of reactants, respectively. In addition, one gas inlet and an outlet with a pressure monitoring facility (pressure gauge) should also be included/attached to obtain and/or to maintain the desired experimental pressure. To avoid any metal or metal-related contamination due to autoclave materials, the inner part of the electrochemical cell is normally made up of glass/quartz or, alternatively, the inner surface of the cell is made inert by the use of Teflon coatings. The quantitative information about the working electrochemical cell has been obtained from the potential difference between counter- and working electrodes as measured against reference electrode (in many cases a calomel electrode is used as reference electrode), while external current is applied to the cell.

In electrochemistry, the reduction products yields are expressed in terms of faradaic efficiency and energy efficiency. The fuel cell by definition is a device that converts chemical energy from a fuel into electricity through chemical reaction with oxygen or another oxidizing agent. In such cases, the desired electrochemical (faradaic) utilization of the reactants can be <100% than that of the theoretical coulombic capacity, which is defined as faradaic efficiency (η_F) [104]:

$$\eta_F = \frac{\text{Coulombs obtained}}{(\text{Total molecules consumed}) \times nF} \qquad (2.8)$$

Where F represents Faraday constant (96,500 coulombs/mol), and n is the number of electrons. Electrochemical reduction of CO$_2$ leads to different products such as HCOOH, HCHO, CO, and CH$_4$. The energy efficiency, ε, for HCOOH can be expressed [105] as:

$$\varepsilon = \frac{\Delta H \times (\text{percent faradaic efficiency})}{nF \, (E_{rev} + \eta c + \eta a + \eta IR)} \qquad (2.9)$$

Where $\Delta H = 64.4$ kcal/mol; $E_{rev} = 1.43$V, for the reaction

$$CO_2(g) + H_2O(1) \rightarrow HCOOH(aq) + O_2(g)$$

and n = number of electrons transferred; F = faraday constant; η_c, η_a, and η_{IR} represent concentration, activation, and ohmic overpotentials, respectively.

As discussed above, CO$_2$ has been considered as a viable energy storage medium [39]. A method for the production of methanol through electrochemical reduction of CO$_2$ is given, which in fact can be fed to a fuel cell or a decomposer to generate hydrogen. The methanol generation by this process is believed to be superior to a hydrogen energy system because of its low cost and convenience of storage and transport. Many investigations of electrochemical reduction of CO_2 have been carried out to resolve the environmental problems, in addition to exploring for an alternate fuel [38–42,106–108].

Transformation of CO_2 to different products is a promising long-term objective in the preparation of fuels and/or raw materials/chemicals. It is important to have information regarding the redox potentials of the various couples involving CO_2 and its reduction products; accompanying reactions with redox potential values are shown below.

$$CO_2 + 2H^+ + 2e^- \rightarrow HCOOH \qquad (E^{\circ\prime} = -0.61 \text{ V}) \qquad (2.10)$$

$$CO_2 + 2H^+ + 2e^- \rightarrow CO + H_2O \qquad (E^{\circ\prime} = -0.52 \text{ V}) \qquad (2.11)$$

$$CO_2 + 4H^+ + 4e^- \rightarrow HCHO + H_2O \qquad (E^{\circ\prime} = -0.48 \text{ V}) \qquad (2.12)$$

$$CO_2 + 6H^+ + 6e^- \rightarrow CH_3OH + H_2O \qquad (E^{\circ\prime} = -0.38 \text{ V}) \qquad (2.13)$$

$$CO_2 + 8H^+ + 8e^- \rightarrow CH_4 + 2H_2O \qquad (E^{\circ\prime} = -0.24 \text{ V}) \qquad (2.14)$$

Although the value for $CO_2/CO_2^{\bullet-}$ redox potential is -2.21 V vs. SCE [109,110], $E^{\circ\prime}$ (standard electrode potential) becomes less and less negative as the reaction uses more and more electrons for reduction.

In electrochemical processes, the products obtained through reduction reactions depend on the electrode material/electrode type, supporting electrolyte, etc. Products reported are, for instance, CO, HCOOH, CH_4, C_2H_4, $(CO_2)_2^{2-}$, CH_3OH, C_2H_5OH, and $(CH_3)_2CO$. Extensive studies on electrochemical reduction of CO_2 has been carried out in the recent past to produce different products such as hydrocarbons, CO, and formic acid [38–42,106–108], which are summarized below.

2.2.2.1 *Effects of Electrode Materials and Electrode Types.* Noda et al.

[107] have carried out electrochemical reduction of CO_2 using various metal electrodes (square plate in shape) in aqueous $KHCO_3$ electrolyte. Among the electrodes, the highest catalytic activity for the production of hydrocarbons, alcohols, and aldehydes has been observed when Cu metal was employed as an electrode [41,107]. Similarly, high yields of $HCOO^-$ have been reported when Hg, In, Sn, and Pb metal were used as electrodes under 1 atm CO_2 pressure in aqueous solutions. Likewise, CO has been found as major product when Ag metal was used as an electrode. For metals such as Ti, V, Nb, Ta, Cr, Mo, W, Re, Fe, Co, Ni, Ru, Ir, and Pt, the faradaic efficiencies for the evolution of hydrogen (H_2) are approximately 100%, in contrast to Pd (29%). This decrease in H_2 in the case of Pd inferred that a portion of H_2 has been adsorbed on the electrode itself [107]. CO_2 has also been reduced to formic acid electrolytically at an amalgamated cathode [38]. Data listed in Table 2.2 show how the electrode material is important for the production of

TABLE 2.2 Electrochemical Reduction of CO_2 under a Pressure of 30 atm on Various Electrodes at 163 mA · cm^{-2}

Group	Elcetrode	ε^3/V	CH$_4$	C$_2$H$_6$	C$_2$H$_4$	CO	HCOOH	H$_2$	CO$_2$red.[b]	Total	PCD(CO$_2$ red.)[a]/mA cm^{-2}
							Faradaic efficiency/%				
4	Ti	−1.57	0.18	0.01	0.08	Trace	4.6	80.8	4.9	85.7	8.0
	Zr	−1.73	0.13	0.01	0.01	32.5	7.6	44.2	40.3	84.5	65.7
5	Nb	−1.45	0.56	0.05	0.01	n[4]	3.5	81.4	4.1	85.5	6.7
	Ta	−1.51	0.55	0.05	Trace	Trace	7.6	74.4	8.2	82.6	13.4
6	Cr	−1.49	0.53	0.05	0.07	11.8	8.2	68.6	20.7	89.3	33.7
	Mo	−1.34	0.40	0.05	0.03	n	6.5	83.3	7.0	90.3	11.4
	W	−1.61	0.38	0.04	0.01	Trace	31.9	53.1	32.3	85.4	52.6
7	Mn	−1.69	0.68	0.10	0.06	2.8	2.8	78.8	6.5	85.3	10.6
8	Fe	−1.63	2.03	0.40	0.16	4.2	28.6	51.6	35.4	87.0	57.7
9	Co	−1.54	3.09	0.17	0.38	15.8	21.9	46.9	41.5	88.4	67.6
	Rh	−1.41	0.26	0.03	0.01	61.0	19.5	13.1	80.8	93.9	131.7
	Ir	−1.55	0.62	0.05	0.05	17.5	22.3	48.3	40.5	88.8	66.0
10	Ni	−1.59	0.72	0.08	0.11	33.5	31.3	26.0	65.7	91.7	107.1
	Pd	−1.56	0.13	0.01	Trace	46.1	35.6	12.8	81.3	94.6	133.3
	Pd[c]	−1.76	0.21	0.01	0.02	35.2	44.0	13.8	79.4	93.2	397.0
	Pt	−1.48	0.22	0.02	Trace	6.1	50.4	33.6	56.7	90.3	92.4
11	Cu	−1.64	9.95	0.06	3.74	20.1	53.7	2.5	87.6	90.1	142.8
	Ag	−1.48	0.20	0.01	Trace	75.6	16.8	3.9	92.6	96.5	150.9
	Au	−1.30	0.21	0.02	0.11	64.7	11.8	15.4	76.8	92.2	125.2
12	Zn	−1.70	0.31	0.03	Trace	48.7	40.5	2.8	89.5	92.3	145.9
13	AJ	−1.97	0.66	0.01	n	n	1.3	86.5	2.0	88.5	3.3
	In[f]	-	0.28	Trace	0.04	3.8	90.1	5.6	90.5	99.1	147.5

(continued)

41

TABLE 2.2 *(Continued)*

Group	Elcetrode	ε^3/V	CH$_4$	C$_2$H$_6$	C$_2$H$_4$	CO	HCOOH	H$_2$	CO$_2$red.[b]	Total	PCD(CO$_2$ red.)[a]/mA cm^{-2}
							Faradaic efficiency/%				
14	C^g	−1.68	0.45	0.03	0.04	44.0	30.2	15.6	74.7	90.3	37.4
	C	−2.14	0.66	0.02	0.05	3.6	6.8	75.5	11.2	86.7	18.3
	π-Si	−2.04	0.87	0.01	0.02	2.0	46.3	40.6	49.2	89.8	80.2
	Sn	−1.39	0.06	Trace	Trace	8.0	92.3	1.3	100.4	101.7	163.0
	Pb	−1.57	0.20	0.01	Trace	Trace	95.5	1.2	95.7	96.9	156.0
15	Bi^h	−1.42	0.17	0.01	Trace	3.3	82.7	6.3	86.2	92.5	140.5

Reaction temperature, 25°C; electrolyte, 0.1 mol dm^{-3} KHCO$_3$; charge passed, 300 C.
[a] Corrected with an *IR* compensation instrument (vs. AgIAgCl).
[b] Total Faradaic efficiency for CO$_2$ reduction.
[c] Partial current density for CO$_2$ reduction.
[d] Not detected.
[e] Current density, 500 mA cm^{-2}.
[f] Current density, 200 mA cm^{-2}.
[g] Current density, 50 mA cm^{-2}.
[h] Current density, 150 mA cm^{-2}.
Reprinted with permission from ref. 103.

various products, especially in electrochemical reduction of CO_2 in high-pressure conditions. For example, at high-pressure CO_2, the faradaic efficiencies for the formation of hydrocarbon and HCOOH are higher when a Cu electrode is used (see Table 2.2).

Russell et al. [105] have also pointed out how modification of the electrode can enhance the yields of reduction products. The faradaic efficiency for the yield of CH_3OH on HCOOH reduction using a Pb electrode and a 0.1 M formic acid and 0.1 M perchloric acid mixed electrolyte is 12%. On replacing Pb with electro-etched Sn, the efficiency can be achieved to 100%, showing the effectiveness of the electrode and electrolyte. Ikeda et al. [42] studied electrochemical reduction of CO_2 on a sintered ZnO electrode in aqueous solution and compared the result with that observed in the case of a Zn electrode system.

A higher catalytic activity for CO formation (faradaic efficiency 70% at -1.4 V Ag/AgCl in 0.1 mol dm^{-3} KHCO$_3$ solutions) has been observed at a ZnO electrode as compared to Zn electrode system. More recently, Le et al. [111] have examined the yield behavior of an electrodeposited cuprous oxide thin film, and they explored a relationship between surface chemistry and reaction behavior comparative to air-oxidized and anodized Cu electrodes. CH_3OH yields (43 µmol·cm^{-2} h^{-1}) and faradaic efficiencies (38%) reported at cuprous oxide electrodes are remarkably higher than air-oxidized or anodized Cu electrodes in the reduction of CO_2. They have suggested a critical role of Cu(I) species in selectivity to CH_3OH formation with a higher yield [111]. In addition, single-crystal electrodes such as Cu(111), Cu(100), Cu(S)-[n(100) × (111)], and Cu(S)-[n(100) × (110)] show significant catalytic activities on electrochemical reduction of CO_2 for generation of divergent products, especially C2 compounds. For example, the Cu(111) electrode yields mainly CH_4 from CO_2, and the Cu(100) favorably gives C_2H_4. Introduction of (111) steps to Cu(100) basal plane, leading to Cu(S)-[n(100) × (111)] orientations, significantly promoted C_2H_4 (C2 compound) formation, suppressing CH_4 yield [112].

2.2.2.2 *Effects of Supporting Electrolytes.* Various electrolytes such as Na_2SO_4, K_2SO_4, $KHCO_3$, K_2CO_3, $HClO_4$, $NaClO_4$, $NaCl$, Na_2SO_4, H_3PO_4, and tetraethyl ammonium perchlorate (TEAP) [38–41,105] have been used for CO_2 reduction to realize better selectivity of product formation. Many studies in this direction have been carried out to show the significance of individual electrolytes for selective products with high yield efficiency. As reported, with modification of the electrolyte at ambient temperature and pressure, the formation of CO_2 reduction products can be enhanced with respect to faradaic efficiency. For example, on changing the electrolyte from K_2CO_3 to $KHCO_3$, HCOOH has been detected as a predominant product in the working potential ranges (from -1.5 to -1.8 V vs. SCE) for both Pb and Sn electrodes [113].

2.2.2.3 *Effect of Reaction Media/Solvents.* The electrochemical reduction of CO_2 has also been reported on application of nonaqueous electrolytes such as propylene carbonate, acetonitrile, dimethyl sulfoxide (aprotic solvents) [41,114].

The reason behind the use of aprotic solvents is their high solubility for CO_2 [115] and also the objective of realizing different chemistry if possible with such solvents because they possess different properties such as dielectric constant, viscosity, and density compared with water. Oxalic acid, an important product, has been observed when Pb, Hg, and Tl were used as working electrodes. On the other hand, CO was formed when the electrode material was changed to In, Zn, Au, or Sn in TEAP/propylene carbonate electrolyte. The reaction pathways suggested that the formation of different products in the reduction of CO_2 depends on the electrode materials and the electrolytes of the electrolysis cell. Therefore, the experimental conditions, the nature of the electrode materials, as well as the applied current density are the most responsible parameters for controlling the selectivity and productivity of specific reduction products.

It is useful to note that Hori et al. [40] have investigated the reduction of CO using Cu metal electrode and correlated the results with those observed in CO_2 reduction under identical conditions. The products on CO reduction at the Cu electrode have suggested that the electrochemical reduction of CO_2 might have proceed via CO or CO-derived intermediates, viz., CH_4.

2.2.2.4 *Effects of CO_2 Pressure.* The faradaic efficiency for the formation of formic acid on Ag and Au electrodes at 1 atm CO_2 pressure is very small, ~0.8% [116]. However, it increases to 12% when CO_2 pressure is increased to 30 atm. The main reduction products at Cu electrodes are formic acid and CO, with faradaic efficiencies of 54% and 20%, respectively [103]. The total faradaic efficiency for the hydrocarbon formation is 14%. It is well known that Cu produces hydrocarbons such as methane and ethylene efficiently. The selectivity of the reduction products on Cu electrodes depends strongly on current density and CO_2 pressure[117], which is clearly understandable from Table 2.3.

2.2.2.5 *Effect of Temperature.* Azuma et al. [118] have reported that at a low temperature (2°C) the reduction efficiency increased dramatically on a Ni electrode compared with that at room temperature on Ni and Fe metal, where H_2 evolution occurs predominantly. At 2°C, the amounts of CO and CH_4 increase linearly with the charge on the Ni electrode during CO_2 reduction. In this case, the total current efficiency for CO_2 reduction exceeds 30%.

2.2.2.6 *Effect of Reactors.* Recently, several reactors with different designs have been used for CO_2 reduction to HCOOH [119–126]. Many of these designs have been based on a fuel cell in which mostly polymer electrolyte membranes were used to separate the anode and cathode. More recently, Whipple et al. [125] have designed a microfluidic reactor successfully, which enables rapid evaluation of catalysts under different operating conditions to reduce CO_2 to HCOOH.

Last but not least, all the processes described above for the generation of different products during chemical reduction of CO_2 through free radical-induced reactions are summarized in Figure 2.6. The scheme shows three different reaction pathways. All three reaction mechanisms lead to methane as a final product starting

TABLE 2.3   Effect of Pressure on Electrochemical Reduction of CO$_2$ on Cu Electrode Without Stirring Electrolyte

CO$_2$ (atm)	E^a (V)	Faradaic efficiency (%)									
		CH$_4$	C$_2$H$_6$	C$_2$H$_4$	C$_2$H$_4$OH	CO	HCOOH	H$_2$	HC^b	CO$_2$ red^c	Total
1	−1.57	Trace	Trace	0.01	nd^d	nd	nd	98.7	0.0	0.0	98.7
10	−1.61	2.5	0.02	0.56	nd	nd	0.8	91.8	3.1	3.9	95.7
20	−1.62	25.0	0.04	2.33	0.9	nd	3.1	58.7	28.3	31.4	90.1
30	−1.61	48.4	0.04	3.62	1.1	Trace	3.3	31.8	53.2	56.5	88.3
40	−1.63	54.4	0.08	3.34	1.1	Trace	9.5	15.9	58.9	68.4	84.3
50	−1.60	43.5	0.13	7.26	2.3	nd	8.9	11.1	53.2	62.1	73.2
60	−1.61	32.9	0.13	5.40	1.7	Trace	13.7	6.5	40.1	53.8	60.3

The reaction was carried out galvanostatically without stirring electrolyte at 25°C. Current density: 163 mA cm^{-2}. Electrolyte: 0.1 mol dm^{-3} KHCO$_3$. Working electrode: Cu wire (surface area: 0.16 cm^{-2}). Passed charge: 300 C.

^{a} vs. Ag/AgCl.
^{b} Total faradaic efficiency for hydrocarbons and alcohol formation.
^{c} Total faradaic efficiency for CO$_2$ reduction.
^{d} Not detected.
Reprinted with permission from ref. 117.

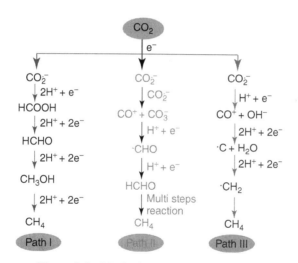

Figure 2.6   Mechanism of formation of CH$_4$.

with CO$_2$ as raw material. However, the intermediates generated are different, for example:

- Path I generates HCOOH, HCHO, CH$_3$OH
- Path II generates CO, $^{\bullet}$CHO, HCHO
- Path III generates CO, $^{\bullet}$C, $^{\bullet}$CH$_2$ after common initial intermediate CO$_2$$^{\bullet -}$ produced through radical-induced reactions between e$^-$ and CO$_2$.

Moreover, the success of the products' selection with high yield of any specific intermediate reported in this chapter from different research groups shows the impact of development in the subject. Furthermore, products such as oxalate, ethylene, and ethane. (mostly C2 and/or higher C atom-containing compounds) are formed through multistep secondary reactions, mainly dimeric and disproportionate reaction(s), on the surface of either electrodes or semiconductor surfaces.

2.3   CONCLUSIONS

It is understood that the solar energy is the Earth's unlimited power supply. The ultimate goal is to demonstrate artificial photosynthesis that may be replicated via photoreduction of CO_2 to produce hydrocarbons, methanol, etc. through free radical-induced reactions. In this way, solar energy can be transformed into chemical energy. Furthermore, methanol is the most promising reduced product from CO_2 because it can be transformed into other useful chemicals with conventional chemical technologies and can be transported and/or used easily as fuel like renewable energy. In other way, this (photocatalytic) process can contribute to the fixation of CO_2 and prevent the increase in atmospheric concentration of CO_2.

In electrochemical reduction of CO_2, there are various ways to generate different reduction products by varying the electrolytes, electrode materials, cell designs, current density, ambient, etc. Under this process also the free radical-induced reactions as discussed above are primary routes for the formation of various reduction products. Development of excellent electro-catalysts is necessary at this juncture for future growth in this field.

ACKNOWLEDGMENTS

The author expresses his sincere thanks to Rightslink, Elsevier, Cambridge University Press, The Electrochemical Society, The American Chemical Society, The American Institute of Physics, and The Materials Research Society for granting copyright permissions. He also thanks his wife, Mrs. Chaitali Dey, for her help during the preparation of this article.

REFERENCES

1. Herzberg, G. (1966). In *Molecular Spectra & Molecular Structure, III Electronic Spectra and Electronic Structure of Polyatomic Molecules*, D. Van Nostrand Co. Inc., New York, p. 500.
2. Inn, E.C.Y.; Watanabe, K.; Zelikoff, M. (1953). *J. Chem. Phys.*, 21: 1648.
3. Chatwal, G.; Anand, S. (1985). In *Spectroscopy (Atomic and Molecular)*, Eds.: M. Arora, and S. Puri, Himalaya Publishing House, Bombay, India, p. 110.

4. Dey, G.R.; Kishore, K. (2005). Carbon dioxide reduction: a brief review, in *Photo/ Electrochemistry & Photobiology in the Environment, Energy and Fuel (PE&PB in EEF)*, (Ed: S. Kaneco), Research Signpost, Kerala, India, p. 357.

5. IR spectrum of CO_2 (1996). Available from: http://www.science.widener.edu/svb/ftir /ir_co2.html

6. Martin, P.E.; Barker, E.F. (1932). *Phys.Rev.*, 41: 291.

7. Orchin, M.; Jaffe, H.H.. (1971). in *Symmetry, Orbitals and Spectra*, Wiley, New York, p. 242.

8. Paso, R.; Kauppinen, J.; Anttila, R. (1980). *J. Mol. Spectrosc.*, 79: 236.

9. Carbon dioxide (2011). Available from http://en.wikipedia.org/wiki/Carbon_dioxide

10. Chemical of the week, carbon dioxide (2008). Available from http://scifun.chem.wisc .edu/chemweek/CO_2/CO_2.html

11. Siegenthaler, U.; Sarmiento, J.L. (1993). *Nature*, 365: 119.

12. Wiebe, R.; Gaddy, V.L. (1939). *J. Am. Chem. Soc.*, 61: 315.

13. Wiebe, R.; Gaddy, V.L. (1940). *J. Am. Chem. Soc.*, 62: 815.

14. Raven, J.A.; Falkowski, P.G. (1999). *Plant Cell Environ*, 22: 741.

15. Teng, H.; Yamasaki, A.; Chun, M.-K.; Lee, H. (1997). *J. Chem. Thermodynamics*, 29: 1301.

16. Berner, R.A. (1991). *Am. J. Sci.*, 291: 339.

17. Hansen, J.E.; Lacis, A.A. (1990). *Nature*, 346: 713.

18. Berner, R.A. (1997). *Science*, 276: 544.

19. Global warming (2003). Available from http://www.elmhurst.edu/~chm/vchembook /global-warmA3.html

20. Earth system research laboratory (2011). Available from http://www.esrl.noaa.gov/ gmd/ccgg/trends/global.html

21. Kuo, C.; Lindberg, C.R.; Thornson, D.J. (1990). *Nature*, 343: 709.

22. Schneider, S.H. (2001). *Nature*, 411: 17.

23. Schneider, S.H. (2002). *Climatic Change*, 52: 441.

24. Schneider, S.H.; Chen, R.S. (1980). *Ann. Rev. Energy*, 5: 107.

25. Friedli, H.; Lotscher, H.; Oeschger, H.; Siegenthhaler, U.; Stauffer, B. (1986). *Nature*, 324: 237.

26. Dey, G.R.; Belapurkar, A.D.; Kishore, K. (2004). *J. Photochem. Photobiol. A:Chem.*, 163: 503.

27. Dey, G.R. (2009). *Res. Chem. Intermed.*, 35: 573.

28. Inoue, T.; Fujishima, A.; Konishi, S.; Honda, K. (1979). *Nature*, 277: 637.

29. Kaneco, S.; Kurimoto, H.; Ohta, K.; Mizuno, T.; Saji, A. (1997). *J. Photochem. Photobiol., A:Chem.*, 109: 59.

30. Kaneco, S.; Shimizu, Y.; Ohta, K.; Mizuno, T. (1998). *J. Photochem. Photobiol. A: Chem.*, 115: 223.

31. Mizuno, T.; Adhachi, K.; Ohta, K.; Saji, A. (1996). *J. Photochem. Photobiol. A:Chem*, 98: 87.

32. Usubharatana, P.; McMartin, D.; Veawab, A.; Tontiwachwuthikul, P. (2006). *Ind. Eng. Chem. Res.*, 45: 2558.

33. Fujita, N.; Matsuura, C.; Ishigure, K. (1990). *Corrosion*, 46: 804.

34. Fujita, N.; Matsuura, C. (1994). *Radiat. Phys. Chem.*, 43: 205.

35. Getoff, N. (1962). *Int. J. Appl. Radiat. Isotop.*, 13: 205.

36. Wang, S.; Lu, G.Q. (1999). *Ind. Eng. Chem. Res.*, 38: 2615.

37. Woods, R.J.; Pikaev, A.K. (1994). in *Applied Radiation Chemistry*, Wiley, New York.

38. Frese, K.W. Jr.; Leach, S. (1985). *J. Electrochem. Soc.*, 132: 259.

39. Hori, Y.; Suzuki, S. (1982). *Bull. Chem. Soc. Jpn.*, 55: 660.

40. Hori, Y.; Murata, A.; Takahashi, R.; Suzuki, S. (1987). *J. Am. Chem. Soc.*, 109: 5022.

41. Ikeda, S.; Takagi, T.; Ito, K. (1987). *Bull. Chem. Soc. Jpn.*, 60: 2517.

42. Ikeda, S.; Hattori, A.; Maeda, M.; Ito, K.; Noda, H. (2000). *Electrochem.*, 68: 257.

43. Harada, H. (1998). *Ultrasonics Sonochemistry*, 5: 73.

44. Henglein, A. (1985). *Z. Naturforsch*, 40b:100.

45. Ohta, K.; Suda, K.; Kaneco, S.; Mizuno, T. (2000). *J. Electrochem. Soc.*, 147: 233.

46. Dey, G.R.; Das, T.N. (2006). *Plasma Chem. Plasma Process.*, 26: 495.

47. Dey, G.R.; Ganguli, R.; Das, T.N. (2006). *Plasmonics*, 1: 95.

48. Dey, G.R.; Singh, B.N.; Kumar, S.D.; Das, T.N. (2007). *Plasma Chem. Plasma Process.*, 27: 669.

49. Eliasson, B.; Egli, W.; Kogelschatz, U. (1994). *Pure Appl. Chem.*, 66: 1275.

50. Inui, T.; Takeguchi, T. (1991). *Catal. Today*, 10: 95.

51. Glasstone, S. (1981). In *Textbook of Physical Chemistry*, M.I. Press, Madras, India, p. 112.

52. Compton, R.N.; Reinhardt, P.W.; Cooper, C.D. (1975). *J. Chem. Phys.*, 63: 3821.

53. Tennakone, K.; Jayatissa, A.H.; Punchihewa, S. (1989). *J. Photochem. Photobiol. A: Chem.*, 49: 369.

54. Anpo, M. (1989). *Res. Chem. Intermed.*, 11: 67.

55. Bahnemann, D.W. (1991). in *Photochemical Conversion and Storage of Solar Energy* (Eds.: E. Pelizzetti, and M. Schiavello), Kluwer Academic Publishers, Dordrecht, p. 251.

56. Fujishima, A.; Rao, T.N.; Tryk, D.A. (2000). *J. Photochem. Photobiol. C: Photochem. Rev.*, 1: 1.

57. Hoffmann, M.R.; Martin, S.T.; Choi, W.; Bahnemann, D.W. (1995). *Chem. Rev.*, 95: 69.

58. Memming, R. (1988). Photo electrochemical solar energy conversion, *Topics Curr. Chem*, 143: 79.

59. Mills, G.; Hoffmann, M.R. (1993). *Environ. Sci. Technol.*, 27: 1681.

60. Mills, A.; Hunte, S. Le (1997). *J. Photochem. Photobiol. A: Chem.*, 108: 1.

61. Rabani, J.; Matheson, M.S. (1964). *J. Am. Chem. Soc.*, 86: 3175.

62. Tabata, Y. (1991). in *Pulse Radiolysis*, CRC Press, Boca Raton, p. 399.

63. Bieski, B.H.J.; Cabelli, D.E.; Arudi, R.L.; Ross, A.B. (1985). *J. Phys. Chem. Ref. Data*, 14: 1041.

64. Rothenberger, G.; Moser, J.; Gratzel, M.; Serpone, N.; Sharma, D.K. (1985). *J. Am. Chem. Soc* 107: 8054.

65. Kamat, P.V. (1993). *Chem. Rev*, 93: 267.

66. Harbour, J.R.; Hair, M.L. (1979). *J. Phys. Chem.*, 83: 652.

67. Jaeger, C.D.; Bard, A.J. (1979). *J. Phys. Chem.*, 83: 3146.

68. Paz, Y.; Luo, Z.; Rabenberg L.; Heller, A. (1995). *J. Mater. Res.*, 10: 2842.

69. Zang, Z.; Wang, C.-C.; Zakaria, R.; Ying, J.Y. (1998). *J. Phys. Chem. B*, 102: 10871.

70. Anpo, M.; Yamashita, H.; Ichihashi, Y.; Ehara, S. (1995). *J. Electroanal. Chem.*, 396: 21.

71. Dey, G.R.; Pushpa, K.K. (2006). *Res. Chem. Intermed.*, 32: 725.

72. Dey, G.R.; Pushpa, K.K. (2007). *Res. Chem. Intermed.*, 33: 631.

73. Kuwabata, S.; Uchida, H.; Ogawa, A.; Hirao, S.; Yoneyama, H. (1995). *J. Chem. Soc., Chem. Commun.*, p. 829.

74. Thampi, K.R.; Kiwi, J.; Gratzel, M. (1987). *Nature*, 327: 506.

75. Tseng, I.-H.; Cheng, W.-C.; Wu, J.C.S. (2002). *Appl. Catal.*, 37: 37.

76. Hirano, K.; Inoue, K.; Yatsu, T. (1992).*J. Photochem. Photobiol., A: Chem.*, 64: 255.

77. Ishitani, O.; Inoue, C.; Suzuki, Y.; Ibusuki, T. (1993). *J. Photochem. Photobiol. A: Chem.*, 72: 269.

78. Chen, J.; Ollis, D.F.; Rulkens, W.H.; Bruning, H. (1999). *Wat. Res.*, 33: 1173.

79. Spinks, J.W.T.; Woods, R.J. (1990). In *An Introduction to Radiation Chemistry*, 3rd edition, John Wiley & Sons, New York.

80. Dey, G.R. (2007). *J. Nat. Gas Chem.*, 16: 217.

81. Kawai, T.; Sakata, T. (1980). *J. Chem. Soc. Chem. Commun.*, p. 694.

82. Hirano, K.; Asayama, H.; Hoshino, A.; Wakatsuki, H. (1997). *J. Photochem. Photobiol. A: Chem.*, 110: 307.

83. Kantrowitz, E.R.; Hoffman, M.Z.; Endicott, J.F. (1971). *J. Phys. Chem.*, 75: 1914.

84. Thomas, J.K. (1967). *J. Phys. Chem.*, 71: 1919.

85. Kraeutler, B.; Bard, A. J. (1977). *J. Am. Chem. Soc.*, 99: 7729.

86. Kraeutler, B.; Bard, A. J. (1978). *J. Am. Chem. Soc.*, 100: 5985.

87. Buxton, G.V.; Greenstock, C.L.; Helman, W.P.; Ross, A.B. (1988).*J. Phys. Chem. Ref. Data*, 17: 513.

88. Sehested, K.; Christensen, H. (1990). *Radiat. Phys. Chem.*, 36: 499.

89. Elliot, A.J.; McCracken, D.R.; Buxton, G.V.; Wood, N.D. (1990). *J. Chem. Soc. Faraday Trans.*, 86: 1539.

90. Ilan, Y.; Rabani, J. (1976). *Int. J. Radiat Phys Chem.*, 8: 609.

91. Christensen, H.; Sehested, K. (1988). *J. Phys. Chem.*, 92: 3007.

92. Gordon, S.; Hart, E.J.; Matheson, M.S.; Rabani, J.; Thomas, J.K. (1963). *Discuss. Faraday Soc.*, 36: 193.

93. Raef, Y.; Swallow A.J. (1963). *Trans Faraday Soc.*, 59: 1631.

94. Mulazzani, Q.G.; D'Angelantonio, M.; Venturi, M.; Hoffman, M.Z.; Rodgers, M.A.J. (1986). *J. Phys. Chem.*, 90: 5347.

95. Wu, L.-M.; Fischer, H. (1984). *Int. J. Chem. Kinet.*, 16: 1111.

96. Butler, J.; Jayson, G.G.; Swallow, A.J. (1974). *J. Chem Soc., Faraday Trans. I*, 70: 1394.

97. Ershov, B.G.; Janata, E.; Henglein, A.; Fojtik, A. (1993). *J. Phys Chem.*, 97: 4589.

98. Marchaj, A.; Kelley, D.G.; Bakac, A.; Espenson, J.H. (1991). *J. Phys. Chem.*, 95: 4440.

99. Getoff, N. (1989). *Appl. Radiat. Isot.*, 40: 585.

100. Liu, B.-J.; Torimoto, T.; Yoneyama, H. (1998). *J. Photochem. Photobiol. A: Chem.*, 115: 227.

101. Dey, G.R.; Nair, K.N.R.; Pushpa, K.K. (2009). *J. Nat. Gas Chem.*, 18: 50.

102. Schwarz, H. A.; Dodson, R. W. (1989). *J. Phys. Chem.*, 93: 409.

103. Hara, K.; Kudo, A.; Sakata, T. (1995). *J. Electroanal. Chem.*, 391:141.

104. Tobias, C.W. (1961). in *Advances in Electrochemistry and Electrochemical Engineering*, Interscience, John Wiley & Sons, New York, p. 262

105. Russell, R.G.; Kovac, N.; Srinivasan, S.; Steinberg, M. (1977). *J. Electrochem. Soc.*, 124; 1329, and references therein.

106. Jitaru, M. (2007). *J. Uni. Chem. Tech. Metal.*, 42: 333.

107. Noda, H.; Ikeda, S.; Oda, Y.; Imai, K.; Maeda, M.; Ito, K. (1990). *Bull. Chem. Soc. Jpn.*, 63: 2459.

108. Scibioh, M.A.; Viswanathan, B. (2004). *Proc. Indian. Nat. Sci. Acad.*, 70: 407.

109. Lamy, E.; Ladjo, L.; Saveant, J.M. (1977). *J. Electroanal. Chem.*, 78: 403.

110. Bard, A.J.; (Eds.), *Encyclopedia of Electrochemistry of Elements*, Dekker, New York 1976, vol. 7.

111. Le, M.; Ren, M.; Zhang, Z.P.; Sprunger, T.; Kurtz, R.L.; Flake, J. C. (2011). *J. Electrochem. Soc.*, 158: E45.

112. Hori, Y.; Takahashi, I.; Koga, O.; Hoshi, N. (2002). *J. Phys. Chem. B*, 106: 15.

113. Köleli, F.; Atilan, T.; Palamut, N. A.; Gizir, M.; Aydin, R.; Hamann, C. H. (2003). *J. Appl. Electrochem.*, 33: 447.

114. Ito, K.; Ikeda, S.; Yamauchi, N.; Iida, T.; Takagi, T. (1985). *Bull.Chem.Soc.Jpn.*, 58: 3027.

115. Sánchez-Sánchez, C. M.; Montiel, V.; Tryk, D. A.; Aldaz, A.; Fujishima, A. (2001). *Pure Appl. Chem.*, 73: 1917.

116. Hori, Y.; Wakebe, J.; Tsukamoto, T.; Koga, O. (1994). *Electrochim. Acta*, 39: 1833.

117. Hara, K.; Tsuneto, A.; Kudo, A.; Sakata, T. (1994). *J. Electrochem. Soc.*, 141: 2097.

118. Azuma, A.; Hashimoto, K.; Hiramoto, H.; Watanabe, M.; Sakata, T. (1989). *J. Electroanal. Chem.*, 260: 441.

119. Akahori, Y.; Iwanaga, N.; Kato, Y.; Hamamoto, O.; Ishii, M. (2004). *Electrochemistry (Tokyo, Japan)*, 72: 266.

120. Bidrawn, F.; Kim, G.; Corre, G.; Irvine, J. T. S.; Vohs, J. M.; Gorte, R. J. (2008). *Electrochem. Solid-State Lett.*, 11: B167.

121. Innocent, B.; Liaigre, D.; Pasquier, D.; Ropital, F.; Leger, J. M.; Kokoh, K. B. (2009). *J. Appl. Electrochem.*, 39: 227.

122. Delacourt, C.; Ridgway, P. L.; Kerr, J. B.; Newman, J. (2008). *J. Electrochem. Soc.*, 155: B42.

123. Li, H.; Oloman, C. (2007). *J. Appl. Electrochem.*, 37: 1107.

124. Subramanian, K.; Asokan, K.; Jeevarathinam, D.; Chandrasekaran, M. (2007). *J. Appl. Electrochem.*, 37: 255.

125. Whipple, D.T.; Finke, E.C.; Kenis, P.J.A. (2010). *Electrochem. Solid-State Lett.*, 13: B109.

126. Xie, K.; Zhang, Y.; Meng, G.; Irvine, J.T.S. (2011). *J. Mater. Chem.*, 21: 195.

CHAPTER 3

Synthesis of Useful Compounds from CO$_2$

BOXUN HU and STEVEN L. SUIB*

3.1 INTRODUCTION

Our daily energy sources are largely dependent on nonrenewable petroleum, natural gas, and coal, which account for about 84% of total energy consumed. Meanwhile, man-made carbon dioxide (CO$_2$) emission causes climate and ecosystem changes, especially global warming and ocean acidification. The

*Indicates the corresponding author.

Green Carbon Dioxide: Advances in CO$_2$ Utilization, First Edition.
Edited by Gabriele Centi and Siglinda Perathoner.
© 2014 John Wiley & Sons, Inc. Published 2014 by John Wiley & Sons, Inc.

atmospheric CO_2 concentration has increased from about 280 parts per million (ppm) in preindustrial times to 382 ppm in 2006 according to the National Oceanic and Atmospheric Administration (NOAA). Currently, the atmospheric CO_2 concentration is still steadily increasing at a rate of about 1.9 ppmv/year. Surface ocean pH is estimated to have dropped from near 8.25 to near 8.14 between 1751 and 2004 [1], representing an increase approaching 30% in "acidity" (H^+ ion). This interweaving energy-environment dilemma is one of the biggest problems for the development of sustainable energy in the twenty-first century. We cannot escape from this problem and need to find viable solutions.

Looking into our future energy supply, we will have many challenges as the fossil fuel reserve is being decreased. Solar, biofuel, wind power, nuclear power, and water power can be alternative energy sources, and they alleviate CO_2 emission. Despite the emergence of these alternative energy sources, fossil fuels will still be the main energy sources in the next decade. CO_2 conversion to liquid fuels and other useful compounds can store renewable solar, wind power, and other kinds of energies in a chemical form with a higher energy density than supercapacitors and batteries. Liquid fuels and useful compounds from CO_2 can be readily used for transportation systems and the chemical industry. This is an ideal solution for the persistent conflicts between increasing global energy and the demand for chemicals and environmental harm.

Nature has developed photosynthesis (Eq. 3.1) via land plants and ocean phytoplankton to synthesize carbohydrates using CO_2, water, and solar energy:

$$6nCO_2\,(g) + 6n\,H_2O\,(l) \xrightarrow{-h\nu} (C_6H_{12}O_6)_n\,(s) \qquad (3.1)$$

This photosynthesis process provides the Earth's ecosystem and human civilization with foods, fossil fuel energy, biofuel, and other daily necessities directly or indirectly. Artificial CO_2 conversion dates back to early in the twentieth century. The Sabatier process invented by Paul Sabatier (Nobel Prize, 1913) converted CO_2 and H_2 into CH_4 and H_2O, using a nickel catalyst at high temperatures and high pressures [2]. A gas-to-liquid process, named the Fischer–Tropsch synthesis, was invented in the 1920s and is still used [3]. This industrial process produces liquid fuel, using carbon oxides/hydrogen mixtures, typically from natural gas, coal, and biomass. Mimicking the natural photosynthesis process, formaldehyde, carbon monoxide, and methanol have been produced by photochemical reduction of CO_2 [4]. Electrochemical reduction of CO_2 using H^+ in aqueous electrolytes has produced numerous products [5]. Low faradaic efficiency limits its practical application. Recently, nanocatalysts, transitional metal complexes, and enzymes have been used as electrocatalysts for CO_2 reduction. A number of products have been produced with high selectivity and high CO_2 conversion [6]. Hydrogenation of CO_2 has produced light olefins, jet fuel, carboxylic acids, and specialty chemicals [7]. Urea, polycarbonate, cyclic carbonates, oxazol 2-carboxylic acids, and propiolic acids have been synthesized by chemical fixation of CO_2.

This chapter shows six reduction pathways for synthesis of useful compounds from CO_2. Each pathway follows specific reaction mechanisms, and its experimental design is unique. Previous reviews [8,9] may have assessed specific pathways, and readers are referred to these. In this chapter, the aims are not only to offer the readers a general overview of the recent progress in the synthesis of useful compounds from CO_2 but also to focus on understanding of structure-component-activity. Useful products are listed by the reduction methods for comparison. The key experimental conditions, selectivities, and energy efficiencies are given in the tables. Readers are referred to specific references for further understanding of these reactions.

A combined approach to maximize CO_2 conversion is more favorable for CO_2 reduction. Intensive interest on CO_2 conversion has recently been focused on the electrocatalytic reduction using heterogeneous catalysts. This process provides the possibility of large-scale storage of electrical energy and thermal energy in a chemical form. Gas-phase CO_2 conversion is a green process without solvents and other chemicals except heterogeneous catalysts. High conversion rates and high selectivities to valuable products are the merits and advantages over other methods.

In these CO_2 (electro)catalytic reduction reactions, understanding of structure-component-activity (selectivity) relationships gives insight for chemical design of the catalysts. Characterization of the catalysts and the products provides evidence to reveal the reaction mechanisms. New catalyst materials and catalysis concepts have led to progress in CO_2 conversion.

3.2 PHOTOCHEMICAL REDUCTION

Mimicking the photosynthesis process, artificial photoreduction of CO_2 has been considered one of the most alluring methods for utilization of CO_2 because of its abundance and free access of sunlight. Total solar irradiance upon earth (TSI) was earlier measured by satellite to be roughly 1.366 kW/m^2 [10]. Photoreduction of CO_2 to useful organic compounds (Table 3.1) has been reported with the use of semiconductor electrodes, homogeneous catalysts, and the enzyme dehydrogenase. The total energy efficiency of photoreduction cells is about 10%, and most energy is lost because of light scattering and energy loss during electron transfer. A single p-n junction semiconductor exhibits the highest solar conversion efficiency, about 33.7% based on the Shockley–Queisser limit. In this section, we focus on photoreduction of CO_2 using semiconductors.

A typical photoreduction system of CO_2 is composed of semiconductor electrodes and photocatalysts. Semiconductors absorb photons, leading to excitation of electrons that transfer from valence bands (VB) to conduction bands (CB); then the electron transfers from the conduction bands of a photoexcited semiconductor to a photocatalyst complex, and, finally, the photocatalyst complex reduces CO_2 to useful organic compounds.

In semiconductors, electrons are confined to a number of bands of energy and forbidden from other regions. The term "band gap" refers to the energy difference between the top of the VB and the bottom of the CB. With the development of the semiconductor industry, band gap engineering is widely applied to control or alter the band gap of a semiconductor by controlling the compositions of certain semiconductor alloys (GaP, InP, and others). Band gaps of the n type of semiconductors

TABLE 3.1   Useful Compounds Synthesized from CO_2 by Photochemical and Photoelectrochemical Reduction

Year	Electrode/Catalysts	Light Sources/Efficiency	Product/Selectivity	Ref.
2010	p-type silicon/Re(bipy − But) $(CO)_3Cl$	Polychromatic (10%)/661 nm (9.3%)	CO (100%)	20
2010	N-Ta_2O_5/ $[Ru(dcbpy)_2(CO)_2]^{2+}$	405 nm (1.9%)	HCOOH (>75%)	18
2008	p-GaP/pyridine −0.5 V	365 nm (Φ_{MeOH}, 44%)	CH_3OH (100%)	22
2006	Metal doped p-InP LiOH/methanol, −0.75 V	>300 nm, faradaic efficiency, 80.4%	CO > HCOOH (Ag, Au, and Cu) > CH_4 > C_2H_4 (Ni)	21
1987	Pd colloid/ P-cyclodextrin	>400 nm, (Φ_{HCO2-}, 1.1%)	Formate	12

(Si, Se, ZnO, and Ta_2O_5) tend to decrease with an increase of temperature or of doping density.

The Shockley–Queisser limit gives the maximum efficiency of a single-junction SC cell under unconcentrated sunlight as a function of the semiconductor band gap. If the band gap is too high or too low, most daylight photons cannot be absorbed or lost. To obtain near-maximum efficiency of energy adsorption, the optimal band gap is 1.1–1.6 V (for example, Si: 1.1 eV and CdTe: 1.5 eV). The Shockley–Queisser limit can be exceeded by tandem cells or by concentrating sunlight onto the cell [11]. The spectrum of sunlight radiation is divided into the ultraviolet (UV; <380 nm), visible (380–700 nm), and infrared (>700 nm) ranges.

In aqueous solutions, because of the competition of photocatalytic splitting of water with photocatalytic CO_2 reduction, the selectivity for hydrocarbon compounds is low. Kinetically, photocatalytic CO_2 reduction yielding useful chemicals is more difficult than H_2 production. Specific photocatalysts, which favor CO_2 reduction, have been selected. Heterogeneous Pd colloid stabilized by β-cyclodextrin (β-CD) was reported as a selective photocatalyst for the conversion of CO_2/HCO_3^- to HCO_2^- (formate) with a high quantum yield ($\Phi = 1.1$) by visible light in 1987 [12]. Transition metal (mainly Co and Ru) complexes are well-known photocatalysts for CO_2 reduction because of their excellent quantum efficiencies and product selectivity [13–16]. Fac-[Re (bpy) $(CO)_3$ {P(OEt)$_3$}]$^+$ (bpy: 2,2′-bipyridine, Et: C_2H_5) is such a metal complex. The quantum yield of conversion of CO_2 to CO was up to 38%, and only a very small amount of hydrogen was produced even in the presence of water [17]. The photoreduction system is composed of p-type semiconductor photosensitizer, N-Ta_2O_5, and a reducing catalyst, a Ru complex such as [Ru-bpy] and [Ru-(dcbpy)(bpy)] (dcbpybpy: 4,4′-dicarboxy-2,2′-bipyridine), in an acetonitrile/triethanolamine solution, and has achieved a selectivity of more than 75% for HCOOH and a quantum efficiency of 1.9% at 405 nm [18].

Silicon of the p-type is a proven photocathode for water reduction to hydrogen using sunlight [19]. Silicon is becoming a cheaper, widely used material. Most recently, the photoreduction of CO_2 to CO on p-type silicon using $Re(bipy-Bu^t)(CO)_3Cl$ (bipy-Bu^t: 4,4'-di-*tert*-butyl-2,2-bipyridine) as an electrocatalyst has achieved a faradaic efficiency of 97 ± 3%, an overall efficiency of about 10%, and a short-circuit quantum efficiency of 61% for light-to-chemical energy conversion [20]. This developed method for CO_2 photoreduction may be feasible to produce synthesis gas (CO/H_2) using sunlight on p-type Si, and then liquid fuel could be synthesized from this synthesis gas via the Fischer–Tropsch synthesis process.

To achieve higher-value products and a high faradaic efficiency, a combined approach of photochemical and electrochemical reduction, called photoelectrochemical reduction, has emerged since the 1980s [21,22]. An electrochemical reduction process involves CO_2 gas and protons mainly from aqueous electrolytes and few from aprotic solvents such as dimethylformamide (DMF), dimethyl sulfoxide (DMSO), propylene carbonate, and acetonitrile. The products of photoelectrochemical reduction of CO_2 using metal-modified p-InP photoelectrodes were dependent on the metals [21]. Carbon monoxide and formic acid were produced on lead-, silver-, gold-, and copper-modified p-InP photocathodes, and a silver-modified p-InP photoelectrode achieved maximum current efficiency (80.4%).

3.3 ELECTROCHEMICAL REDUCTION

Electrochemical reduction of CO_2 (ERC) is the conversion of CO_2 to more reduced chemical species on electrodes using electricity as the energy source. The history of electrochemical reduction of CO_2 began in the nineteenth century, when carbon dioxide was reduced to formic acid with a zinc cathode. The equilibrium potentials (E vs. SCE, pH 7.0) of aqueous CO_2 reduction reaction involving multielectron reduction are shown in Eqs. 3.2–3.8 [23]:

$$CO_{2(ad)} + e^- \rightarrow CO_2^-{}_{(aq)} \qquad (E^\circ = -2.14 \text{ V}) \qquad (3.2)$$

$$CO_{2(ad)} + 2e^- + 2H^+{}_{(aq)} \rightarrow HCOOH_{(l)} \qquad (E^\circ = -0.85 \text{ V}) \qquad (3.3)$$

$$CO_{2(ad)} + 2e^- + 2H^+{}_{(aq)} \rightarrow CO_{(g)} + H_2O_{(l)} \qquad (E^\circ = -0.76 \text{ V}) \qquad (3.4)$$

$$CO_{2(ad)} + 4e^- + 4H^+{}_{(aq)} \rightarrow H_2CO_{(g)} + H_2O_{(l)} \qquad (E^\circ = -0.72 \text{ V}) \qquad (3.5)$$

$$CO_{2(ad)} + 6e^- + 6H^+{}_{(aq)} \rightarrow CH_3OH_{(l)} + H_2O_{(l)} \qquad (E^\circ = -0.62 \text{ V}) \qquad (3.6)$$

$$CO_{2(ad)} + 8e^- + 8H^+{}_{(aq)} \rightarrow CH_{4(g)} + 2H_2O_{(l)} \qquad (E^\circ = -0.48 \text{ V}) \qquad (3.7)$$

$$2H^+{}_{(aq)} + 2e^- \rightarrow H_{2(g)} \qquad (E^\circ = 0 \text{ V}) \qquad (3.8)$$

In aqueous solutions, the selectivity for the above carbon species is low because of the competition of evolution of hydrogen with CO_2 reduction. Part of the electrical energy is converted to heat energy because of the existence of overpotential on

TABLE 3.2   The Overpotential for the Evolution of H_2 and O_2 Gases on Various Electrode Materials at 25°C

Material of Electrode	H_2 Overpotential (V)	O_2 Overpotential (V)
Pt (platinized)	−0.05	+0.77
Pt (shiny)	−0.68	+1.49
Cu	−0.48	+0.79
Zn	−1.23	—
Iron	−1.29	—
Au	−0.24	+1.63
Ni	−0.56	+0.85
Ag	−0.48	+1.13
Graphite	−0.60	+0.95

the metal electrodes, therefore decreasing the faradaic efficiency of electrochemical reduction. The overpotential of different metal electrodes varies from metal to metal. Table 3.2 lists the activation overpotential for the evolution of selected gases on various electrode materials at a current density of 1.0 A/cm^2 and at 25°C [24].

Hydrogen will form first rather than carbon species in aqueous solution because of the low hydrogen overpotential on platinum electrodes. In the gas phase, electrochemical reduction of CO_2 over a Pt-containing gas diffusion electrode under high pressure (<50 atm) has been successfully performed [25]. Methane was produced at a faradaic efficiency of 35% with a current density of 313 mA/cm^2. Ethanol was also produced at a faradaic efficiency of 2.2% with a partial current density of 19.8 mA/cm^2. With increasing CO_2 pressure, the faradaic efficiency for methane formation increased while hydrogen formation decreased. Because of the existence of the higher hydrogen overpotential on copper electrodes, hydrogen does not form until the cell voltage is above 0.48 V; except for Zn, Pd, graphite, iron, nickel, and gold, metals cannot be used as electrodes for electrochemical reduction of CO_2 in aqueous solution.

Low-cost copper-based electrodes have been developed and tested in various electrolytes. The Cu(I) species in copper halides [Cu(I)X] acted as active catalysts in the selective conversion of CO_2 to C_2H_4 [26–28], while Cu(I) species in Cu_2O played a critical role in selectivity to CH_3OH (a yield of 43 μmol/cm^2 h) with a faradaic efficiency of 38%. The activity of the Cu_2O species was remarkably higher than that of air-oxidized or anodized Cu electrodes [29].

The interfaces of electrode-gas–liquid are critical for the electrochemical reduction of CO_2. In ERC reactions, gas diffusion electrodes (GDE) provided a conjunction of a solid, liquid, and gaseous interface, and an electrically conducting catalyst supported an electrochemical reaction between the liquid and the gaseous phase. This GDE design increased the current efficiency.

One unique electrode design for the ERC reaction is that of high-surface-area nickel electrocatalysts supported on activated carbon fibers, which contain slit-shaped pores with widths on the order of nanometers. The current efficiency for CO_2 reduction to CO reached a value of 70%, while on the same type of high-area nickel catalyst supported on nonactivated carbon fibers much smaller amounts of CO were generated. The enhancement of selectivity for CO_2 reduction with the microporous support is due to a nanospace effect, which gives rise to high pressure-like effects at ambient pressure [30]. Another novel electrode for ERC is using microbes as electrochemical CO_2 conversion catalysts. Two acetogenic bacteria [*Moorella thermoacetica* (Mt) and faradaic *Clostridium formicoaceticum* (Cf)] of five tested microorganisms efficiently converted CO_2 to formate. The current efficiency was 80% for Mt and 100% for Cf upon electrolysis in CO_2-saturated phosphate buffer solution (0.1 M, pH 7.0) at a pressure of 1 atm [31]. The applied voltage of -0.58 V (vs. NHE) is near the equilibrium potential of CO_2/formate.

This catalyst design has advantages over enzyme catalysts and provides more choices of catalysts for ERC with microorganisms that are already developed and optimized in nature. The ERC reaction proceeded not only in the aqueous and gas phases but also in molten carbonate phases. CO_2 was reduced to CO in molten Li_2CO_3/K_2CO_3 in the range of 923–973K. The maximum current efficiency of 95–100% occurred at about -1.4 V. The activation energy of the reduction of CO_2 to CO calculated from an Arrhenius plot was about 70 kJ/mol [32].

In summary, the advantage of the ERC method is that renewable energy sources, such as hydropower, wind power, solar power, tidal power, or nuclear power can be taken from electric power grids for ERC. Various products, including CO, HCOOH, alcohols, and light hydrocarbons like methane, have been reduced from CO_2 (Table 3.3) [27–40]. In general, lots of improvements need to be accomplished in the areas of thermodynamic efficiency (overcome high overpotential), current efficiency, selectivity, and stability.

3.4 ELECTROCATALYTIC REDUCTION

Electrocatalytic reduction of CO_2 involves an electrocatalyst in the electrochemical reduction reaction. The electrocatalyst participates in and modifies electrochemical reactions and increases the rate of chemical reactions without being consumed in the process. The electrocatalyst functions at electrode surfaces or may be the electrode surface itself. The electrocatalyst promotes the transfer of electrons between the electrode and reactants/intermediates and/or accelerates an intermediate transformation described by overall half-reactions. Specifically for the CO_2 reduction electrocatalysts, these electrocatalysts exist in various formations, such as platinum nanoparticles (NPs), coordination complexes, and enzymes. Such materials are heterogeneous and homogeneous catalysts. Each electrocatalyst involves a characteristic mechanism. These catalysts may work at totally different reaction conditions and show differences in activity, selectivity, and products.

Table 3.4 lists recently developed electrocatalysts for CO_2 reduction [41–52]. Further studies of structure-component-activity relationships may lead to a better understanding of the mechanisms involved in CO_2 electrocatalytic reduction.

Here we classify three types of CO_2 reduction electrocatalysts: transition metal nanoparticles, coordination complexes, and enzymes.

TABLE 3.3 **Useful Compounds Synthesized by the Electrochemical Method**

Year	Electrode/Electrolyte	Current Efficiency	Products	Ref.
2011	Microbes/0.1 M H$_3$PO$_4$	80% (Mt) 100% (Cf)	Formate	31
2011	Cu$_2$O/Cu	38%	CH$_3$OH	29
2008	TiO$_2$/[EMIBF$_4$]	8-14%	Polyethylene	33
2005	Cu, Ag, Ni/CuX,	Cu: 64%, Ag: 42.9%	C$_2$H$_4$ (69.4%), CO (7.1%)	27
2004	Cu mesh/CuX, KX	49.9% (C$_2$H$_4$)	C$_2$H$_4$ (75%), CO$_2$ (90%)	28
2003	Cu mesh/CuCl	70% (−1.8V)	C$_2$H$_4$ (75%)	26
2002	MP/carbon fiber	70%	CO	34
2000	Au, Cu, Ni, Co/molten carbonate	98% (−1.52V)	CO	32, 35
1999	Co, Fe, Cu, Zn-phthalocyanine (TPP)	97.4% (Co)	CO, HCOOH	36–39
1998	Ni/carbon fibers	70%	CO	30
1995	Pt/GDE, 50 atm	46%	Methane and ethanol	25
1993	Perovskite/GDE	40%	Alcohols	40

3.4.1 Transition Metal Nanoparticle Catalysts

3.4.1.1 Aqueous Electrolyte. Transition metal nanoparticles (NPs) are extensively used as catalysts in many reactions from organic synthesis to carbon nanotube synthesis and from fuel cells to CO$_2$ electrocatalytic reduction. Supported metal NP catalysts have practical or potential applications in many industries, such as the biopharmaceutical, environmental protection, energy, and chemical industries. The performance of these NP catalysts in the reactions depends on which of their crystal faces are exposed. The sizes and shapes of the NP catalysts are key factors for these catalytic reactions. In modern molecular catalysis, it is crucial to keep the NP catalysts at the nanoscale and thereby increase selectivity and efficiency in the catalysis reaction. These NPs are clusters containing from a few tens to several thousands of metal atoms, stabilized by surfactants, polymers, and ligands to protect their surfaces. Their sizes vary from 1 nm to several tens or hundreds of nanometers. The most active NPs in catalysis are only one or a few nanometers in diameter, equal to a few to a few hundred atoms only. The transition metal NPs have been synthesized with different isolated shapes including cubes, spheres, wires, octopod-cubes, stars, rods, bilobes, tetrahedra, and multipods.

Many synthetic methods, such as colloid, sol−gel, photochemical synthesis, and metal organic chemical vapor deposition, have been reported for successful control of the size and shape of transition metal NPs. Commercial NP catalysts, such as carbon-supported Pt NPs, have been produced with these methods. The particle size

TABLE 3.4   Typical Useful Compounds from Electrocatalytic Reduction of CO_2

Year	Electrocatalytic Cell	Reaction Condition	Products	Ref.
2010	RuPd/$KHCO_3$/Pt Gas diffusion electrodes (GDE)	Microfluidic reactor, pH: 4–7, 2.5–4 V	Formic acid with 45% energy efficiency (E. E.)	41
2010	Pt(Pd)/Pyridines, KCl	Galvanostatic mode, 50 μA cm^{-2}, pH: 4.7–5.6	HCOOH, HCHO, CH_3OH, 33% E. E.	42
2010	Benzimidazole, pyridine/bipyridine polymers in acetonitrile, 1% H_2O	1.2–1.7 V	Form adducts with CO_2	43
2010	CO_2, H_2O, Pt/yttria-stabilized zirconia (YSZ)/Pt CO_2, H_2O, Pt/OMS-2/Pt	600–900°C, 0.5–1.5 V, 250–450°C, 0.5–1.5 V	HCHO (8% conv.) HCHO (0.5–5% conv.)	44
2009	H_2O-CO_2-H_2, Ni-YSZ/$La_{0.6}Sr_{0.4}Co_{0.2}Fe_{0.8}O_3$ (LSCF)-GDC, air	800°C, 1.3 V	CO/H_2, 7 sccm/cm^2	45
2009	Pt/CNT or Fe/CNT on carbon cloth/$KHCO_3$/Pt	60°C, 20 mA	Oxygenates	46
2008	Ag, Nifion/$KHCO_3$, Pt, polytetrafluoroethylene (PTFE)	−1.7 to −2 V, 80 mA/cm^2	CO/H_2, 30–50% E. E.	47
2007	Nano Pt, GDE/0.5M $KHCO_3$,	20 mA/cm^2, CO_2: 20 cm^3/min	C_{5+}: 90–96%	48
2004	CO_2, Pt/YSZ/Pt	750–850°C, 1.2–1.95 V	CO, <50% at 1.95 V	49
2003	Electrodes/carbon monoxide dehydrogenase (CODH), methyl viologen (MV)	0.1 M phosphate buffer, pH 6.3, −0.57 V	CO, 60–100%	50, 51
1984	Co, Cu, Ni, and Ru phthalocyanine-modified carbon electrode	0.3 V, turnover number: 105	CO	39, 52

of the Pt NPs was 2–5.4 nm (from E-TEK). By using these Pt NP electrocatalysts, long-carbon-chain hydrocarbons (>C_5) have been first converted from CO_2 at room temperature and atmospheric pressure in a continuous CO_2 flow cell. CO was the main product. The amount of CO produced was 10^3 times higher than that of hydrocarbons. The evaluation of long hydrocarbons is a step-by-step process as shown

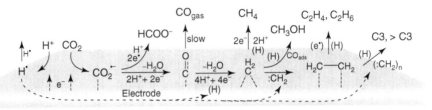

Figure 3.1   Stepwise electrocatalytic reduction of CO_2 to higher-molecular-weight hydrocarbons at the surface of the Pt NPs/C electrode. Reproduced from ref. 48 by permission of The Royal Society of Chemistry.

in Figure 3.1. These electrocatalytic reduction experiments showed that the surface reaction-chain growth was very slow. Desorption, readsorption, and consecutive transformation of the lower-molecular-weight hydrocarbons at room temperature and pressure are reactions that occurred. It took about 20 min for the conversion of $=CH_2$ to ethane, and another 20 min for the conversion of ethane to propane. This method provided an opportunity for tailoring the production of hydrocarbon products. The carbon chain growth is largely diverted from the expected ASF model with a probability of chain growth of 0.85 [48].

In a microfluidic reactor (Fig. 3.2), the same E-TEK type Pt NPs electrocatalysts demonstrated different activity and selectivity. CO_2 was reduced to formic acid at higher faradaic and energetic efficiencies of 89% and 45%, respectively. The current density increased from 20 mA/cm^2 to 100 mA/cm^2. Operating at acidic pH of 4 resulted in a significant increase in performance [53].

Pt, Co, and Fe electrocatalysts on carbon nanotube supports have produced oxygenates (isopropanol, methanol, ethanol, acetone, and acetaldehyde) [54]. Isopropanol was the main product. The continuous mode and batch mode showed different products on the carbon black-supported nano Pt electrocatalysts. Mainly C_{5+} hydrocarbons and alkyl-substituted aromatic (toluene and xylenes) products and small amounts of alcohols (ethanol and butanol) were produced in the semi-half continuous cell; while n-butanol was produced in the full continuous configuration with one-third of the productivity in the semi-half continuous cell.

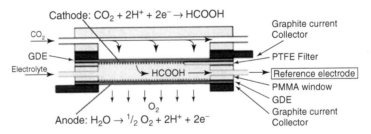

Figure 3.2   A microfluidic reactor for electrocatalytic reduction of CO_2 using Pt GDE electrodes. Reproduced from ref. 53, Copyright 2010, with permission from The Electrochemical Society.

This indicated that the products were kinetically controlled, and the chain growth is slow. These catalysts are unstable and deactivated very fast in 4 h. The Fe catalysts were more active than the Pt catalysts. The deactivation of the Fe catalysts was due to crossover of electrolyte, particularly of K^+ ions. The Pt catalysts may lose activity, since K^+ ions covered active Pt sites.

Long hydrocarbons and alcohols were first synthesized at room temperature with transition metal NP electrocatalysts in the above syntheses. However, these reaction rates were extremely slow at such a low temperature, and a long residence time was critical for the production of long-chain compounds. These obstacles restricted the practical application of the electrocatalytic CO_2 reduction using this process.

3.4.1.2 Solid Oxide Electrolyte.

Electroreduction of CO_2 using a solid oxide fuel cell (SOFC) technology is another pathway to synthesize useful compounds from CO_2. This technology is based on relatively well-developed SOFC technology, and these electrolyzers are operated in a reversed fuel cell mode. Compared to aqueous electrocatalysis, solid oxide electrodes (SOEs) work at high temperatures (600–900°C), and less electrical energy is required because part of the energy required for splitting H_2O and reduction of CO_2 is thermal energy. A short residence time and a fast reaction rate can be achieved at high temperatures. This makes it possible to apply this process for practical applications of the reduction of CO_2.

Solid oxide electrolysis (SOE) proof-of concept experiments were initially designed for the production of O_2 and fuel for long-term human exploration of Mars missions using SOE CO_2 electrolysis, and this work was published in 2004 [49]. This strategy of in situ resource utilization (ISRU) could provide substantial savings in mission costs and launch/landing masses and significantly reduce risks. In a CO_2 (gas), Pt | YSZ | Pt, O_2 (gas) electrolysis cell, active electrolysis of CO_2 to form CO and oxygen anions occurs under conditions of a temperature between 750 and 850°C and applying a cell voltage of 1–2 V. The maximum current density of 92 mA/cm^2 occurred at a temperature of 850°C while applying a cell voltage of 2 V. No further analyses of reaction mechanisms and formation of O_2 and CO were provided.

A simultaneous high-temperature electrolysis of steam and CO_2 using SOFC technology was sponsored by the Idaho National Laboratory (INL) and Ceramatec Inc. (Salt Lake City, UT) for large-scale nuclear-powered syngas production. The simulation FLUENR model based on an electrolysis cell design showed that the outlet mole fraction of CO and H_2 could reach 0.06 and 0.14, respectively, under a current density of 0.4 A/cm^2 and a temperature of 800°C [55].

Formaldehyde, ethylene, and methanol had not been produced from CO_2 and H_2O in SOFC cells in a continuous gas phase reaction until electrocatalysts were applied to the SOFC cells [56]. Electrocatalysis plays an important role in the activation of CO_2 and splitting of water. Both steps are key rate-determining steps for the synthesis of organic compounds from CO_2 and H_2O.

This work initially started with nano-sized platinum catalysts, because nano-sized platinum catalysts are well known as one of the most efficient catalysts in fuel cell reactions [57–61]. The CO_2 activation on precious metal-coated

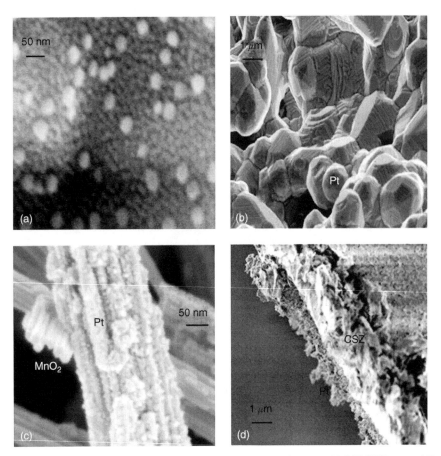

Figure 3.3 Field emission scanning electron microscopy images. (a) MOCVD coated Pt on CSZ, (b) calcined Pt paste, (c) MOCVD coated Pt on OMS-2, (d) the cross-section of Pt/CSZ. Reprinted from ref. 56, Copyright 2010, with permission from Elsevier.

electrodes has been performed with low-temperature AC plasma methods, and thermal dissociation to CO was observed with a CO$_2$ conversion of about 30% [62]. Water splitting has also been observed in a fan-type Pt reactor and on Pt electrodes [63]. But nano-sized platinum catalysts on solid oxide electrolyte supports have not been reported. In the latter experiment, nano-sized platinum was coated by a metal organic chemical vapor deposition (MOCVD) method. A thin fluffy platinum layer was deposited on reticulated calcia-stabilized zirconia (CSZ) or cryptomelane-type octahedral molecular sieve manganese oxide (OMS-2) [Fig. 3.3 (c) and (d)].

This electrocatalytic reaction showed that the particle size is very important for catalytic activity. The size of MOCVD-coated platinum can be controlled at a few nanometers to 20 nm (Fig. 3.3a). With atomic layer deposition equipment, the coating layer can be precisely controlled at a thickness of a few angstroms, and the cost of electrocatalysts could be further reduced. The particle sizes of the calcined platinum paste are a few micrometers in size, and platinum particles were aggregated (Fig. 3.3b). The surface area of the calcined platinum paste was much smaller than the MOCVD-coated platinum nanoparticles. Therefore, fewer catalytic sites of platinum paste are exposed to the reactants. The electrocatalytic effect of the platinum paste is very low. Another problem of a thick electrocatalyst layer is that such thick layers increase the difficulty of charge transfer. This is discussed in the following section about reaction mechanisms.

The electrocatalytic cells were placed in a setup connected with electronic control devices. In a gas-phase electrocatalysis reaction of CO_2 and H_2O, the reaction temperature (250–900°C) was varied with different catalysts and electrolytes. The reaction works at a pressure of 1 atm. The flow rates of the reactants were controlled by mass flow controllers. The moderate potentials are applied in the electrocatalyts/solid ion electrolyte cell to enhance charge transfer in the solid oxide electrolytes.

The mechanisms of electrocatalytic reduction of CO_2 and H_2O are still not clear. Chemically adsorbed hydrogen species (H^*) form on the Pt catalyst surfaces via agostic interactions and lead to dissociation in the first layer of water on Pt (111) [64]. These H * species reduced the OPt^+CO species [65] and play an important role in the formation of HCHO. Oxygen ions in the solid oxide electrolyte act as "oxygen vehicles" to remove oxygen ions at three phase boundaries [66–68] (gas, catalyst, and electrolyte) in the electrical fields. O_2 gas is produced by oxidation of O_2^- at the counterelectrode in the third step as shown in Figure 3.4. A summary of reactions is shown in Eqs. 3.9–3.13.

$$H_2O_{(g)} \rightarrow H_{ad} + O^{2-}{}_{(aq)} \tag{3.9}$$

$$CO_{2(g)} \rightarrow CO_{ad} + O^{2-}{}_{(aq)} \tag{3.10}$$

$$PtO^+{}_{(s)} + CO_{2(g)} + 2e^- \rightarrow OPt^+ - CO_{(s)} + O_2^-{}_{(s)} \tag{3.11}$$

$$2H_{ad} + CO_{ad} \rightarrow HCHO_{(g)} \tag{3.12}$$

$$O^{2-}{}_{(aq)} - 2e^- \rightarrow 1/2O_{2(g)} \tag{3.13}$$

The HCHO products can react with water and form methylene hydrate, $HO-CH_2-OH$. The methylene hydrate molecules react with one another, forming paraformaldehyde [$HO(CH_2O)_n$ H, n = 8–30]. In these electrocatalytic reduction reactions of CO_2 and H_2O, the product is present in the formation of paraformaldehyde and not in the formation of small HCHO molecules. This polymer can be reversibly converted to formaldehyde by heating or addition of acid. Paraformaldehyde has the same uses as formaldehyde. Formaldehyde is a common building block for the synthesis of numerous polymers.

High selectivity to formaldehyde (up to 100%) is one of the characteristics of electrocatalytic reduction of CO_2 and H_2O using the Pt/CSZ catalysts. The selectivity of the gas-phase electrocatalytic reduction is greatly improved compared with the liquid-phase electrochemical reduction. In these continuous high-temperature

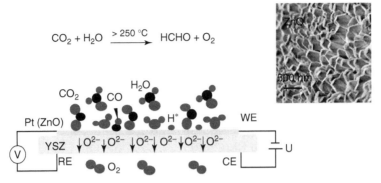

Figure 3.4 Illustration of electrocatalytic reduction of CO_2 and H_2O using Pt (or ZnO)/YSZ catalysts and applying a DC voltage at the interfaces. A thin catalyst layer (inset) was deposited on the electrolyte (YSZ or other solid ion electrolyte).

gas-phase reactions, a short residence time (<0.05 s) was needed for a CO_2 conversion up to 8%. The reaction turnover frequency TOF, defined as [(moles of converted CO_2/s)/moles of Pt catalyst used] at 900°C was 5.2 ± 0.5 s^{-1}. This process shows an apparent advantage over the CO_2 electrochemical reduction reaction in aqueous electrolytes.

Endothermic reduction reactions of CO_2 to useful compounds require an energy input. The energy may come from different formations, such as photochemical energy, electrical energy, and thermal energy. The energy of these gas-phase electrocatalytic reduction reactions came from electrical energy and also thermal energy. Electrical energy inputs are calculated from Eq. 3.14:

$$G_f = tV^2/R = I^2Rt \tag{3.14}$$

where V is the DC voltage, R is the resistance, I is the current, and t is the reaction time. Heat energy inputs were calculated based on the analytical data, the mass balance, and thermodynamic data. The energy balances for different electrocatalysts are shown in Table 3.5. In a typical electrocatalytic conversion of CO_2 with Pt/CSZ and Pt/OMS-2 catalysts, the heat energy demand (at 900°C) for the Pt/CSZ catalyst is 1000 times more than the electrical energy demand, and the heat energy

TABLE 3.5 Energy Balance of Electrocatalytic Reactions with Different Electrocatalysts

Catalysts (Reaction Temp.)	Electronic Energy Input (J/H)	Heat Energy Input (kJ/H)	Chemical Energy Stored (kJ/H)	Ref.
Pt/CSZ (900°C)	0.21 ± 0.02	10.5 ± 0.5	10.5 ± 0.6	56
Pt/OMS-2 (400°C)	230 ± 12	10.1 ± 0.5	10.3 ± 0.5	56
ZnO/YSZ (700°C)	0.83 ± 0.04	3.1 ± 0.2	3.1 ± 0.2	This work

demand (at 400°C) for the Pt/OMS-2 catalyst is about 43 times more than the electrical energy demand. Solid oxide electrolyte affects the electrical energy input. The thermal dissociation of CO_2 and splitting of water favor high temperature. The reduction reaction proceeds faster, and more thermal energy is stored in a chemical form. These thermal reactions demonstrate that these CO_2 reduction reactions are catalytic reactions and are different from coelectrolysis of CO_2 and H_2O.

The effects of DC voltages and currents are critical to the CO_2 electrocatalytic reduction reaction. Figure 3.5 shows that the yields of the CO_2 reduction products were controlled by DC voltages and currents at the same reaction temperature. The product yield is proportional to V^2 (V = DC voltages). Although thermal energy provides most of the energy, the reduction reaction is controlled by the applied DC voltage. Without applying a DC voltage, the control experiment showed no products found by GC and NMR analyses (a detection limit of $2-5$ μg/ml for paraformaldehyde).

Electrochemical impedance spectra (EIS) and simulated circuits are two important tools for studies of electrocatalytic reactions. A simulated circuit is based on EIS spectra. EIS spectra were monitored in real time as the activation reaction proceeded a polarizing step with a DC voltage. The effects of DC voltages and currents are observed in EIS spectra. Figure 3.6b shows the Nyquist spectrum when a -1.0 V DC voltage at 900°C polarized the Pt–CSZ interface. Based on this Nyquist spectrum, an equivalent circuit was proposed to model the interface of the Pt/CSZ catalysts with Echem analyst software (Echem Analyst 3000, Gamry). The values of the circuit components are calculated as shown in Figure 3.6a. The current flow from the electrolyte to the reaction sites, as presented by the half-cell

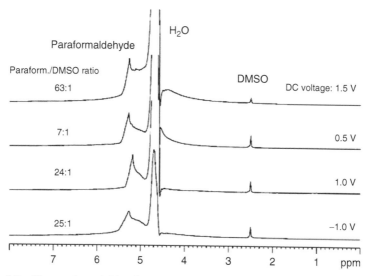

Figure 3.5 The product yields when applying different DC voltages on the Pt/CSZ catalytic cell under conditions of a temperature of 900°C and a CO_2/H_2O molar ratio of 2. Reprinted from ref. 56, Copyright 2010, with permission from Elsevier.

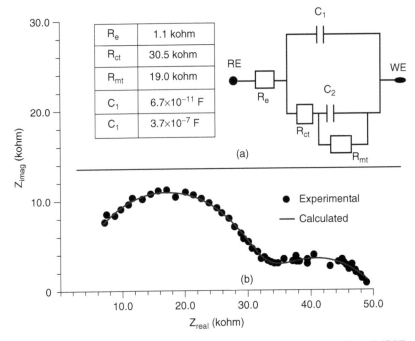

R_e	1.1 kohm
R_{ct}	30.5 kohm
R_{mt}	19.0 kohm
C_1	6.7×10^{-11} F
C_1	3.7×10^{-7} F

Figure 3.6 Typical Nyquist spectrum (b) and its equivalent circuit (a) of the Pt/CSZ catalytic cell polarized with a DC voltage of -1.0 V at 900°C. Reprinted from ref. 56, Copyright 2010, with permission from Elsevier.

(CO_2,H_2O),Pt|CSZ, is modeled by the resistor (R_{ct}) representing the charge transfer, in series with the parallel (R_{mt},C_2) circuit, where R_{mt} models the mass transfer and C_2 models the double layer capacitance. The C_1 capacitance models the capacitive behavior associated with the nano-sized Pt coating. The resistance (R_{mt}) is parallel with a constant phase element (CPE). CPE behavior is associated with fractal electrode geometries, the fraction of exposed edge plane orientations, and inhomogeneous reaction rates.

The polarization effect is significant as shown in the real-time EIS spectra (Fig. 3.7). EIS spectra were also monitored in real time as the activation reaction proceeded without applying a DC voltage (or current) and with different DC voltages (or currents). This study revealed how an electric field affected the electrochemical reduction reaction and what kinds of mechanisms were involved. Figure 3.7 shows Nyquist spectra of different catalysts polarized on different catalysts with different DC voltages and currents. The elementary values of the equivalent circuit are shown in Table 3.6.

Looking at the charge transfer resistance (R_{ct}) change when applying 0 V and 1.5 V on the Pt/CSZ catalysts, the R_{ct} value significantly dropped from 1.0 M Ω to 0.5 MΩ (Fig. 3.7a). A significant drop of the R_{ct} value from 2.0 kΩ to 1.3 kΩ and 1.0 kΩ was also observed when the galvanostatic polarization current across the

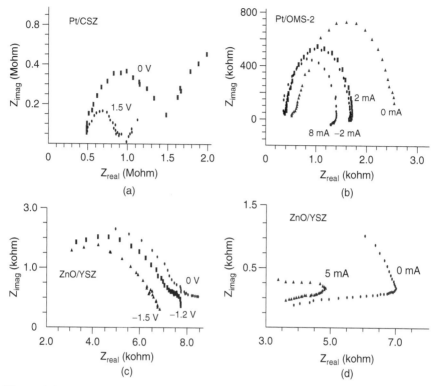

Figure 3.7 Real time Nyquist spectra of the polarized ZnO/YSZ catalysts. (a): poten-tiostatic polarized Pt/CSZ catalysts, (b): galvanostatic polarized Pt/OMS-2 catalysts. (c): potentiostatic polarization, (d): galvanostatic polarization. Reprinted from ref. 56, Copyright 2010, with permission from Elsevier.

Pt/OMS-2 interfaces increased from 0 mA to 2 mA and 8 mA (Fig. 3.7b), respec-tively. For the ZnO/YSZ catalysts, the R_{ct} values decreased from 6.6 kΩ to 5.7 kΩ and 4.1 kΩ when the potentiostatic polarizations on the ZnO/YSZ catalysts changed from 0 V to −1.2 V and −1.5 V, respectively (Table 3.6). The electrolyte resistance R_e slightly increased and the mass transfer resistance R_{mt} slightly decreased for the

TABLE 3.6   Element Values of Equivalent Circuit with Different DC Voltages and Catalysts

Catalyst	DC Voltage, V	R_e, K ohm	R_{ct}, K ohm	C_a, F	R_{mt}, K ohm	C_b, F
ZnO/YSZ	0	1.2	6.6	7.5×10^{-9}	0.50	5×10^{-5}
ZnO/YSZ	−1.2	1.4	5.7	3.0×10^{-9}	0.47	1×10^{-5}
ZnO/YSZ	−1.5	1.5	4.1	1.4×10^{-9}	0.44	4×10^{-6}
Pt/CSZ	−1.0	1.1	30.5	6.7×10^{-11}	19	3.7×10^{-7}

ZnO/YSZ catalysts. The charge transfer resistances followed the Arrhenius-type temperature dependence as shown below (Eq. 3.15):

$$T/R_{ct} = A \exp(-E_a/RT) \tag{3.15}$$

$$E_a = RT(\ln A + \ln R_{ct} - \ln T) \tag{3.16}$$

where T is temperature, R_{ct} is charge transfer resistance, A is the preexponential factor, E_a is the activation energy, and R is the gas constant. Equation 3.16 shows that the activation energy E_a of the CO_2 reduction decreases when R_{ct} is decreased by applying DC voltages. This explains why applying a DC voltage promotes the reduction of CO_2 and H_2O.

Electrocatalysts determine the selectivity to different products at applied voltages. The Pt/CSZ catalysts produced mainly paraformaldehyde and small amounts of methanol; the ZnO/YSZ catalysts produced ethylene and paraformaldehyde; the SiC/CSZ catalysts produced methane, hydrogen, and carbon monoxide. The activity of these electrocatalysts is determined by several factors, such as size, morphology, thickness, and the nature of the electrocatalysts. The size and morphology could change under conditions of high temperature, electrical modification of the surfaces, and chemical reactions as shown in Figure 3.8. Ethylene was produced with the ZnO/YSZ catalysts in the first stage (<1 h) with high CO_2 conversion. The activity decreased fast when the electrocatalytic reaction proceeded. The surface morphology was changed from nanorod arrays to nanosheet flowers. The stability of these electrocatalysts is critical for practical applications. Doping other metal ions in the electrode may improve the stability and activity [69]. The Pt/CSZ catalysts show good stability for >100 h. The activity did not show much decrease (<10%) under low voltage (or small current) polarization. Currently, doping methods with other transition metals are used for stabilizing the electrocatalysts [70,71]. These effects are believed to be due to electron transfer from the support to the metal

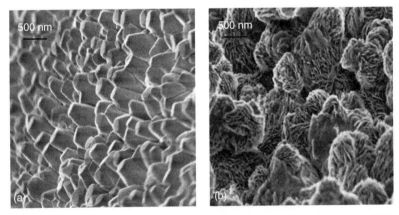

Figure 3.8 FESEM images of ZnO before and after electrocatalytic reduction of CO_2 and H_2O at 700°C.

crystallites originating from an electronic type of interaction induced by doping at the metal-support interface. Other promoters (K^+ ions) are also being considered for the growth of long carbon chains [72].

SOEs play an important role in charge and mass transfer. Oxygen ions are responsible for the main contribution for the conductivity of CSZ and YSZ; much lower electronic conductivity exists in CSZ and YSZ. These types of SOEs need to work at a high temperature to have a low charge transfer resistance and mass transfer resistance. OMS-2 is a mixed conductor. Oxygen ions and electrons are responsible for the conductivity of OMS-2. The OMS-2 electrolyte has excellent ionic and electrical conductivity at room temperature [73]. This characteristic of OMS-2 provides an opportunity for low-temperature CO_2 reduction. Activation of CO_2 requires an increase in the working temperature. The lowest temperature for electrocatalytic reduction of CO_2 in these experiments is 250°C with the Pt/OMS-2 catalysts. Low temperature reduces the equipment and operation cost. Low-temperature operation using low-cost electrocatalysts for CO_2 reduction is a current strategy for CO_2 utilization toward practical applications.

3.4.2 Coordination Complexes

Transition metal (Cu, Zn, Ru, and Pd) complexes play important roles in the stoichiometric transformation of CO_2 to carbonate salts [36,39,74–76]. Complex agents such as glycol-amines, porphyrins [37], polyamines, pyridines, and benzimidazole bind CO_2 reversibly through the formation of carbamates in this transformation. The complex agents around the metal are carefully chosen, and then the metal aids in stoichiometric or catalytic transformations of molecules. Electron transfer redox reactions between coordination complexes involve bridging ligands and coordination centers (metals). For example, in the cobalt-porphyrin-catalyzed electrochemical reduction of CO_2 in water, the two-electron, multistep electrochemical reduction reactions of CO_2 to CO in water can be summarized in several steps as shown in Figure 3.9 [76]. The electron transfer steps were determined by density functional theory (DFT) calculations with hybrid functions and dielectric continuum solvation. The CO_2 adsorption

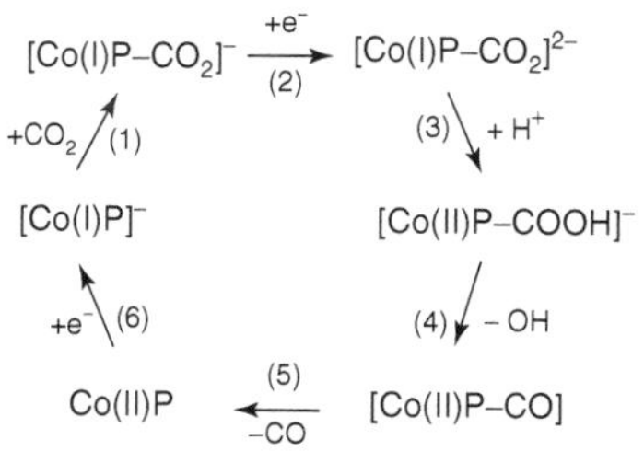

Figure 3.9 Possible mechanism of CO_2 reduction with electron addition deduced from hybrid DFT plus dielectric continuum redox potential calculations. Adapted with permission from ref. 76, Copyright 2010 American Chemical Society.

(step 1) and the first electron insertion (step 2) likely occurred simultaneously and cooperatively. The electron transfer between the gas diffusion electrode and the polymerized porphyrin catalyst is the rate-limiting step of the CO_2 to CO reduction reaction in water. The protonation of $[Co(I)PCO_2]^{2-}$ (step 3) is downhill at the bicarbonate buffer experimental conditions (pH: 7). The subsequent cleavage of the C–OH bond (step 4) is also exothermic($\Delta G° = -8.5 \pm 1.1$ kcal/mol). C–OH cleavage may occur on a nanosecond time scale at 27°C. The cobalt porphyrin has drastically reduced the C–OH cleavage barrier. Hence two key steps (steps 3 and 4) in the multistep reaction may proceed readily.

Without a metal-based multielectron transfer, pyridinium-based catalysts are effective and stable catalysts for the reduction of CO_2 to various products operating at low overpotential [42]. A pyridinium radical is a one-electron charge-transfer mediator and efficiently transfers six electrons necessary for transforming CO_2 to methanol. In these processes, the pyridinium radical reacts with the various species and intermediates through a coordinative interaction that stabilizes the intermediate species. A single pyridinium catalyst performing as an electron shuttle has the ability to reduce multiple species. High faradaic yields were obtained with both metal and semiconductor electrode materials. This homogeneous reaction pathway was dominant in solution processes in the absence of an electrocatalytic interface.

3.4.3 Enzymes

Enzyme-based electrocatalysts provide critical solutions to improving selectivity, lowering the overpotential, and increasing the reaction kinetics of CO_2 conversion. This enzyme-based catalysis involves dehydrogenase [77] as an electrocatalyst and a mediator. Isocitrate dehydrogenase [78], carbonate dehydratase [79], and carbon monoxide dehydrogenase (CODH) [80,81] are the typical enzymes for CO_2 conversion [50]. These CODH enzymes have been optimized by nature for the equilibration of CO_2 and CO (Fig. 3.10). Recently, mimicking the natural process, these artificial electrocatalysts have provided a design for highly efficient homogeneous catalysts for CO_2 conversion to CO.

A process similar to CODH electrocatalysts was found in electrochemical CO_2 conversion using microbes. The acetogenic bacteria *Moorella thermoacetica* and *Clostridium formicoaceticum* converted CO_2 into formate with a current efficiency of 80–100% in phosphate buffer solution (pH 7.0) at −0.58 V vs. NHE; near the

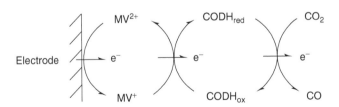

Figure 3.10 Electrocatalytic reduction of CO_2 to CO by CODHs. Reprinted with permission from ref. 50. Copyright 2003 American Chemical Society.

equilibrium potential of CO$_2$/formate [31]. This high energy efficiency is comparable to the conversion of CO$_2$ to CO at -0.57 V (NHE) by CODH from *M. thermoacetica* (the thermodynamic CO$_2$/CO redox potential is -0.48 V at pH $= 6.3$).

3.5 CO$_2$ HYDROGENATION

CO$_2$ hydrogenation is similar to Fischer–Tropsch synthesis, which converts a mixture of CO and hydrogen into liquid hydrocarbons. CO$_2$ is much more inert compared to CO. In CO$_2$ hydrogenation, the first step is CO$_2$ conversion to CO via the reverse water-gas shift (RWGS) reaction (Eq. 3.17). This step needs energy, hydrogen, and catalysts. These catalysts function as RWGS catalysts [82–86] and as FTS catalysts.

$$CO_{2(g)} + H_{2(g)} \leftrightarrow CO_{(g)} + H_2O_{(g)} \tag{3.17}$$

Typical low-temperature RWGS catalysts are Fe$_3$O$_4$, Co$_3$O$_4$, Fe$_x$C, and Co$_x$C [87]. Raney copper catalysts were also active in the RWGS reaction [88–90]. But for CO$_2$ conversion catalysts, converting CO$_2$ to CO is just the first step. The next step follows Fischer–Tropsch synthesis to convert CO to valuable industrial feedstocks, such as lower olefins and liquid hydrocarbons. In CO$_2$ conversion, these steps proceed simultaneously. FTS catalysts can be used for CO$_2$ hydrogenation, but these catalysts must tolerate CO$_2$ and also function as a catalyst in the RWGS reaction.

Table 3.7 shows that products such as methane, ethanol, hydrocarbons, and oxygenates were also selectively produced from CO$_2$ hydrogenation using different catalysts [91–97].

3.5.1 Active Phases

Methanol has been produced industrially from syngas containing CO and CO$_2$ using Cu/ZnO/Al$_2$O$_3$ catalysts [98]. The mechanisms of selective hydrogenation of CO$_2$ to methanol have been extensively studied [99–101]. The Cu/ZnO-based catalysts have been found to be the most efficient catalysts for methanol production from CO$_2$ hydrogenation. Aurichalcite [(Zn,Cu)$_5$(CO$_3$)$_2$(OH)$_6$] exhibits high activity and selectivity for methanol synthesis [102]. Morphology, size, and shape-dependent interactions of ZnO with Cu nanoparticles also play an important effect in selective hydrogenation of CO$_2$ to methanol [103–106].

CO$_2$ hydrogenation to hydrocarbon via Fischer–Tropsch synthesis is another pathway for CO$_2$ hydrogenation. Although Fischer–Tropsch catalysts have extensively been developed based on Co, Fe, Mo, Ru, and Ni catalysts on SiO$_2$, Al$_2$O$_3$, MgO, zeolites, and carbon (carbon nanotubes or carbon fibers) [107–112], the development of CO$_2$ hydrogenation catalysts has lagged behind. The main reason is that CO$_2$ hydrogenation is much more difficult than CO hydrogenation because of the inertness of CO$_2$. Generally, CO conversion (high as 87%) is much higher than CO$_2$ conversion (high as 45%) in hydrogenation. CO$_2$ could be a poison for CO hydrogenation catalysts.

TABLE 3.7 Useful Compounds Synthesized by CO_2 Hydrogenation Method

Year	Catalysts	Reaction Conditions	CO_2 Conversion & Products	Ref
2011	Cu/ZnO,	4.5 Mpa, CO_2/H_2: 1: 2.2–2.5, 40 sccm, 270–280°C	CO_2 Conv.: 11–18% CH_3OH: 72.7–32%	91
2009	Zn/ZnO	335–648°C, 13.7–31.2 MPa, supercritical CO_2	CO, H_2 >95% Other HC, oxygenates	92
2009	Co/Al$_2$O$_3$	220°C, 20 bar, gas hourly space velocity (GHSV): 4800 cm^3/h/gcat. H_2/CO_2: 2.45–4.9	CH_4 (>90%), CO	93
2007	Ionic liquids,	60°C, 9Mpa H_2, 18Mpa CO_2	Formic acid, 103 h^{-1}	94
2002	Fe-Co/magnetite	CO_2/H_2 (1/4), 1 MPa, GHSV: 3000 h^{-1}, 260°C	CO_2 conversion: 28%, CO: 52%, C_2-C_4: 65.5%	95
1996	Rh/SiO$_2$	5 Mpa, 240°C, 6000cm^3/h.g cat.	CO_2 conversion: 7%, CH_4: 64%, CO: 18%, alcohols: 21%	96
1993	Ball milling Fe and C (cementite)	0.1 Mpa, 250°C, GHSV: 15,000–20,000 h^{-1}	1.4×10^3 mol CO/g cat.h, less CH_4 and C_2-C_4 (23%)	97

There is some progress in CO_2 hydrogenation. Tihay et al. [95] reported a CO_2 conversion of 28.4% and an olefin-to-paraffin ratio of 1.2 under CO/H$_2$ (1/1) at a pressure of 1 MPa, a GHSV of 3000 h^{-1}, and a temperature of 250°C using Fe-Co alloy/magnetite catalysts for CO_2 hydrogenation. Kishan et al. [113] reported CO_2 conversion of 27.5% and a selectivity to C_{2+} hydrocarbons of 58.5% under a total pressure of 10 atm, a GHSV of 1.9 l/(g·h), and a temperature of 300°C using Fe(10%)K(5%) catalysts supported on Al$_2$O$_3$ –MgO. We recently developed new CO_2 conversion catalysts by a chemical design method. The CO_2 conversion was increased to 45%, and the olefin-to-paraffin ratio was improved to above 2.0 under a total pressure of 13.6 atm, a GHSV of 7.2 l/(g·h), and a temperature of 320°C using OMS-2-supported iron or cobalt catalysts [114]. A low selectivity to CH_4 (6%) and a high selectivity to C_{2+} hydrocarbons (75%) were achieved.

CO_2 hydrogenation catalysts undergo dynamic phase transformations under high-pressure and high-temperature conditions. These changes made it difficult to identify which phases, metallic phases (Fe and Co) or carbide phases (Fe_2C, $Fe_{2.5}C$, Fe_3C, Co_2C, and Co_3C), are the actual active phases in CO_2 hydrogenation [115]. In situ characterization techniques, such as scanning transmission X-ray microscopy [116], X-ray diffraction [117], X-ray absorption fine structure spectroscopy, and Raman spectroscopy [118], have been used to characterize working catalysts under complex conditions. Although these metallic phases and carbide phases have been identified in ex situ XRD experiments, they may be oxidized or decomposed when exposed to air and moisture. Recently, in situ characterization experiments have identified the phases of Co/Al$_2$O$_3$ [117] and unsupported Fe$_2$O$_3$ catalysts [118] under realistic reaction conditions. More complex supported catalysts, such as OMS-2-supported Co, Cu, and Fe catalysts, have been characterized by in situ XRD under CO$_2$ hydrogenation conditions and have been analyzed concomitantly with online gas chromatography to measure activity. Structure-activity relationships have been determined with this procedure.

When OMS-2-supported iron catalysts were reduced in H$_2$ (or CO), Fe NPs were formed on the surfaces of manganese oxide support. Manganese oxides are mixed-valent oxides. Temperature programmed reduction-mass spectra data show that reduction changed the average oxidation states of manganese in the support. Therefore, the oxidation property of the manganese oxide supports change. Different products were produced. Light olefins and jet fuels were produced when the OMS-2-supported catalysts were reduced at about 450°C; carboxylic acids were formed when these catalysts were reduced at about 350°C.

Metallic particles convert to metal carbides in CO- or CO$_2$-containing gases at medium or high temperatures. In situ XRD patterns (Fig. 3.11a) show that Fe was initially carburized to Fe$_5$C$_2$ at 200°C and then converted to Fe$_3$C at 320°C in CO/H$_2$. K$_2$O promoters also gradually migrated to the MnO$_x$ surface and then formed KHCO$_3$ in CO/H$_2$. Similar phase transformation occurred for the reduced Co$_3$O$_4$ (CuO)/K-OMS-2 catalysts. Hexagonal-closed-packed (hcp) Co and K$_2$O were formed on the MnO$_x$ surface. hcp Co converted into fcc Co at high temperatures (300–450°C) because of a small energy gap (0.03 eV) [119], and both hcp Co and fcc Co transformed into Co$_2$C at 200°C and then converted to Co$_3$C at 320°C in CO/H$_2$. The carburization in CO$_2$/H$_2$ is different from that in CO$_2$/H$_2$. Metallic Fe NPs partially converted Fe$_5$C$_2$ in CO$_2$/H$_2$. Metallic Co NPs partially converted Co$_2$C in CO$_2$/H$_2$. Metallic particles are the main phases.

These different metallic phases and metal carbide phases were tested in CO$_2$ hydrogenation. The activity tests showed that metal carbide phases (Co$_3$C and Fe$_3$C) are more active in CO$_2$ hydrogenation. These supported Co$_3$C/MnO/KHCO$_3$ and Fe$_3$C/MnO/KHCO$_3$ catalysts have 3 and 2.5 times the conversion of CO$_2$ compared with supported Co/MnO/KHCO$_3$ and Fe/MnO/KHCO$_3$, respectively. Metallic Co and Fe phases accounted for the high selectivity to CO and CH$_4$. This is consistent with reported results that metallic Co/Al$_2$O$_3$ catalysts have a selectivity of >90% to methane in CO$_2$ hydrogenation [93,120].

Potassium promotion is evident in CO$_2$ hydrogenation using these OMS-2-supported Fe and Co catalysts. Up to 4% of potassium was naturally doped into the tunnel sites of KMn$_8$O$_{16}$ octahedral molecular sieve (OMS-2). Compared with the support without potassium, the CO$_2$ conversion increased 2–3%. The selectivity to C$_{2+}$ hydrocarbon increased from 35% to 65%, and the selectivity to CH$_4$ decreased from 52% to 15%. More valued products were produced.

Overall, the high-efficiency CO$_2$ hydrogenation catalysts have optimal components and nanostructure. These components include a RWGS catalyst (Fe$_3$O$_4$ or

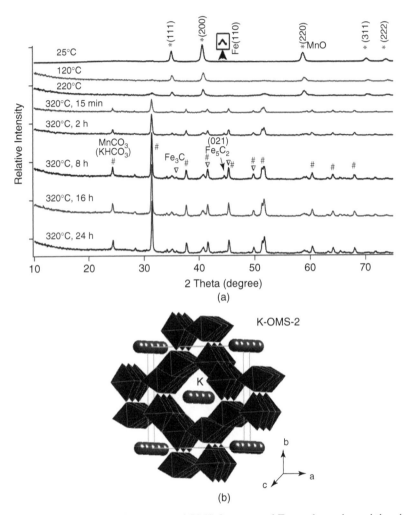

Figure 3.11 (a) In situ XRD patterns of OMS-2 supported Fe catalysts showed the phase transformation with the changes of temperature and time. #: MnCO$_3$, Δ: Fe$_3$C. (b) Illustration of KMn$_8$O$_{16}$ octahedral molecular sieve.

Co$_3$O$_4$), iron (cobalt) carbides, KHCO$_3$ promoters, and nanostructured supports. Manganese oxide can increase light olefin production. The catalysts must locate on the support surfaces and interact with other promoters and the supports. The synergistic effects of these components contribute to the high activity and selectivity of the catalysts.

3.5.2 Products of CO$_2$ Hydrogenation

3.5.2.1 Methanol. Methanol is used as a solvent, chemical material, and fuel for fuel cells and internal combustion engines. The Lurgi MegaMethanol process [121] is a process for commercial methanol production using CO_2, CO, and hydrogen. The methanol production capacity with this method was 0.67–2.3 million tons/year in 2009. Another method developed by ICI Co. Ltd. used CO_2 and H_2 initially under extreme reaction conditions, 250–300°C and 50–100 bar with copper–zinc-based oxide catalysts [122]. Low-pressure methanol (LPM) production was developed by ICI in the late 1960s and is now owned by Johnson Matthey [123]. Typical operating conditions for the LPM process are a pressure range from 5 to 10 MPa and a temperature range from 220 to 260°C. Copper-based catalysts supported on ZnO exhibit high performance for these reactions. Methanol fuel is corrosive to some metals (Al) and plastic components. Anticorrosive additives and special design are needed for methanol fuel applications. Ethanol is substituted for methanol as a fuel because of its lower degree of corrosion. Large amounts of ethanol are produced from biomass [124–128].

3.5.2.2 *Fine Chemicals.*

The products of CO_2 hydrogenation vary over different catalysts. Cu-based catalysts produced alcohol products, mainly methanol. High-molecular-weight alcohol synthesis was performed by using Co–Cu-based perovskites as catalysts and alkali additives as promoters [129]. Co and alkali additives contributed to the carbon chain growth toward long chains. But manganese oxides restricted the chain growth and converted the formed intermediates to carboxylic acids. Co–Cu catalysts supported on manganese oxides produced acetic acid, propanoic acid, butanoic acid, 4-hydroxy, pentanoic acid, 5-hydroxy, 2*H*-pyran-2-one, tetrahydro, and 2*H*-pyran-2-one, tetrahydro-4-methyl (molar ratio 63:18:4:8:5:2) in CO_2 hydrogenation [114].

Multireaction sites existed on the catalyst surfaces in these complex reactions. Copper catalysts contributed to the formation of methanol. Manganese oxide-supported cobalt catalysts led to formation of carboxylic acids. Manganese oxides function as coupling agents [130] in the intermolecular and intramolecular condensations. Methanol reacted with carboxylic acids to form α-hydroxylic acids via intermolecular condensation. For example, CH_3OH and propanoic acid form butanoic acid, 4-hydroxy via the intermolecular (aldol) condensation (Eq. 3.18):

CH₃OH   Propanoic acid   butanoic acid, 4-hydroxy

$$H_2C=O + CH_3-CH_2-\underset{\underset{OH}{|}}{C}=O \xrightarrow[280°C]{\text{aldol condensation}} H_2C-CH_2-CH_2-\underset{\underset{OH}{|}}{C}=O$$

(3.18)

Pentanoic acid, 5-hydroxy was also produced from methanol and butanoic acid via similar intermolecular condensation. 2*H*-pyran-2-one tetrahydro was formed from pentanoic acid, 5-hydroxy converted to tetrahydro-2*H*-pyran-2-one via the intramolecular condensation (Eq. 3.19). Similarly, 5-hydroxy-pentanoic acids converted to tetrahydro-4-methyl-2*H*-pyran-2-one via similar intramolecular condensation.

$$
\text{5-hydroxy-pentanoic acid} \xrightarrow[\text{280°C, MnO}_x]{\text{Intramolecular condensation}} \text{tetrahydro-2H-pyran-2-one} + H_2O
$$

(3.19)

The synergistic effects of Co, Cu, MnO, and K provide pathways for the synthesis of various fine chemicals from CO_2 and H_2. Alpha hydroxyl acids are natural acids derived from fruit and milk sugars. These synthetic fine products from CO or CO_2 hydrogenation can be used for medicines and perfume.

OMS-2-supported cobalt catalyst produced carboxylic acids (liquid products) in CO_2 hydrogenation without copper in the catalyst. The liquid products of CO_2 hydrogenation are ethanoic acid (53 molar%), propanoic acid (27%), butanoic acid (14%), and pentanoic acid (6%). Carboxylic acids are colorless, and they have an unpleasant pungent smell. These carboxylic acids react with methanol or ethanol to form esters. Esters have a pleasant smell, and they are extensively used for perfumes.

3.5.2.3 Light Olefins. Light olefins include C_2H_4, C_3H_6, C_4H_8, C_5H_{10}, and C_6H_{12}. Global production of ethylene rose to 119 million tons in 2010 because of high demand growth [Chemical Industry News & Intelligence (ICIS)]. Worldwide propylene production has increased to 80.3 million tons in 2010 from 57.6 million tons in 2003 [about 5%/year, Chemical Market Associates Inc. (CMAI)]. Ethylene and propene are mainly used for polyethylene and polypropene production. C_4H_8, C_5H_{10}, and C_6H_{12} are used as monomers for the production of copolymers. These monomers are more expensive than ethylene and propene because of low yield and difficulty of production [131–134]. Ethylene and propene are mainly produced by cracking of ethane and liquefied petroleum gas. 1-Butene and 1-hexene are produced from ethylene. Linear alpha-olefins, especially 1-hexene and 1-octene, are key components for the production of linear low-density polyethylene (LLDPE), and the demand for 1-hexene and 1-octene has increased enormously in recent years. Millions of tons of LLDPE are produced from a Ziegler–Natta-type catalyst and alpha-olefin comonomers every year. For example, LLDPE is produced by the gas-phase EXPOL- or UNIPOL-type processes, where 1-butene and 1-hexene are employed [135–137]. Solution-type processes such as the Dow/DuPont process use 1-octene and 1-hexene comonomers [138,139].

Light olefins were produced from CO_2 hydrogenation using manganese oxide-supported iron catalysts as shown in Figure 3.12. In these catalysts, iron is located at the surfaces of manganese oxide supports. Iron exists in the form of Fe_3O_4, Fe_5C_2, and Fe_3C. Fe_3O_4 functions as catalysts for the RWGS reaction. Fe_5C_2 and Fe_3C are very active and selective Fischer–Tropsch catalysts. Manganese oxide restricts the long carbon chain growth and increases the formation of light olefins. In the manganese oxides, 3–5 atomic% of potassium

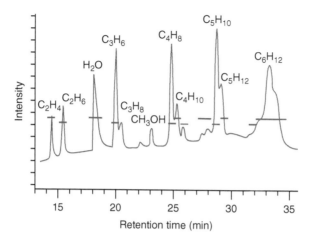

Figure 3.12 GC spectra of the gaseous products of CO$_2$ hydrogenation using OMS-2 supported Fe catalysts.

ions exist and act as promoters for carbon chain growth. The synergistic effects of Fe components, manganese oxide, and potassium contribute to the high activity and selectivity of these catalysts. CO$_2$ conversion is up to 45%, and the selectivity to C$_{2+}$ is up to 75%. About 65% of C$_{2+}$ hydrocarbons are light olefins.

3.5.2.4 *Fuels.*

Synthetic fuel is a liquid fuel obtained from natural gas, coal, biomass, or oil shale. "Synthetic fuel" is most often used to describe fuels manufactured via Fischer–Tropsch conversion, methanol to gasoline conversion, or direct coal liquefaction. Jet fuel is a type of aviation fuel designed for use in aircraft powered by gas-turbine engines. A standardized international specification has been set for jet fuel production. The carbon numbers, freezing point, and smoke point are restricted by the requirements of the product. Most commonly used kerosene-type jet fuel (including Jet A and Jet A-1) has a carbon number distribution between about 8 and 16. Wide-cut or naphtha-type jet fuel (including Jet B) used for cold weather has a carbon number distribution between about 5 and 15. Synthetic fuels from syngas contain no sulfur, and they are environmentally friendly products.

Synthesis of fuels from CO$_2$ hydrogenation is much more difficult than that from CO hydrogenation. However, it is possible for CO$_2$ hydrogenation to convert CO$_2$ to liquid fuel. Figure 3.13a shows that it is possible to produce Jet A type fuel with a carbon number distribution between about 8 and 16. This synthetic fuel has a small amount of aldehydes and aromatics as shown in Figure 3.13b. Aldehydes and aromatics components can be reduced by optimizing the catalysts and the molar ratio of H$_2$/CO$_2$.

3.5.3 Deactivation and Regeneration

The challenge for practical fuel production from CO$_2$ hydrogenation is low CO$_2$ conversion and low selectivity for C$_5$–C$_{15}$ hydrocarbons. The RWGS reaction

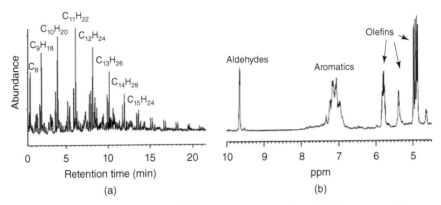

Figure 3.13 (a) Liquid products of CO_2 hydrogenation using OMS-2 supported Fe catalysts. (b) NMR spectrum of the liquid products of CO_2 hydrogenation with OMS-2 supported iron catalysts.

forms water, and water is a poison to most FTS catalysts; CO_2 is also a poison to some FTS catalysts. Fe, Co, and Cu catalysts formed interfaces of metal supported on $MnO_x (1 \leq x \leq 2)$ as well as a metal ion-doped MnO_x matrix. This complex was very active in the RWGS reaction at 300°C. OMS-2-supported iron catalysts formed a higher water concentration (1.23 mol%) than OMS-2-supported cobalt catalysts (0.58 mol%) in CO_2 hydrogenation (Fig. 3.14A). Compared with CO hydrogenation, CO_2 hydrogenation produces more water. Therefore, water reduces the catalytic activity of these catalysts. Higher water concentrations may

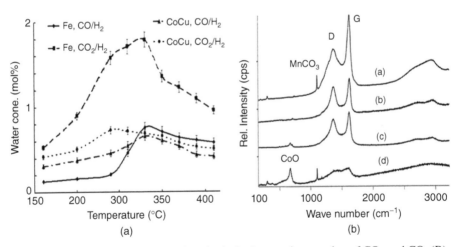

Figure 3.14 (A) Water concentrations in the hydrogenation reaction of CO_2 and CO. (B) Raman spectra of the post-reaction catalysts. Note: (a) OMS-2 supported Fe catalysts in CO_2 hydrogenation at 360°C for 1 h and 320°C for 144 h; (b) Co/Al_2O_3 catalysts in CO hydrogenation at 320°C for 120 h; (c) $CoCu/MnO/KHCO_3$ catalysts in CO_2/H_2 at 320°C for 144 h; (d) $CoCu/MnO/KHCO_3$ catalysts in CO_2/H_2 at 320°C for 2 h.

kill the reaction because metallic Fe and Co components, as well as metal carbides (Fe$_5$C$_2$, Fe$_3$C, Co$_2$C, and Co$_3$C), are vulnerable to water. When these active FTS active phases convert to metal oxides, the FTS reactions decrease and leave only the RWGS reaction.

Another deactivation factor is due to formation of carbon species on the catalyst surface. Disordered carbon (D) and graphitic carbon (G) species have been identified by Raman spectra as shown in Figure 3.14B. These carbon species are formed from dissociation of CO species and decomposed metal carbides. Disordered carbon can be dehydrogenated to form CH$_x$ species. Graphitic carbon species deposit on the catalyst surfaces and block the adsorption of reactants, therefore deactivating the catalysts. High temperatures increase the amount of carbon deposits, and the carbon deposits increase with reaction time.

Typical regeneration of catalysts was generally performed in three steps. The first step is to purge the catalysts by using hot nitrogen to remove adsorbed surface hydrocarbons at 350°C. The second step is to oxidize the adsorbed carbon species in air. The oxidation temperature is controlled to prevent the sintering of iron or cobalt catalysts. Like many industrial reactors, thermocouples are used to monitor the temperature of the catalyst beds. Diluted gases can be used for decreasing the temperature. Thermogravimetric analysis (TGA) and differential scanning calorimetry (DSC) analysis of the postreaction catalysts are performed to optimize the regeneration temperature. The postreaction CO$_2$ hydrogenation catalysts were regenerated at about 400°C, and surface carbon species are totally oxidized as shown by DSC and TGA data. The third step is to reduce the catalysts by using CO/He at 300–350°C to form active metal carbide phases.

3.5.4   Mechanisms of CO$_2$ Hydrogenation

The pathways of CO$_2$ hydrogenation using transition metal-catalyzed reactions are generally considered as follows (Fig. 3.15): A metal carbonyl complex ([M]-CO) reacts with hydroxide to form a metallocarboxylic acid ([M]-COOH$^-$);

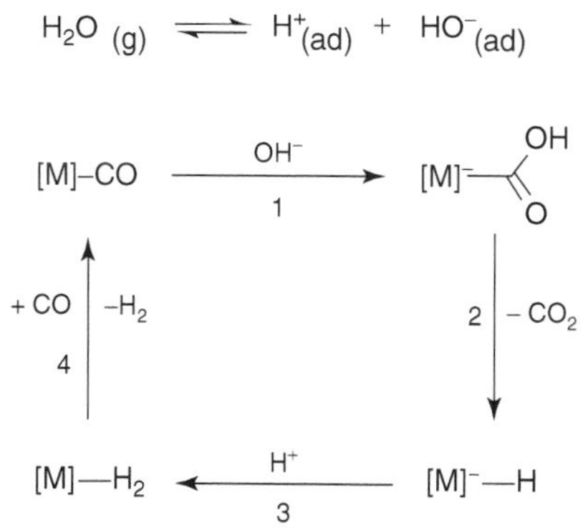

Figure 3.15 Possible mechanism of the reverse water-gas shift reaction. Adapted from ref. 140.

then this acid decarboxylates with metal hydride ([M]-H$^-$) form. The metal carbonyl complex is finally regenerated via the reaction with hydronium from water and carbon monoxide [140]. However, this mechanism of decarboxylation is debated, and the regeneration of the carbonyl complex may follow other pathways. With the understanding of the catalyst surface and intermediate species, new mechanisms have been developed based on modern characterization techniques, such as real-time time-of-flight mass spectrometry (DART TOF MS) [141] and time-resolved step-scan FT-IR [142,143].

Metal carbonyl complexes ([M]-CO) can form on different transition metal surfaces [144–148]. In situ XRD experiments (Fig. 3.16) showed that Co NPs reacted with CO and H$_2$ at room temperature to form tetracarbonylhydrocobalt [HCo(CO)$_4$]. This compound was decomposed to Co–H and CO as the temperature increased to 300°C. Metal hydrides (M–H$^-$) are the products of hydrogen dissociation on metal NPs or metal carbides. Metallocarboxylic acid ([M]-COOH$^-$) species have been detected by direct analysis in DART TOF MS.

The high throughput of DART coupled with high mass accuracy (5 ppm or 2 mmu) obtained with the TOF analyzer was used for the rapid identification of unknown adsorbed species in solid materials. DART TOF MS has been performed on postreaction catalysts with an AccuTOF instrument in an open-air environment.

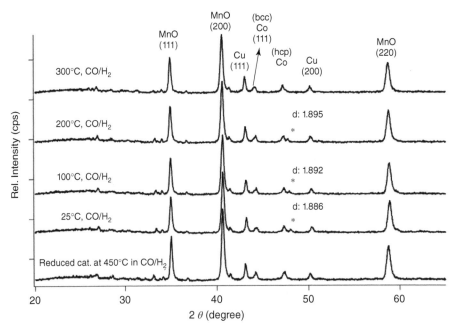

Figure 3.16 *In situ* XRD experiment of OMS-2 supported CoCu catalysts reduced in H$_2$/He and then exposed to CO/H$_2$ at temperatures of 25–300°C. The peak marked "*" is assigned to HCo(CO)$_4$. The peak marked (hcp) is assigned to hexagonal-closed-packed (hcp) cobalt.

The positive mode was used for the analysis of oxygenate intermediates. The corona discharge voltage was 3000 V; the exit electrode was held at 350 V; and the orifice of the atmospheric pressure interface was set to 30 eV. Helium (2.8 lL/min) was used as the corona discharge gas and was heated at 350–450°C.

In the mass range of 55–62 (the first group of peaks in Fig. 3.17), the four adsorbed species [$CH_3C \equiv CCH$ (1), $CH_3CH_2CH = CH_2$ (2), CH_3CH_2CHO (3), and CH_3COOH (4)] on three catalysts were identified by their masses. The experimental masses of these peaks are 54.0554, 56.0699, 58.0482, and 60.0281, respectively, which closely correspond to their theoretical masses (54.0908, 56.1067, 58.0491, 60.0520). The other groups of peaks show a mass increment of 14, indicating the addition of a CH_2 group to these compounds. These DART MS data indicated that the metallocarboxylic acid ([M]–COOH⁻) not only was

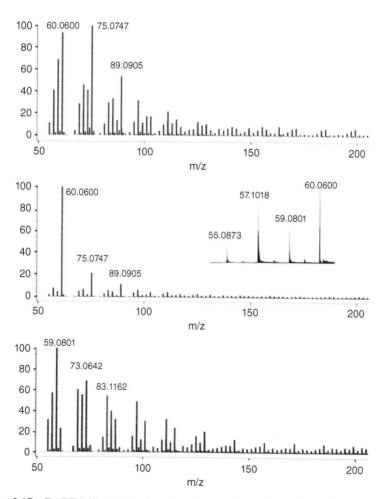

Figure 3.17 DART MS spectra of postreaction catalysts show adsorbed species on the catalysts.

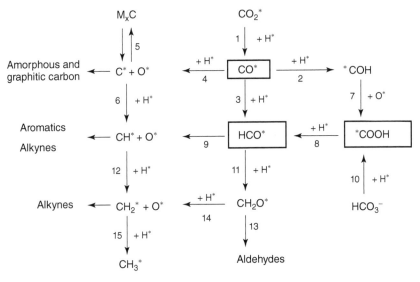

Figure 3.18   Possible CO_2 hydrogenation mechanisms.

adsorbed on the catalyst but also was involved in a series of carbon chain growths and reactions.

Current CO hydrogenation (Fischer–Tropsch synthesis) mechanisms of carbide phases, CO insertion, H-assisted CO dissociation, and oxygenate formation have theoretically elucidated how CO hydrogenation proceeds. These theories can be applied to CO_2 hydrogenation. Direct desorption, ionization, and identification of intermediates on these postreaction catalysts with the DART TOF MS method provided useful mechanistic information for the study of CO_2 hydrogenation. These experimental data (Fig. 3.17) suggested that CO_2 hydrogenation involves several mechanisms. Figure 3.18 is proposed to explain the pathways of the selective formation of carboxylic acids, alkenes, and other intermediates as shown in Figure 3.17.

For the CO_2 hydrogenation on the OMS-2-supported Fe catalysts, the chain initiation starts with CO_2 adsorbed on Fe-H, and adjacent adsorbed methylene groups are formed via a carbide mechanism [149] and then react with A species to form B species (Eq. 3.20):

$$(3.20)$$

The oxygenate mechanism [150] explains the high percentage of C_2–C_4 carboxylic acids on the Fe catalysts (Fig. 3.17). The B species or other species formed from B species are protonated by DART ion sources (Eq. 3.21) to form C_2–C_4

carboxylic acids.

$$\underset{\substack{\text{Fe}}}{\overset{\substack{\text{CH}_3 \\ | \\ \text{C}}}{O \diagdown \diagup O}} + \text{H}^+ \longrightarrow \underset{\substack{\text{OH}}}{\overset{\substack{\text{CH}_3 \\ | \\ \text{C}=\text{O} \\ |}}{}} \tag{3.21}$$

CO$_2$ converts to CO in the RWGS reaction. In the following CO hydrogenation, the carbide mechanism involves direct CO dissociation and subsequent hydrogenation of chemisorbed carbon (C*). For active iron carbides, such as Fe$_3$C (001), H$_2$ can easily cleave into adsorbed H and CH species on the Fe-C hybrid sites [151]. CH$_x$* (x = 1–3) species grow to hydrocarbon chains (steps 1, 4, 5, 6, 10, 13, 14, and 15 of Fig. 3.18). Alkynes, alkenes, and aromatics form on the catalyst surfaces. The combination of Fe = CH$_2$ and adsorbed CO$_2$ on the Fe catalysts contributes to the formation of carboxylic acids.

The H-assisted CO dissociation mechanism [152] also involves the dissociation of CO on the iron (or cobalt) catalyst surfaces. Chemisorbed HCO* and CHOH* species form because of the reaction of chemisorbed hydrogen atoms with chemisorbed CO* on saturated surfaces (steps 2, 3, 9, and 12). The combinations of HCO* and Fe = CH$_2$ on the Fe catalysts lead to the formation of aldehydes on the catalyst surfaces.

Manganese oxides are dehydrogenation catalysts. They not only restricted further hydrogenation of CH* and CH$_2$* to form CH$_3$* (step 13) but also converted alcohols to alkenes by removing a H$_2$O molecule. Carboxylic acids were reduced to aldehydes by dissociated H* and were further reduced to alkenes via loss of an H$_2$O molecule. These synergistic effects led to the high selectivity to light olefins up to 90%. Other nontraditional CO$_2$ hydrogenation pathways have recently been developed. Ionic liquids [153–158] (ILs) are organic salts with a melting point below 100°C. Hydrogenation of CO$_2$ in IL has advantages of high selectivity, catalyst recycling, and product recovery. Low-temperature (25–100°C) ILs and supercritical carbon dioxide as alternative solvents have combined to provide a wide range of homogeneously catalyzed reaction conditions for CO$_2$ hydrogenation. Liu et al. [159] reported that N,N-dialkylformamides have formed from hydrogenation of CO$_2$ with a turnover number of 820 mol product/mol ruthenium in the presence of dialkylamines and RuCl$_2$(PMe$_3$)$_4$ (Me: CH$_3$) catalysts. Zhang et al. [94] report the formation of formic acid with a turnover frequency of 103 h^{-1} using Si-(CH$_2$)$_3$ NH (CSCH$_3$)-RuCl$_3$-PPh$_3$ catalysts and ILs called 1-(N,N-dimethyl-aminoethyl)-2,3-dimethylimidazolium trifluoromethanesulfonate. Vostrikov et al. [92] reported the formation of carbon and CO by reactions of bulk Zn with H$_2$O and CO$_2$ at sub- and supercritical conditions (514–600°C, 7.1–24.4 MPa, 0.041–0.158 g/cm^3). These nontraditional CO$_2$ hydrogenation processes demonstrated the different approaches of CO$_2$ hydrogenation, but most of these processes are not economically feasible because of high cost and low product value.

3.6 CO_2 REFORMING

The above approaches focus on the direct conversion of CO_2 to useful compounds. Another CO_2 utilization approach is chemical fixation of CO_2 into organic compounds as shown in Table 3.8, in which CO_2 is used as a carbon resource in organic synthesis. Electrochemical activation of CO_2 in ILs has been extensively used for CO_2 reduction because of increased solubility and overcoming the thermodynamic stability and kinetic inertness of CO_2. These electrochemical reduction reactions proceed under mild conditions without additional supporting electrolytes. Cyclic carbonates (Eq. 3.22) have been synthesized through cyclo addition of CO_2 to epoxides in a series of ionic liquids consisting of 1-butyl-3-methyl-imidazolium ($BMIm^+$), 1-ethyl-3-methylimidazolium ($EMIm^+$), BPy^+ cations and BF_4^-, PF_6^- anions using a constant potential (-2.4 V vs. Ag/AgCl) [160].

$$R\text{-epoxide} + CO_2\,(1\,\text{atm}) \xrightarrow[\text{Ionic liquids, RT}]{-2.4\,\text{V vs. Ag/AgCl}} \text{cyclic carbonate} \qquad (3.22)$$

This reaction has high selectivity up to 100%, and the current efficiency and conversion varied from 27% to 90%. Since the report of the reaction of Eq.3.22, more applications of electrocatalytic reduction of CO_2 in ILs have been reported [161–164].

   The copolymerization of CO_2 and epoxides is a particularly promising route to activate and use CO_2 as a renewable C-1 source. Since polycarbonate was synthesized by S. Inoue and co-workers using CO_2 and propylene oxide in the presence of $ZnEt_2$ and water in 1969 [165], considerable efforts have been put into developing a copolymer synthesis based on CO_2 [166–168], especially polycarbonates (Eq. 3.23). More active and controllable catalysts based on Co, Zn, and Cr complexes show very high turnover number (TON) up to 26,000, as well as excellent control for the copolymerization of CO_2 and cyclohexene oxide or CO_2 and propylene oxide (Eq. 3.24) [169,170].

$$R\text{-epoxide} + CO_2 \longrightarrow \underset{\text{Cyclic carbonate}}{\text{cyclic carbonate}} \xrightarrow{\text{Polymerization}} \text{Polycarbonates} \qquad (3.23)$$

$$\text{epoxide} + O{=}C{=}O \xrightarrow{\text{cat.}} \left[\!\!\left(\!\! \begin{array}{c} \end{array}\!\!\right)_{\!n}\right] \left(+ \text{cyclic carbonate}\right) \qquad (3.24)$$

   The formation of oxazole 2-carboxylic acid in carboxylation of oxazole (Eq. 3.25) has been reported using [(NHC)AuOH] complexes and 1.05 mmol of KOH in THF at 45°C [171]. Various propiolic acids were synthesized through copper- and copper–N-heterocyclic carbene (NHC)-catalyzed transformation of CO$_2$ to carboxylic acid through C–H bond activation and carboxylation of terminal alkyne under mild conditions (Eq. 3.26) [172].

$$(3.25)$$

$$(3.26)$$

Except for the fixation of CO$_2$ onto polymer compounds, CO$_2$ can go through an electrocatalytic reaction with PhBr to form benzoic acid in dipolar aprotic solvent and 0.1 M Et$_4$NBF$_4$ [173]. CO$_2$ can react with PhCCH catalyzed by P(NHC-CO$_2$)$_2$ (NHC-Cu)/Cs$_2$CO$_3$ to form PhCCCOOH products in DMF solvent at room temperature. CO$_2$ also can be used as a mild oxidant to react with ethane to produce ethylene at 600–800°C with a high selectivity [174]. Table 3.8 reports some selected examples of the use of CO$_2$ as a carbon resource in organic synthesis [170–176].

TABLE 3.8 Synthesis of Useful Compounds by Chemical Fixation

Year	Raw Materials and Catalysts	Reaction Conditions	Products	Ref
2011	PhBr + CO$_2$ in dipolar aprotic solvents, 0.1 M Et$_4$NBF$_4$	25°C, Ag cathode, Al anode, electrocatalysis	benzoic acid (~80)	173
2010	PhCCH + CO$_2$ catalyzed by P (NHC-CO$_2$)$_2$ (NHC-Cu)/Cs$_2$CO$_3$	Dimethylformamide (DMF) solvent 25°C, 24 h	PhCCCOOH (52%)	172
2009	cyclohexene oxide + CO$_2$ catalyzed by [L^{1}Zn$_2$(OAc)$_2$]	10 atm for 24 h	Copolymer TON: 3300	170
2009	Ethane + CO$_2$ catalyzed by OMS-2	600-800°C	C$_2$H$_4$ + CO	174
2008	Propylene oxide + CO$_2$ catalyzed by [Co(salen)]/PPNCl PPN: bis(triphenylphosphine) iminium	2.0–1.7 MPa, 80°C, 1–3 h	Copolymer TON: 26000	169
2007	Benzene + CO$_2$ catalyzed by mesoporous C$_3$N$_4$	150°C, 10 bar, 12–48 h	Phenol (100%) +CO	175
2002	Epoxides + CO$_2$ + ionic liquids	−2.4 V, 25°C	100% cyclic carbonates	176

3.7 PROSPECTS IN CO_2 REDUCTION

Current catalytic technologies for CO_2 conversion have achieved high energy efficiency, high reaction rates, and high value products, although these are not achieved at the same time. An integration of several strategies may simultaneously achieve practical production of high-value products from CO_2. Electrocatalytic reduction combined with photochemical reduction, hydrogenation, and chemical fixation are possible new directions for CO_2 conversion to useful compounds. Energy costs, catalysts, and product value largely determine the feasibility of the CO_2 conversion process. The cost of CO_2 sequestration directly affects the cost of products. Favorable environmental and financial policies play an important positive role in CO_2 conversion.

The electrocatalytic approach can make a compact operation unit flexible for storage of renewable solar, wind, wave, and water power and thermal energy simultaneously stored in a chemical form, ideally to produce clean liquid fuels. The reactor can take electricity from the grid or a solar electrical station. The CO_2 conversion reactor would be operated on demand, using intermittent renewable energy and low-cost and stable electrocatalysts tolerant to small amounts of oxygen and moisture. The CO_2 gases captured from power plants and chemical plants contain small amounts of oxygen and moisture.

New nanostructured functional materials will play an important role in photosynthesis and electrocatalytic reduction. Thin film photocatalysts and electrocatalysts could be obtained by atomic layer deposition at a thickness of a few atoms. Size, shape, and morphology of the catalysts can be tuned for better catalytic performance. Dopants in the lattice structure could be controlled for an enhanced catalytic activity. Modern in situ characterization [177] of the real-time reaction by X-ray diffraction, extended X-ray absorption fine structure (EXAFS), time-resolved Fourier transform infrared (FT-IR), and X-ray absorption near edge structure (XANES) may lead to new fundamental information about these reactions and catalysts. These surface characterization technologies help us understand the heterogeneous catalysis reaction mechanisms and give insight for the design of cost-effective catalysts.

ACKNOWLEDGMENTS

We acknowledge the support of the U. S. Department of Energy, Office of Basic Energy Sciences, Division of Chemical, Biological, and Geological Sciences for this work.

REFERENCES

1. Jacobson, M. Z. (2005). Studying ocean acidification with conservative, stable numerical schemes for nonequilibrium air-ocean exchange and ocean equilibrium chemistry. *J. Geophys. Res.-Atmos.*, 110:D07302.

2. Taylor, H., Paul Sabatier (1944). *J. Amer. Chem. Soc.*, 66:1615–1617.

3. Khodakov, A. Y.; Chu, W.; Fongarland, P. (2007). Advances in the development of novel cobalt Fischer-Tropsch catalysts for synthesis of long-chain hydrocarbons and clean fuels. *Chem. Rev.*, 107:1692–1744.

4. Grimes, C. A.; Roy, S. C.; Varghese, O. K.; Paulose, M. (2010). Toward Solar Fuels: Photocatalytic conversion of carbon dioxide to hydrocarbons. *ACS Nano*, 4:1259–1278.

5. Gattrell, M.; Gupta, N.; Co, A. (2006). A review of the aqueous electrochemical reduction of CO_2 to hydrocarbons at copper. *J. Electroanal. Chem.*, 594:1–19.

6. Kenis, P. J. A.; Whipple, D. T. (2010). Prospects of CO_2 utilization via direct heterogeneous electrochemical reduction. *J. Phys. Chem. Lett.*, 1:3451–3458.

7. Jessop, P. G.; Ikariya, T.; Noyori, R. (1995). Homogeneous hydrogenation of carbon-dioxide. *Chem. Rev.*, 95:259–272.

8. Song, C. (2002). CO_2 conversion and utilization: an overview. *ACS Symp. Series*, 809: 2–30.

9. Song C. (2006). Global challenges and strategies for control, conversion and utilization of CO_2 for sustainable development involving energy, catalysis, adsorption and chemical processing. *Catal. Today*, 115:2–32.

10. Torr, M. R. (1993). The scientific objectives of the Atlas-1 shuttle mission. *Geophys. Res. Lett.*, 20:487–490.

11. Werner, J. H.; Kolodinski, S.; Queisser, H. J. (1994). Novel optimization principles and efficiency limits for semiconductor solar-cells. *Phys. Rev. Lett.*, 72: 3851–3854.

12. Mandler, D.; Willner, I. (1987). Effective photoreduction of CO_2 HCO_3^- to formate using visible-light. *J. Am. Chem. Soc.*, 109:7884–7885.

13. Carballo, R.; Castineiras, A.; Hiller, W.; Strahle, J. (1996). Photochemical activation of carbon dioxide by visible light mediated by cobalt(II) and nickel(II) complexes of 1,6-bis(benzimidazol-2-yl)-2,5-dithiahexane (BBDH). *J. Coord. Chem.*, 40:253–271.

14. Grodkowski, J.; Neta, P. (2000). Cobalt corrin catalyzed photoreduction of CO_2. *J. Phys. Chem. A*, 104:1848–1853.

15. Hartl, F.; Aarnts, M. P.; Nieuwenhuis, H. A.; van Slageren, J. (2002). Electrochemistry of different types of photoreactive ruthenium(II) dicarbonyl alpha-diimine complexes. *Coord. Chem. Rev.*, 230:107–125.

16. Takeda, H.; Ishitani, O. (2009). Highly efficient CO_2 reduction using metal complexes as photocatalyst. *J. Synthetic Org. Chem. Jpn*, 67: 486–493.

17. Hori, H.; Johnson, F. P. A.; Koike, K.; Ishitani, O.; Ibusuki, T. (1996). Efficient photocatalytic CO_2 reduction using $[Re(bpy)(CO)_3\{P(OEt)_3\}]^+$. *J. Photochem. and Photobiol. A Chem.*, 96:171–174.

18. Sato, S.; Morikawa, T.; Saeki, S.; Kajino, T.; Motohiro, T. (2010). Visible-light-induced selective CO_2 reduction utilizing a ruthenium complex electrocatalyst linked to a p-type nitrogen-doped Ta_2O_5 semiconductor. *Angew. Chem.*, 49:5101–5105.

19. Bookbinder, D. C.; Bruce, J. A.; Dominey, R. N.; Lewis, N. S.; Wrighton, M. S. (1980). Synthesis and characterization of a photosensitive interface for hydrogen generation-chemically modified P-type semiconducting silicon photo-cathodes. *Proc. of the Natl. Acad. of Sciences of the United States of America-Phys. Sciences*, 77:6280–6284.

20. Kumar, B.; Smieja, J. M.; Kubiak, C. P. (2010). Photoreduction of CO_2 on p-type silicon using Re(bipy-But)(CO)$_3$Cl: photovoltages exceeding 600 mV for the selective reduction of CO_2 to CO. *J. Phys. Chem. C*, 114:14220–14223.

21. Kaneco, S.; Katsumata, H.; Suzuki, T.; Ohta, K. (2006). Photoelectrocatalytic reduction of CO_2 in LiOH/methanol at metal-modified p-InP electrodes. *Appl. Catal. B: Env.*, 64:139–145.

22. Barton, E. E.; Rampulla, D. M.; Bocarsly, A. B. (2008). Selective solar-driven reduction of CO_2 to methanol using a catalyzed p-GaP based photoelectrochemical cell. *J. Am. Chem. Soc.*, 130:6342–6344.

23. Tanaka, K.; Ooyama, D. (2002). Multi-electron reduction of CO_2 via Ru-CO_2, -COOH, -CO, -CHO, and -CH$_2$OH species. *Coord. Chem.Rev.*, 226:211–218.

24. Dean, J. A., Editor, *Lange's Handbook of Chemistry* (15th Edition). McGraw-Hill: 1999; p 8.140.

25. Hara, K.; Kudo, A.; Sakata, T.; Watanabe, M. (1995). High efficiency electrochemical reduction of carbon dioxide under high pressure on a gas diffusion electrode containing Pt catalysts. *J. Electrochem. Soc.*, 142:57–59.

26. Ogura, K. (2003). Electrochemical and selective conversion of CO_2 to ethylene. *Electrochemistry* (Tokyo, Japan), 71:676–680.

27. Ogura, K.; Oohara, R.; Kudo, Y. (2005). Reduction of CO_2 to ethylene at three-phase interface. effects of electrode substrate and catalytic coating. *J. Electrochem. Soc.*, 152: 213–219.

28. Ogura, K.; Yano, H.; Tanaka, T. (2004). Selective formation of ethylene from CO_2 by catalytic electrolysis at a three-phase interface. *Catal. Today*, 98:515–521.

29. Le, M.; Ren, M.; Zhang, Z.; Sprunger, P. T.; Kurtz, R. L.; Flake J. C. (2011). Electrochemical reduction of CO_2 to CH$_3$OH at copper oxide surfaces. *J. Electrochem. Soc.*,158:45–49.

30. Yamamoto, T.; Hirota, K.; Tryk, D. A.; Hashimoto, K.; Fujishima, A.; Okawa, M. (1998). Electrochemical reduction of CO_2 in micropores. *Chem. Lett.*, 8:825–826.

31. Song, J.; Kim, Y.; Lim, M.; Lee, H.; Lee, J. I.; Shin, W. (2011). Microbes as Electrochemical CO_2 Conversion Catalysts. *ChemSusChem*, 4:587–590.

32. Ishihara, A.; Fujimori, T.; Motohira, N.; Ota, K.-I.; Kamiya, N. (2000). High temperature electrochemical heat pump using water gas shift reaction. Electrolytic reduction of CO_2 in molten carbonate. *Proc. Electrochem. Soc.*, 99–41, (Molten Salts XII):744–751.

33. Chu, D.; Qin, G.; Yuan, X.; Xu, M.; Zheng, P.; Lu, J. (2008). Fixation of CO_2 by electrocatalytic reduction and electropolymerization in ionic liquid-H$_2$O solution. *ChemSusChem*, 1:205–209.

34. Magdesieva, T. V.; Butin, K. P.; Yamamoto, T.; Tryk, D. A.; Fujishima, A. (2003). Lutetium monophthalocyanine and diphthalocyanine complexes and lithium naphthalocyanine as catalysts for electrochemical CO_2 reduction. *J. Electrochem. Soc.*, 150:608–612.

35. Dunks, G. B.; Stelman, D. (1983). Electrochemical studies of molten sodium carbonate. *Inorg. Chem.*, 22:2168–77.

36. Sonoyama, N.; Kirii, M.; Sakata, T. (1999). Electrochemical reduction of CO_2 at metal-porphyrin supported gas diffusion electrodes under high pressure CO_2. *Electrochem. Comm.*, 1:213–216.

37. Behar, D.; Dhanasekaran, T.; Neta, P.; Hosten, C. M.; Ejeh, D.; Hambright, P.; Fujita, E. (1998). Cobalt porphyrin catalyzed reduction of CO_2. Radiation chemical, photochemical, and electrochemical studies. *J. Phys. Chem. A*, 102:2870–2877.

38. Bockris, J. O.; Minevski, Z. S. (1992). Electrocatalysis: a futuristic view. *Int. J. Hydrogen Energy*, 17:423–444.

39. Lieber, C. M.; Lewis, N. S. (1984). Catalytic reduction of carbon dioxide at carbon electrodes modified with cobalt phthalocyanine. *J. Am. Chem. Soc.*, 106:5033–4.

40. Schwartz, M.; Cook, R. L.; Kehoe, V. M.; MacDuff, R. C.; Patel, J.; Sammells, A. F. (1993). Carbon dioxide reduction to alcohols using perovskite-type electrocatalysts. *J. Electrochem. Soc.*, 140:614–18.

41. Whipple, D. T.; Kenis, P. J. A., Prospects of CO_2 utilization via direct heterogeneous electrochemical reduction. *Journal of Physical Chemistry Letters* 2011, 1:3451–3458.

42. Barton Cole, E.; Lakkaraju, P. S.; Rampulla, D. M.; Morris, A. J.; Abelev, E.; Bocarsly, A. B. (2010). Using a one-electron shuttle for the multielectron reduction of CO_2 to methanol: kinetic, mechanistic, and structural insights. *J. Am. Chem. Soc.*, 132:11539–11551.

43. Smith, R. D. L.; Pickup, P. G. (2010). Nitrogen-rich polymers for the electrocatalytic reduction of CO_2. *Electrochem. Comm.*, 12:1749–1751.

44. Hu, B.; Stancovski, V.; Morton, M.; Suib, S. L. (2010). Enhanced electrocatalytic reduction of CO_2/H_2O to paraformaldehyde at Pt/metal oxide interfaces. *Appl. Catal., A*, 382:277–283.

45. Barnett, S. A.; Zhan, Z.; Kobsiriphat, W.; Wilson, J. R.; Pillai, M.; Kim, I. (2009). Syngas production by coelectrolysis of CO_2/H_2O: the basis for a renewable energy cycle. *Energy & Fuels*, 23:3089–3096.

46. Gangeri, M.; Perathoner, S.; Caudo, S.; Centi, G.; Amadou, J.; Begin, D.; Pham-Huu, C.; Ledoux, M. J.; Tessonnier, J. P.; Su, D. S.; Schloegl, R. (2009). Fe and Pt carbon nanotubes for the electrocatalytic conversion of carbon dioxide to oxygenates. *Catal. Today*, 143:57–63.

47. Delacourt, C.; Ridgway, P. L.; Kerr, J. B.; Newman, J. (2008). Design of an electrochemical cell making syngas ($CO+H_2$) from CO_2 and H_2O reduction at room temperature. *J. Electrochem. Soc.*, 155:42–49.

48. Wine, G.; Centi, G.; Perathoner, S.; Gangeri, M. (2007). Electrocatalytic conversion of CO_2 to long carbon-chain hydrocarbons. *Green Chem.*, 9:671–678.

49. Tao, G.; Sridhar, K. R.; Chan, C. L. (2004). Study of carbon dioxide electrolysis at electrode/electrolyte interface: Part I. Pt/YSZ interface. *Solid State Ionics*, 175:615–619.

50. Shin, W.; Lee, S. H.; Shin, J. W.; Lee, S. P.; Kim, Y. (2003). Highly selective electrocatalytic conversion of CO_2 to CO at −0.57 V (NHE) by carbon monoxide dehydrogenase from Moorella thermoacetica. *J. Am. Chem. Soc.*, 125:14688–14689.

51. Sugimura, K.; Kuwabata, S.; Yoneyama, H. (1989). Electrochemical fixation of carbon dioxide in oxoglutaric acid using an enzyme as an electrocatalyst. *J. Am. Chem. Soc.*, 111:2361–2.

52. Angamuthu, R.; Byers, P.; Lutz, M.; Spek, A. L.; Bouwman, E. (2010). Electrocatalytic CO_2 conversion to oxalate by a copper complex. *Science* (Washington, DC, U. S.), 327:313–315.

53. Kenis, P. J. A.; Whipple, D. T.; Finke, E. C. (2010). Microfluidic reactor for the electrochemical reduction of carbon dioxide: the effect of pH. *Electrochemical and Solid State Lett.*, 13:109–111.

54. Perathoner, S.; Gangeri, M.; Caudo, S.; Centi, G.; Amadou, J.; Begin, D.; Pham-Huu, C.; Ledoux, M. J.; Tessonnier, J. P.; Su, D. S.; Schlogi, R. (2009). Fe and Pt carbon nanotubes for the electrocatalytic conversion of carbon dioxide to oxygenates. *Catal. Today*, 143:57–63.

55. Hawkes, G.; O'Brien, J.; Stoots, C.; Jones, R. (2007). In 3D CFD model of high temperature H_2O/CO_2 Co-electrolysis, *ANS Summer Meeting* 2007.

56. Suib, S. L.; Hu, B. X.; Stancovski, V.; Morton, M. (2010). Enhanced electrocatalytic reduction of CO_2/H_2O to paraformaldehyde at Pt/metal oxide interfaces. *Appl. Catal. A*, 382:277–283.

57. Chu, D.; Jiang, R. Z. (2002). Novel electrocatalysts for direct methanol fuel cells. *Solid State Ionics*, 148:591–599.

58. Qi, Z. G.; Kaufman, A. (2003). Low Pt loading high performance cathodes for PEM fuel cells. *J. Power Sources*, 113:37–43.

59. Zhou, W. J.; Song, S. Q.; Li, W. Z.; Sun, G. Q.; Xin, Q.; Kontou, S.; Poulianitis, K.; Tsiakaras, P. (2004). Pt-based anode catalysts for direct ethanol fuel cells. *Solid State Ionics*, 175:797–803.

60. Lee, H. I.; Lim, D. H.; Lee, W. D. (2008). Highly dispersed and nano-sized Pt-based electrocatalysts for low-temperature fuel cells. *Catal. Surveys from Asia*, 12: 310–325.

61. Liang, Z. X.; Liu, B.; Liao, S. J. (2011). Core-shell structure: the best way to achieve low-Pt fuel cell electrocatalysts. *Progress in Chem.*, 23:852–859.

62. Luo, J.; Brock, S. L.; Marquez, M.; Matsumoto, H.; Hayashi, Y.; Suib, S. L. (1998). Efficient catalytic plasma activation of CO_2, NO, and H_2O. *J. Phys. Chem. B*, 102:9661–9666.

63. Luo, J.; Suib, S. L.; Hayashi, Y.; Matsumoto, H. (2000). Water splitting in low-temperature AC plasmas at atmospheric pressure. *Res. on Chem. Intermediates*, 26:849–874.

64. Goddard, W. A.; Jacob, T. (2004). Agostic interactions and dissociation in the first layer of water on Pt(111). *J. Am. Chem. Soc.*, 126:9360–9368.

65. Zhang, X. G.; Armentrout, P. B. (2003). Activation of O_2 and CO_2 by PtO^+: The thermochemistry of PtO_2^+. *J. Phys. Chem. A*, 107:8915–8922.

66. Li, W.; Hammer, B. (2005). Reactivity of a gas/metal/metal-oxide three-phase boundary: CO oxidation at the Pt(111)-c(4 x 2)-2CO/alpha-PtO_2 phase boundary. *Chem. Phys. Lett.*, 409:1–7.

67. Xiong, Y. P.; Yamaji, K.; Kishimoto, H.; Brito, M. E.; Horita, T.; Yokokawa, H. (2009). Deposition of platinum particles at LSM/ScSZ/air three-phase boundaries using a platinum current collector. *Electrochem. and Solid State Lett.*, 12:31–33.

68. Ryll, T.; Galinski, H.; Schlagenhauf, L.; Elser, P.; Rupp, J. L. M.; Bieberle-Hutter, A.; Gauckler, L. J. (2011). Microscopic and nanoscopic three-phase-boundaries of platinum thin-film electrodes on YSZ electrolyte. *Adv. Funct. Mat.*, 21:565–572.

69. Ishihara, A.; Ohgi, Y.; Matsuzawa, K.; Mitsushima, S.; Ota, K. (2010). Progress in non-precious metal oxide-based cathode for polymer electrolyte fuel cells. *Electrochimica Acta*, 55:8005–8012.

70. Ioannides, T.; Verykios, X. E. (1995). Modification of the catalytic properties of supported noble-metal catalysts by carrier doping. *Chem. Eng. & Techn.*, 18:25–32.

71. Wang, R. M.; Sun, Q.; Ren, Z.; Wang, N.; Cao, X. (2011). Platinum catalyzed growth of NiPt hollow spheres with an ultrathin shell. *J. Mat. Chem.*, 21: 1925–1930.

72. Goodwin, J. G.; Lohitharn, N. (2008). Effect of K promotion of Fe and FeMn Fischer-Tropsch synthesis catalysts: Analysis at the site level using SSITKA. *J. Catal.*, 260:7–16.

73. Deguzman, R. N.; Awaluddin, A.; Shen, Y. F.; Tian, Z. R.; Suib, S. L.; Ching, S.; Oyoung, C. L. (1995). Electrical-resistivity measurements on manganese oxides with layer and tunnel Structures-birnessites, todorokites, and cryptomelanes. *Chem. Mat.*, 7:1286–1292.

74. Pal, M.; Ganesan, V. (2009). Zinc phthalocyanine and silver/gold nanoparticles incorporated MCM-41 type materials as electrode modifiers. *Langmuir*, 25:13264–13272.

75. Zhang, A.; Zhang, W.; Lu, J.; Wallace, G. G.; Chen, J. (2009). Electrocatalytic reduction of carbon dioxide by cobalt-phthalocyanine-incorporated polypyrrole. *Electrochem. Electrochemical and Solid-State Lett.*, 12:17–19.

76. Leung, K.; Nielsen, I. M. B.; Sai, N.; Medforth, C.; Shelnutt, J. A. (2010). Cobalt-porphyrin catalyzed electrochemical reduction of carbon dioxide in water. 2. Mechanism from first principles. *J. Phys. Chem. A*, 114:10174–10184.

77. Kuwabata, S.; Tsuda, R.; Nishida, K.; Yoneyama, H. (1993). Electrochemical conversion of carbon-dioxide to methanol with use of enzymes as biocatalysts. *Chem. Lett.*, 9:1631–1634.

78. Steen, I. H.; Madsen, M. S.; Birkeland, N. K.; Lien, T. (1998). Purification and characterization of a monomeric isocitrate dehydrogenase from the sulfate-reducing bacterium Desulfobacter vibrioformis and demonstration of the presence of a monomeric enzyme in other bacteria. *FEMS Microbiol. Lett.*, 160:75–79.

79. Wiwanitkit, V., Plasmodium and host carbonic anhydrase: molecular function and biological process—Research article. *Gene Therapy and Molecular Biology* 2006, 10:251–254.

80. Heo, J.; Staples, C. R.; Ludden, P. W. (2001). Redox-dependent CO_2 reduction activity of CO dehydrogenase from Rhodospirillum rubrum. *Biochem.*, 40:7604–7611.

81. Armstrong, F. A.; Parkin, A.; Seravalli, J.; Vincent, K. A.; Ragsdale, S. W. (2007). Rapid and efficient electrocatalytic CO_2/CO interconversions by Carboxydothermus hydrogenoformans CO dehydrogenase I on an electrode. *J. Am. Chem. Soc.*, 129:10328–10329.

82. Andreev, A.; Idakiev, V.; Mihajlova, D.; Shopov, D. (1986). Iron-based catalysts for the water-gas shift reaction promoted by 1st-Row transition-metal oxides. *Appl. Catal.*, 22:385–387.

83. Boisen, A.; Janssens, T. V. W.; Schumacher, N.; Chorkendorff, I.; Dahl, S. (2010). Support effects and catalytic trends for water gas shift activity of transition metals. *J. Mol. Catal. A-Chem.*, 315:163–170.

84. Gines, M. J. L.; Amadeo, N.; Laborde, M.; Apesteguia, C. R. (1995). Activity and structure-sensitivity of the water-gas shift reaction over Cu-Zn-Al mixed-oxide catalysts. *Appl. Catal. A*, 131:283–296.

85. Pettigrew, D. J.; Trimm, D. L.; Cant, N. W. (1994). The effects of rare-earth-oxides on the reverse water-gas shift reaction on palladium alumina. *Catal. Lett.*, 28:313–319.

86. Wang, S. D.; Du, X. R.; Yuan, Z. S.; Cao, L.; Zhang, C. X. (2008). Water gas shift reaction over Cu-Mn mixed oxides catalysts: Effects of the third metal. *Fuel Proc. Techn.*, 89:131–138.

87. Nagai, M.; Matsuda, K. (2006). Low-temperature water-gas shift reaction over cobalt-molybdenum carbide catalyst. *J. Catal.*, 238:489–496.

88. Baronskaya, N. A.; Minyukova, T. P.; Khassin, A. A.; Yurieva, T. M.; Parmon, V. N. (2010). Enhancement of water-gas shift reaction efficiency: catalysts and the catalyst bed arrangement. *Russian Chem. Rev.*, 79:1027–1046.

89. Coville, N. J.; Mellor, J. R.; Sofianos, A. C.; Copperthwaite, R. G. (1997). Raney copper catalysts for the water-gas shift reaction: I. Preparation, activity and stability. *Appl. Catal. A*, 164:171–183.

90. Mellor, J. R.; Coville, N. J. (1997). Alkali metal promoted Raney copper catalysts for the water-gas shift reaction. *South African Journal of Chemistry-Suid-Afrikaanse Tydskrif Vir Chemie*, 50:153–156.

91. Hong, X. L.; Liao, F. L.; Huang, Y. Q.; Ge, J. W.; Zheng, W. R.; Tedsree, K.; Collier, P.; Tsang, S. C. (2011). Morphology-dependent interactions of ZnO with Cu nanoparticles at the materials' interface in selective hydrogenation of CO_2 to CH_3OH. *Angew. Chemie*, 50:2162–2165.

92. Vostrikov, A. A.; Fedyaeva, O. N.; Shishkin, A. V.; Sokol, M. Y. (2009). ZnO nanoparticles formation by reactions of bulk Zn with H_2O and CO_2 at sub- and supercritical conditions: I. Mechanism and kinetics of reactions. *J. Supercritical Fluids*, 48:154–160.

93. Visconti, C. G.; Lietti, L.; Tronconi, E.; Forzatti, P.; Zennaro, R.; Finocchio, E. (2009). *Appl. Catal. A*, 355:61–68.

94. Han, B. X.; Zhang, Z. F.; Xie, E.; Li, W. J.; Hu, S. Q.; Song, J. L.; Jiang, T. (2008). Hydrogenation of carbon dioxide is promoted by a task-specific ionic liquid. *Angew. Chemie*, 47:1127–1129.

95. Tihay, F.; Roger, A. C.; Pourroy, G.; Kiennemann, A. (2002). Role of the alloy and spinel in the catalytic behavior of Fe-Co/cobalt magnetite composites under CO and CO_2 hydrogenation. *Energy & Fuels*, 16:1271–1276.

96. Kusama, H.; Okabe, K.; Sayama, K.; Arakawa, H. (1996). CO_2 hydrogenation to ethanol over promoted Rh/SiO_2 catalysts. *Catal. Today*, 28:261–266.

97. Trovarelli, A.; Matteazzi, P.; Dolcetti, G.; Lutman, A.; Miani, F. (1993). Nanophase iron carbides as catalysts for carbon-dioxide hydrogenation. *Appl. Catal. A*, 95: 9–13.

98. Behrens, M.; Girgsdies, F.; Trunschke, A.; Schlogl, R. (2009). Minerals as model compounds for Cu/ZnO catalyst precursors: structural and thermal properties and IR spectra of mineral and synthetic (Zincian) malachite, Rosasite and Aurichalcite and a catalyst precursor mixture. *Eur. J. Inorg. Chem.*, 10:1347–1357.

99. Beltramini, J.; Liu, X. M.; Lu, G. Q.; Yan, Z. F. (2003). Recent advances in catalysts for methanol synthesis via hydrogenation of CO and CO_2. *Ind. & Eng. Chem. Res.*, 42:6518–6530.

100. Tabatabaei, J.; Sakakini, B. H.; Waugh, K. C. (2006). On the mechanism of methanol synthesis and the water-gas shift reaction on ZnO. *Catal. Lett.*, 110:77–84.

101. Deng, J. F.; Sun, Q.; Liu, C. W.; Pan, W.; Zhu, Q. M. (1998). In situ IR studies on the mechanism of methanol synthesis over an ultrafine $Cu/ZnO/Al_2O_3$ catalyst. *Appl. Catal. A-General*, 171:301–308.

102. Fujita, S.; Kanamori, Y.; Satriyo, A. M.; Takezawa, N. (1998). Methanol synthesis from CO_2 over Cu/ZnO catalysts prepared from various coprecipitated precursors. *Catal. Today*, 45:241–244.

103. Himelfarb, P. B.; Simmons, G. W.; Klier, K.; Joseyacaman, M. (1985). Microanalysis of a Copper-Zinc oxide methanol synthesis catalyst precursor. *ACS Symp. Series*, 288:351–360.

104. Huang, A. S.; Caro, J. (2010). Shape evolution of zinc oxide from twinned disks to single spindles through solvothermal synthesis in binary solvents. *J. Crystal Growth*, 312:2977–2982.

105. Kasatkin, I.; Kniep, B.; Ressler, T. (2007). Cu/ZnO and Cu/ZrO$_2$ interactions studied by contact angle measurement with TEM. *Phys. Chem. Chem. Phys.*, 9: 878–883.

106. Burghaus, U.; Wang, J. (2005). Adsorption of CO on the copper-precovered ZnO(0001) surface: A molecular-beam scattering study. *J. Chem. Phys.*, 123: 184716.

107. Davis, B. H.; Graham, U. M.; Dozier, A.; Khatri, R. A.; Bahome, M. C.; Jewell, L. L.; Mhlanga, S. D.; Coville, N. J. (2009). Carbon nanotube docking stations: a new concept in catalysis. *Catal. Lett.*, 129:39–45.

108. Vitidsant, T.; Hinchiranan, S.; Zhang, Y.; Nagamori, S.; Tsubaki, N. (2008). TiO$_2$ promoted Co/SiO$_2$ catalysts for Fischer-Tropsch synthesis. *Fuel Proc. Techn.*, 89:455–459.

109. Okabe, K.; Takahara, I.; Inaba, M.; Murata, K.; Yoshimura, Y. (2007). Effects of Ru precursors on activity of Ru-SiO$_2$ catalysts prepared by alkoxide method in Fischer-Tropsch synthesis. *J. Jpn Petrol. Inst.*, 50:65–68.

110. Hinchiranan, S.; Zhang, Y.; Nagamori, S.; Vitidsant, T.; Tsubaki, N. (2006). TiO$_2$ promoted Co/SiO$_2$ catalysts for Fischer-Tropsch synthesis. *J. Jpn Petrol. Inst.*, 49:45–46.

111. Das, P. C.; Pradhan, N. C.; Dalai, A. K.; Bakshi, N. N. (2004). Carbon monoxide hydrogenation over zirconia supported Ni and Co-Ni bimetallic catalysts. *Catal. Lett.*, 98:153–160.

112. Das, P. C.; Pradhan, N. C.; Dalai, A. K.; Bakhshi, N. N. (2004). Carbon monoxide hydrogenation over various titania-supported Ru-Ni bimetallic catalysts. *Fuel Proc. Techn.*, 85:1487–1501.

113. Lee, K. W.; Kishan, G.; Lee, M. W.; Nam, S. S.; Choi, M. J. (1998). The catalytic conversion of CO$_2$ to hydrocarbons over Fe-K supported on Al$_2$O$_3$-MgO mixed oxides. *Catal. Lett.*, 56:215–219.

114. Hu, B.; Frueh, S.; Garces, H. F.; Zhang, L.; Aindow, M.; Brooks, C.; Kreidler, E.; Suib, S. L. (2011). Selective hydrogenation of CO$_2$ and CO to useful light olefins over octahedral molecular sieve manganese oxide supported iron catalysts. *Appl. Catal. B*,132: 54–61.

115. Herranz, T.; Rojas, S.; Perez-Alonso, F. J.; Ojeda, M.; Terreros, P.; Fierro, J. G. (2006). Genesis of iron carbides and their role in the synthesis of hydrocarbons from synthesis gas. *J. Catal.*, 243:199–211.

116. de Smit, E.; Swart, I.; Creemer, J. F.; Hoveling, G. H.; Gilles, M. K.; Tyliszczak, T.; Kooyman, P. J.; Zandbergen, H. W.; Morin, C.; Weckhuysen, B. M.; de Groot, F. F. (2008). Nanoscale chemical imaging of a working catalyst by scanning transmission X-ray microscopy. *Nature* (London, U. K.), 456:222–225.

117. Karaca, H.; Hong, J. P.; Fongarland, P.; Roussel, P.; Griboval-Constant, A.; Lacroix, M.; Hortmann, K.; Safonova, O. V.; Khodakov, A. Y. (2010). In situ XRD investigation of the evolution of alumina-supported cobalt catalysts under realistic conditions of Fischer-Tropsch synthesis. *Chem. Comm.*, 46:788–790.

118. de Smit, E.; Cinquini, F.; Beale, A. M.; Safonova, O. V.; van Beek, W.; Sautet, P.; Weckhuysen, B. M. (2010). Stability and reactivity of epsilon-chi-theta iron carbide catalyst phases in Fischer-Tropsch synthesis: controlling mu(c). *J. Am. Chem. Soc.*, 132:14928–14941.

119. Sort, J.; Nogues, J.; Surinach, S.; Baro, M. D. (2003). Microstructural aspects of the hcp-fcc allotropic phase transformation induced in cobalt by ball milling. *Phil. Magazine*, 83: 439–455.

120. Zhang, Y.; Jacobs, G.; Sparks, D. E.; Dryb, M. E.; Davis, B. H. (2002). CO and CO_2 hydrogenation study on supported cobalt Fischer-Tropsch synthesis catalyst. *Catal. Today*, 71:411–418.

121. Pontzen, F.; Liebner, W.; Gronemann, V.; Rothaemel, M.; Ahlers, B. (2011). CO_2-based methanol and DME-Efficient technologies for industrial scale production. *Catal. Today*, 171:242–250.

122. Neophytides, S. G.; Marchi, A. J.; Froment, G. F. (1992). Methanol synthesis by means of diffuse reflectance infrared Fourier-transform and temperature-programmed reaction spectroscopy. *Appl. Catal. A*, 86:45–64.

123. Pinto, A.; Rogerson, P. L. (1977). Optimizing ICI Low-pressure methanol process. *Chem. Eng.*, 84:102–108.

124. Goldman, G. H.; Amorim, H. V.; Lopes, M. L.; Oliveira, J. V. D.; Buckeridge, M. S. (2011). Scientific challenges of bioethanol production in Brazil. *Appl. Microbiol. and Biotechn.*, 91:1267–1275.

125. Zhong, C.; Cao, Y. X.; Li, B. Z.; Yuan, Y. J. (2010). Biofuels in China: past, present and future. *Biofuels Bioproducts & Biorefining-Biofpr*, 4:326–342.

126. Sulak, M.; Smogrovicova, D. (2008). Bioethanol: Current trends in research and practice. *Chemicke Listy*, 102:108–115.

127. Gray, K. A. (2007). Cellulosic ethanol-state of the technology. *Int. Sugar J.*, 109:145–151.

128. Prasad, S.; Singh, A.; Joshi, H. C. (2007). Ethanol as an alternative fuel from agricultural, industrial and urban residues. *Resources Conservation and Recycling*, 50:1–39.

129. Nguyen, T.; M., H. Z.; Alamdari, H.; Kaliaguine, S. (2007). Effect of alkali additives over nanocrystalline Co–Cu-based perovskites as catalysts for higher-alcohol synthesis. *J. Catal.*, 245:348–357.

130. Fatiadi, A. J. (1976). Active Manganese-dioxide oxidation in organic-chemistry I, *Synthesis-Stuttgart*, 2:65–104.

131. Chauvel, A.; Delmon, B.; Holderich, W. F. (1994). New catalytic processes developed in Europe during the 1980s. *Appl. Catal. A*, 115:173–217.

132. Brookhart, M.; Svejda, S. A. (1999). Ethylene oligomerization and propylene dimerization using cationic (alpha-diimine)nickel(II) catalysts. *Organomet.*, 18:65–74.

133. Zhang, Q. L.; Kantcheva, M.; DallaLana, I. G. (1997). Oligomerization of ethylene in a slurry reactor using a nickel/sulfated alumina catalyst. *Ind. & Eng. Chem. Res.*, 36:3433–3438.

134. van Leeuwen, P. W. N. M.; Clement, N. D.; Tschan, M. J. L. (2011). New processes for the selective production of 1-octene. *Coord. Chem. Rev.*, 255:1499–1517.

135. Liu, H. T.; Davey, C. R.; Shirodkar, P. P. (2003). Bimodal polyethylene products from UNIPOL (TM) single gas phase reactor using engineered catalysts. *Macromol. Symp.*, 195:309–316.

136. Chiu, D. Y.; Ealer, G. E.; Moy, F. H.; Buhler-Vidal, J. O. (1999). Unipol II LLDPE-gas phase LLDPE for the shrink market. *J. Plastic Film & Sheeting*, 15:153–178.

137. Kim, Y. M.; Park, J. K. (1996). Effect of short chain branching on the blown film properties of linear low density polyethylene. *J. Appl. Polymer Science*, 61:2315–2324.

138. Puig, C. C.; Aviles, M. V.; Joskowicz, P.; Diaz, A. (2001). On the melting behavior of isothermally crystallized 1-octene linear low-density polyethylene copolymers. *J. Appl. Polymer Science*, 79:2022–2028.

139. Dow delivers metallocene LLDPE (1997). *Chemical Week*, 159:10.

140. Crabtree, R. H. (2005). *Applications of Organometallic Chemistry*. 4th edition; Wiley:; p 360–361.

141. Morlock, G.; Ueda, Y. (2007). New coupling of planar chromatography with direct analysis in real time mass spectrometry. *J. Chromat. A*, 1143:243–251.

142. Jin, B. K.; Liu, P.; Wang, Y.; Zhang, Z. P.; Tian, Y. P.; Yang, J. X.; Zhang, S. Y.; Cheng, F. L. (2007). Rapid-scan time-resolved FT-IR spectroelectrochemistry studies on the electrochemical redox process. *J. Phys. Chem. B*, 111:1517–1522.

143. Akao, K.; Sugiyama, H.; Koshoubu, J.; Kashiwabara, S.; Nagoshi, T.; Larsen, R. A. (2008). Time-resolved step-scan infrared imaging system utilizing a linear array detector. *Appl. Spectroscopy*, 62:17–23.

144. Liu, Z.; Rittermeier, A.; Becker, M.; Kahler, K.; Loffler, E.; Muhler, M. (2011). High-pressure CO adsorption on Cu-based catalysts: Zn-induced formation of strongly bound CO monitored by ATR-IR spectroscopy. *Langmuir*, 27:4728–33.

145. Politano, A.; Marino, A. R.; Chiarello, G. (2010). CO-promoted formation of the alkali-oxygen bond on Ni(111). *J. Chem. Phys.*, 132:044706.

146. Sadique, A. R.; Brennessel, W. W.; Holland, P. L. (2008). Reduction of CO_2 to CO using low-coordinate iron: formation of a four-coordinate iron dicarbonyl complex and a bridging carbonate complex. *Inorg. Chem.*, 47:784–786.

147. Herman, J.; Foutch, J. D.; Davico, G. E. (2007). Gas-phase reactivity of selected transition metal cations with CO and CO_2 and the formation of metal dications using a sputter ion source. *J. Phys. Chem. A*, 111:2461–8.

148. Visart de Bocarme, T.; Kruse, N.; Gaspard, P.; Kreuzer, H. J. (2006). Field-induced CO adsorption and formation of carbonyl waves on gold nanotips. *J. Chem. Phys.*, 125:054704.

149. Craxford, S. R.; Rideal, E. K. (1939). Mechanism of the synthesis of hydrocarbons from water gas. *J. Chem. Soc.*, 0:1604–1614.

150. Davis, B. H. (2009). Fischer-Tropsch Synthesis: Reaction mechanisms for iron catalysts. *Catal. Today*, 141:25–33.

151. Huo, C. F.; Li, Y. W.; Wang, J. G.; Jiao, H. J. (2009). Insight into CH_4 formation in iron-catalyzed Fischer-Tropsch synthesis. *J. Am. Chem. Soc.*, 131:14713–14721.

152. Ojeda, M.; Nabar, R.; Nilekar, A. U.; Ishikawa, A.; Mavrikakis, M.; Iglesia, E. (2010). CO activation pathways and the mechanism of Fischer-Tropsch synthesis. *J. Catal.*, 272:287–297.

153. Bates, E. D.; Mayton, R. D.; Ntai, I.; Davis, J. H. (2002). CO_2 capture by a task-specific ionic liquid, *J. Am. Chem. Soc.*, 124:926–927.

154. Dixon, J. K.; Muldoon, M. J.; Aki, S. N. V. K.; Anderson, J. L.; Brennecke, J. F.; Maginn, E. J. (2006). Designing ionic liquids for CO_2 capture. IEC-221, *231st ACS National Meeting*, Atlanta, GA, United States, March 26–30, 2006.

155. Reddy, R. G. (2008). Reducing CO_2 using room-temperature ionic liquids. *J. Minerals, Metals and Materials Soc.*, 60:33.

156. Schwab, P. F.; Seiler, H. M.; Weyershausen, B. (2006). CO_2 absorption using ionic liquids. *Chimica Oggi-Chemistry Today*, 24:21–23.

157. Zhang, S. J.; Zhang, Y. Q.; Lu, X. M.; Zhou, Q.; Zhang, X. P. (2008). Capture of CO$_2$ by using ionic liquids. I&EC 011, *236th ACS National Meeting*, Philadelphia, PA, United States, August 17–21, 2008.

158. Zhang, S. J.; Chen, Y. H.; Li, F. W.; Lu, X. M.; Dai, W. B.; Mori, R. (2006). Fixation and conversion of CO$_2$ using ionic liquids. *Catal. Today*, 115:61–69.

159. Baker, R. T.; Liu, F. C.; Abrams, M. B.; Tumas, W. (2001). Phase-separable catalysis using room temperature ionic liquids and supercritical carbon dioxide. *Chem. Comm.*, 5:433–434.

160. Yang, H. Z.; Gu, Y. L.; Deng, Y. Q. (2002). Electrocatalytic insertion of carbon dioxide to epoxides in ionic liquids at room temperature. *Chinese J. Org. Chem.*, 22:995–998.

161. Cai, Q. H.; Yuan, D. D.; Yan, C. H.; Lu, B.; Wang, H. X.; Zhong, C. M. (2009). Electrochemical activation of carbon dioxide for synthesis of dimethyl carbonate in an ionic liquid. *Electrochimica Acta*, 54:2912–2915.

162. Compton, R. G.; Barrosse-Antle, L. E. (2009). Reduction of carbon dioxide in 1-butyl-3-methylimidazolium acetate. *Chem. Comm.*, 25:3744–3746.

163. Han, B. X.; Zhao, G. Y.; Jiang, T.; Li, Z. H.; Zhang, J. M.; Liu, Z. M.; He, J.; Wu, W. Z. (2004). Electrochemical reduction of supercritical carbon dioxide in ionic liquid 1-n-butyl-3-methylimidazolium hexafluorophosphate. *J. Supercritical Fluids*, 32:287–291.

164. Snuffin, L. L.; Whaley, L. W.; Yu, L. (2011). Catalytic Electrochemical Reduction of CO$_2$ in Ionic Liquid EMIMBF$_3$Cl. *J. Electrochem. Soc.*, 158:155–158.

165. Inoue, S.; Koinuma, H.; Tsuruta, T. (1969). *J. Polymer Science, Part A: Polymer Chem.*, 7:287–292.

166. Coates, G. W.; Cohen, C. T.; Chu, T. (2005). Cobalt catalysts for the alternating copolymerization of propylene oxide and carbon dioxide: Combining high activity and selectivity. *J. Am. Chem. Soc.*, 127:10869–10878.

167. Fukuoka, S.; Tojo, M.; Hachiya, H.; Aminaka, M.; Hasegawa, K. (2007). Green and sustainable chemistry in practice: Development and industrialization of a novel process for polycarbonate production from CO$_2$ without using phosgene. *Polymer J.*, 39:91–114.

168. Lu, X. B.; Shi, L.; Wang, Y. M.; Zhang, R.; Zhang, Y. J.; Peng, X. J.; Zhang, Z. C.; Li, B. (2006). Design of highly active binary catalyst systems for CO$_2$/epoxide copolymerization: Polymer selectivity, enantioselectivity, and stereochemistry control. *J. Am. Chem. Soc.*, 128:1664–1674.

169. Lee, B. Y.; Sujith, S.; Min, J. K.; Seong, J. E.; Na, S. J. (2008). A highly active and recyclable catalytic system for CO$_2$/propylene oxide copolymerization. *Angew. Chemie*, 47:7306–7309.

170. Williams, C. K.; Kember, M. R.; Knight, P. D.; Reung, P. T. R. (2009). Highly active dizinc catalyst for the copolymerization of carbon dioxide and cyclohexene oxide at one atmosphere pressure. *Angew. Chemie*, 48:931–933.

171. Nolan, S. P.; Boogaerts, I. I. F. (2010). Carboxylation of C-H Bonds using N-heterocyclic carbene gold(I) complexes. *J. Am. Chem. Soc.*, 132:8858–8859.

172. Zhang, Y. G.; Yu, D. Y. (2010). Copper- and copper-N-heterocyclic carbene-catalyzed C-H activating carboxylation of terminal alkynes with CO$_2$ at ambient conditions. *Proc. of the National Academy of Sciences of the United States of America*, 107:20184–20189.

173. Isse, A. A.; Durante, C.; Gennaro, A. (2011). One-pot synthesis of benzoic acid by electrocatalytic reduction of bromobenzene in the presence of CO_2. *Electrochem. Comm.*,13:810–813.

174. Jin, L.; Reutenauer, J.; Opembe, N.; Lai, M.; Martenak, D. J.; Han, S.; Suib, S. L. (2009). Studies on dehydrogenation of ethane in the presence of CO_2 over octahedral molecular sieve (OMS-2) catalysts. *Chem. Cat. Chem.*, 1:441–444.

175. Goettmann, F.; Thomas, A.; Antonietti, M. (2007). Metal-free activation of CO_2 by mesoporous graphitic carbon nitride. *Angew. Chemie.*, 46:2717–20.

176. Deng, Y. Q.; Yang, H. Z.; Gu, Y. L.; Shi, F. (2002). Electrochemical activation of carbon dioxide in ionic liquid: synthesis of cyclic carbonates at mild reaction conditions. *Chem. Comm.*, 3:274–275.

177. Somorjai, G. A.; Frei, H.; Park, J. Y. (2009). Advancing the Frontiers in Nanocatalysis, biointerfaces, and renewable energy conversion by innovations of surface techniques. *J. Am. Chem. Soc.*, 131:16589–16605.

CHAPTER 4

Hydrogenation of Carbon Dioxide to Liquid Fuels

MUTHU KUMARAN GNANAMANI, GARY JACOBS, VENKAT RAMANA RAO PENDYALA, WENPING MA, and BURTRON H. DAVIS*

4.1 INTRODUCTION

The increase in carbon dioxide (CO_2) emissions arguably contributes to an increase in global temperatures and climate change due to the greenhouse effect. The sources of CO_2 include residential, commercial, industrial, and transportation sectors. Countries all over the world have been facing tremendous pressure to reduce CO_2 emissions and develop more efficient ways for capturing and utilizing CO_2 [1–3]. In the past, many strategies have been tried, and among these the following three methods have attracted much attention: (i) reduction in the amount of CO_2 produced, (ii) storage of CO_2, and (iii) conversion of CO_2 to produce fuels and chemicals [4–6]. A number of studies, including reviews, have been published on the topic of capture and sequestration of CO_2 [5–8]. CO_2 is of great interest as a C_1 feedstock because of the vast amount of carbon that exists in this form, and because of the low cost of bulk CO_2. Therefore, it is prudent to identify methodologies for converting such low-value CO_2 to valuable chemicals and/or liquid fuels.

*Indicates the corresponding author.

Green Carbon Dioxide: Advances in CO₂ Utilization, First Edition.
Edited by Gabriele Centi and Siglinda Perathoner.
© 2014 John Wiley & Sons, Inc. Published 2014 by John Wiley & Sons, Inc.

Catalytic hydrogenation of CO_2 to liquid fuels has attracted considerable interest in recent years, as it provides an alternative to the conventional petroleum refining processes of crude oil. Despite the fact that numerous studies have been published on this topic, CO_2 is not being used extensively as a chemical feedstock for any major industrial process, including Fischer–Tropsch synthesis (FTS). This is primarily because of the thermodynamic stability of CO_2, requiring high-energy substances like H_2 to transform CO_2 into valuable products. Hydrogen could play an important role in utilizing CO_2 to produce transportation fuels. However, this is a challenging task to accomplish in an economical way. On the other hand, the growing consumption of diminishing resources of fossil fuels is causing the world to now look at alternative fuels that are based on non-fossil fuel sources. Unfortunately, hydrogen does not exist in its natural state on Earth, and few processes can generate H_2 without producing CO_2 in the process. One possibility is the use of nuclear energy to produce H_2 through the hydrolysis of water.

Depending on conditions and reaction system, hydrogenation of CO_2 mainly produces methanol, dimethyl ether, higher alcohols, methane, and higher hydrocarbons. There have been several excellent reviews regarding hydrogenation of CO_2 [8–15]. However, many reviews are focused mainly on the homogeneous hydrogenation of CO_2 [13,14] and other processes that obtain chemicals from CO_2 [15]. Therefore, the aim of this review is to update the heterogeneously catalyzed processes and reaction mechanisms for hydrocarbon synthesis obtained from hydrogenation of CO_2. The first part of the review focuses on the methanation of carbon dioxide. Second, we review the progress in the elucidation of reaction mechanisms for the syntheses of methanol and higher alcohols from CO_2. Finally, a summary of current understanding on catalytic reactivity and reaction mechanisms for hydrogenation of CO_2 over FTS-based catalysts is provided.

4.2 METHANATION OF CARBON DIOXIDE

Catalytic hydrogenation of carbon oxides was observed by Sabatier and Senderens [16] at the beginning of the twentieth century. Since then, numerous studies have been carried out [17–22]. The group VIIIB metals on various oxide supports have been investigated extensively. Recently, Gong et al. [23] reviewed the catalyst system for methanation of carbon dioxide and in particular the support effect. Halmann and Steinberg [24] reviewed CO_2 methanation using supported Ni catalysts. Methanation of CO_2 on Ni catalysts has been explored under standard catalytic reaction conditions with either the pure or the supported metal in either the presence or the absence of promoters [25–27]. Catalytic systems other than the supported single-component catalyst have also been studied [28–32].

A number of mechanisms have been proposed for CO_2 methanation. In general, they fall into two categories: (i) conversion of CO_2 to CO via the reverse water-gas shift (RWGS) reaction followed by CO methanation and (ii) direct hydrogenation of CO_2 to methane. There are some mechanistic studies that show that after CO_2 adsorption and dissociation methanation apparently proceeds via the same route as

for CO methanation [33,34]. Another viewpoint [35,36] is that CO does not partic-
ipate in the CO_2 methanation reaction, that is, the mechanisms of hydrogenation of
CO and CO_2 are different.

A kinetic study by Weatherbee and Bartholomew [20] suggests that methanation
of CO_2 involves dissociative adsorption of CO_2 to form CO and atomic oxygen and
this is followed by dissociation of adsorbed CO to atomic carbon and atomic oxygen. Methane is then formed by the hydrogenation of atomic carbon. The authors
proposed a sequence of elementary steps involved in CO_2 methanation as given
below (Fig. 4.1). The authors found that dissociation of CO is involved in the
rate-determining step for hydrogenation of CO_2.

Araki and Ponec [37] also suggest that CO_2 was first transformed into CO, and
CO subsequently hydrogenated to methane. Delmon and Martin [33] proposed a
similar pathway for hydrogenation of CO and CO_2 using Ni/SiO_2 through the formation of active carbonaceous species (Ni_3C_{surf}). The authors further demonstrated
that the formation of Ni_4CO_{ads} at lower temperature was responsible for the more
rapid hydrogenation of CO_2 than of CO.

A transient kinetic study was employed to study the mechanism for hydrogenation of CO_2 on Ru/TiO_2 [38]. The authors identified intermediates and reaction
products using DRIFTS-IR. It was proposed that CO_{ads} is the key intermediate
for methanation of CO_2. Interestingly, it was demonstrated that formate species,
adsorbed at the metal-support interface, were identified to be the precursor of CO.

The hydrogenation of C, CO, and CO_2 has been studied by Lahtinen et al. [39]
on polycrystalline cobalt foils with a combination of UHV studies and atmospheric
pressure reactions in the temperature range of 475 to 575 K. The measured activation energies for methane formation from surface C, CO, and CO_2 on cobalt foils
are 57 ± 3 kJ/mol, 86 ± 10 kJ/mol, and 155 ± 5 kJ/mol, respectively. The authors
concluded that the reaction proceeds via dissociation of C–O bonds and formation
of CoO on the surface. The reduction of CoO is the rate-limiting step in the CO and
CO_2 hydrogenation reactions.

It appears from this literature study that there are differences in opinion on
the nature of the intermediate compounds involved in the rate-determining step

$$H_2 + 2S \underset{k_{-1}}{\overset{k_1}{\rightleftharpoons}} 2H\text{—}S$$

$$CO_{2(g)} + 2S \underset{k_{-2}}{\overset{k_2}{\rightleftharpoons}} OC\text{—}S + O\text{—}S$$

$$OC\text{—}S \underset{k_{-3}}{\overset{k_3}{\rightleftharpoons}} C\text{—}S + O\text{—}S$$

$$C\text{—}S + H\text{—}S \underset{k_{-4}}{\overset{k_4}{\rightleftharpoons}} HC\text{—}S + S$$

$$HC\text{—}S + H\text{—}S \underset{k_{-5}}{\overset{k_5}{\rightleftharpoons}} H_2C\text{—}S + S$$

$$H_2C\text{—}S + H\text{—}S \underset{k_{-6}}{\overset{k_6}{\rightleftharpoons}} H_3C\text{—}S + S$$

$$H_3C\text{—}S + H\text{—}S \underset{k_{-7}}{\overset{k_7}{\rightleftharpoons}} H_4C\text{—}S + S$$

$$H_4C\text{—}S \underset{k_{-8}}{\overset{k_8}{\rightleftharpoons}} CH_{4(g)} + S$$

$$O\text{—}S + H\text{—}S \underset{k_{-9}}{\overset{k_9}{\rightleftharpoons}} OH\text{—}S + S$$

$$OH\text{—}S + H\text{—}S \underset{k_{-10}}{\overset{k_{10}}{\rightleftharpoons}} H_2O\text{—}S + S$$

$$H_2O\text{—}S \underset{k_{-11}}{\overset{k_{11}}{\rightleftharpoons}} H_2O_{(g)} + S$$

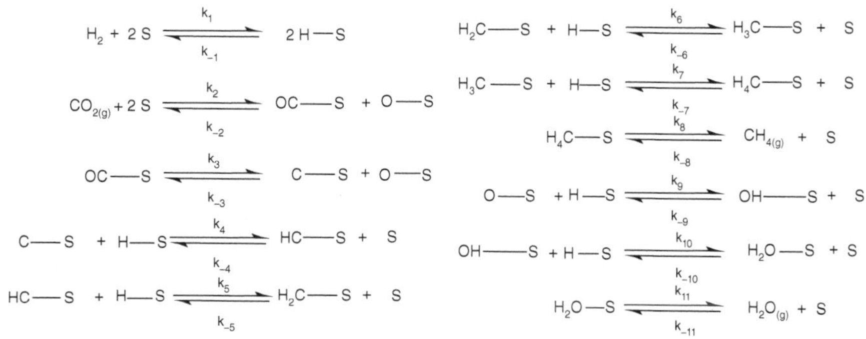

Figure 4.1 Proposed sequence of elementary steps in CO_2 methanation; S refers to a surface site. Adapted from ref. 20.

of the process and on the methane formation scheme. Novel surface characterization techniques, isotopic transient kinetic measurements, and recent computational techniques have advanced our understanding on elementary steps that are involved in hydrogenation of carbon oxides. Reaction intermediates such as formates and CO were identified by using DRIFTS-IR under realistic methanation reaction conditions. Hence, hydrogenation of either one of these species should be involved in the rate-limiting step. However, even for CO methanation, there is still no consensus on the reaction mechanism.

4.3 METHANOL AND HIGHER ALCOHOL SYNTHESIS BY CO_2 HYDROGENATION

Methanol is an alternative fuel and a starting material in many industrial processes. Also, mixed alcohols can be used alone as fuels or as an additive to gasoline for automobiles. Methanol (MeOH) synthesis by CO_2 hydrogenation has been extensively studied, mainly over Cu-based catalysts [40–42]. Many have addressed the disadvantages of using CO_2 in methanol synthesis because the yield of methanol is much lower than that obtained from syngas conversion under the same temperature and pressure [43]. Also, the equilibrium value for methanol obtained from a CO_2-H_2 mixture is about one-third that of CO hydrogenation.

$$CO_2 + 3H_2 \rightarrow CH_3OH + H_2O \tag{4.1}$$

A large number of investigations have addressed the effect of active components, supports, promoters, preparation methods, and surface morphology on the activity and selectivity for methanol synthesis. Although many transition metal-based catalysts have been examined for the synthesis of methanol, Cu remains the main active catalyst component and ZnO the main component used for dispersion and stabilization of copper. Other modifiers such as Zr, Ce, Al, Si, V, Ti, Ga, B, Cr, La, and Fe have also been tried. Wang et al. [23] recently reviewed the various catalytic systems that are used for the synthesis of methanol by CO_2 hydrogenation. In order to improve the activity and selectivity of methanol synthesis catalysts, it is a highly desirable and challenging task to understand the mechanism of methanol formation at an atomic level. Despite the large number of studies devoted to the reaction mechanism [44–46], the role of Cu^0, Cu^+, Cu-Zn alloy, and carrier sites on methanol synthesis is still undefined. Many argue that this depends upon a greater sensitivity of the catalyst morphology, possibly surface composition, to the reaction atmosphere [47–49]. Hydrogenation of CO_2 to methanol occurs by two different pathways: (i) direct hydrogenation of CO_2 via adsorbed formate or (ii) the reduction of CO_2 to carbon monoxide with subsequent hydrogenation to methanol (Fig. 4.2a).

Sun et al. [50] used in situ IR to study the mechanism of methanol formation over Cu/ZnO/Al$_2$O$_3$ catalysts. The experimental results indicate that methanol was formed directly from CO_2 hydrogenation both for CO_2 and CO/CO_2 hydrogenation. $HCOO^-_s$ was the key intermediate for methanol synthesis, and hydrogenation

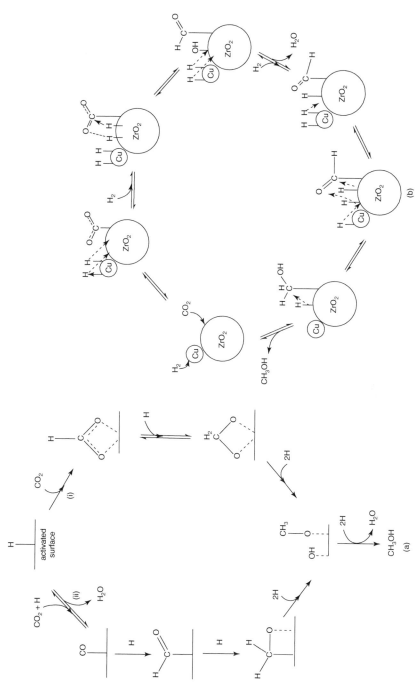

Figure 4.2 (a) Reaction pathway for the formation of methanol from hydrogenation of CO_2. (b) Proposed reaction pathway for the formation of methanol from CO_2 hydrogenation with Cu/ZrO_2 catalyst; adapted from ref. 49.

103

of $HCOO^-{}_s$ was the rate-limiting step for methanol synthesis. A recent DFT study by Grabow and Mavrikakis [51] suggests that under realistic conditions on commercial $Cu/ZnO/Al_2O_3$ catalysts, both the CO and CO_2 pathways are active for methanol synthesis. The authors further argue that hydrogenation of CH_3O species is slow; the product of CH_3O and H qualitatively describes the behavior of overall methanol synthesis rates for a large range of reaction conditions and CO_2-rich feed compositions. However, for CO-rich feeds, the formation of CH_3O can be rate-limiting. Arena et al. [49] proposed a mechanism for the formation of methanol from CO_2 using a Cu/ZrO_2 catalyst. The authors concluded that the metal/oxide interface plays a fundamental role for the CO_2 hydrogenation functionality, enabling the adsorption of CO_2 on basic sites of ZnO and ZrO_2 in the neighborhood of Cu^0 hydrogenation sites (Fig. 4.2b). The formation-hydrogenation of the reactive intermediate on ZnO and ZrO_2 surface sites in the interface with Cu is proposed to be the rate-limiting step.

Hence, much development has been made on understanding the chemistry of methanol formation by using modern surface characterization techniques and in the advancement of theoretical studies as well. However, an efficient method of converting CO_2 to higher alcohols is a subject that has been less studied [52,53]. For example, the production of ethanol by CO hydrogenation using supported Rh catalysts has been investigated extensively; however, direct conversion of CO_2 to ethanol has not been accomplished for a large-scale operation [54]. A catalyst for the formation of higher alcohols requires a combination of RWGS activity along with hydrogenation of syngas.

In general, Fe-based FT catalysts mixed with Cu-based alcohol production catalysts have been utilized. Okamoto and Arakawa [55] and Inui [56] developed combined catalyst systems based on $K/Fe-Cu-ZnO(-Al_2O_3)$. The authors reported that the concept of catalyst design is based on the combination of a C–C bond formation catalyst and a catalyst for oxygenate formation. Several other catalytic systems (Rh-Li-Fe/SiO_2 [57], (Rh/MFI-silicate)-(Fe-Cu-Zn-Al-K) [58]) and physically mixed Fe- and Cu-based catalysts [58] have been tested for producing higher alcohols from hydrogenation of carbon dioxide. Interestingly, Inui [59] obtained highly dispersed Co/SiO_2 catalysts derived from cobalt acetate that exhibited high hydrogenation activity. The author claimed that the selectivity to ethanol was increased up to 8% by adding Na salt to the Ir/Co(A)/ SiO_2 catalyst. Nagata et al. [60] investigated Ir-Mo/SiO_2 for the hydrogenation of CO_2 to $C_2{}^+$ higher alcohols. Among the alcohols, ethanol was the major product obtained from CO_2 hydrogenation irrespective of Fe catalyst and reaction conditions used.

The mechanism for the formation of higher alcohols including ethanol by CO_2 hydrogenation may proceed by either a CO intermediate or via a direct hydrogenation route. Using in situ IR analysis, Arakawa et al. [61] suggested that CO_2 was hydrogenated to ethanol through CO intermediate(s) over Rh/SiO_2. The authors proposed a plausible pathway for the formation of ethanol as shown in Figure 4.3a. By considering the possibility that ethanol formation from CO_2 may proceed through CO, one might expect that the formation of other higher alcohols may also follow a similar mechanism, as shown below (Fig. 4.3b). In summary,

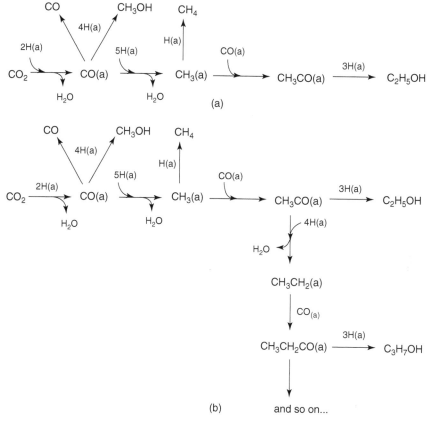

Figure 4.3 (a) Plausible reaction mechanism of CO_2 hydrogenation to ethanol and (b) proposed mechanism for synthesis of higher alcohols from CO_2. Adapted from ref. 61.

the results obtained to date show that higher linear alcohols are formed on most of the catalysts through the reaction of CO insertion with a C_{n-1} hydrocarbon entity. However, a nondissociative direct hydrogenation of CO_2 to ethanol route using $Rh_{10}Se/TiO_2$ catalyst was also proposed [53].

4.4 HYDROCARBONS THROUGH MODIFIED FISCHER-TROPSCH SYNTHESIS

Fischer–Tropsch Synthesis (FTS) is an industrially established process for the production of mainly linear hydrocarbons having a wide range of carbon numbers from synthesis gas [62–65]. One of the main advantages of this process is that the products derived from FTS are endowed with a tremendous environmental value, as they are virtually free of sulfur, nitrogen, and aromatic compounds. Coal, natural gas, and biomass are commercially being used for the production of syngas ($H_2 + CO$).

One drawback from an environmental standpoint is that during this process a significant quantity of CO_2 may be generated. This depends on the nature of the resource used to produce the syngas and, hence, the resulting $CO/H_2/CO_2$ ratios generated. In the case of coal and biomass, the gasifier type determines at what point CO_2 is generated in the overall process. If a Lurgi gasifier is used, considerable CO_2 is produced at the gasifier. However, if an O_2-blown entrained-flow gasifier is used (e.g., Texaco type), the H_2/CO ratio is low (e.g., 0.7), and CO_2 rejected by the gasifier is low. To achieve stoichiometric ratios of reactants to generate diesel fuels, WGS boosts the H_2/CO ratio, and the CO_2 is rejected in this process step, downstream from the gasifier (Fig. 4.4a). CO_2 may be separated out from the syngas stream by solvent-based processes (e.g., Rectisol). Thus, to avoid CO_2 generation, in the case of coal/biomass either FTS would have to generate unsaturated hydrocarbons, which has not been successfully developed, or H_2 would have to be added from another source (e.g., nuclear/water electrolysis). CO_2 might also be collected from CO_2-producing processes and then converted by H_2 produced from nuclear or solar, etc., power-driven electrolysis of water.

Hydrogenation of carbon dioxide to hydrocarbons has been less studied than conventional FTS. It can be divided primarily into two categories (Fig. 4.4b): methanol mediated and non-methanol mediated [66,67]. In methanol-mediated

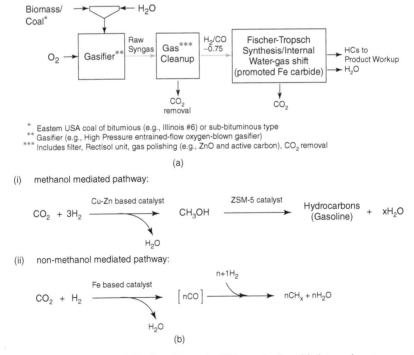

(a)

(i)   methanol mediated pathway:

(ii)   non-methanol mediated pathway:

(b)

Figure 4.4 (a) Iron-based Fischer–Tropsch (FT) synthesis with internal water-gas. (b) Routes for synthesizing hydrocarbons by modified FT synthesis using CO_2.

synthesis, initially, CO_2 and hydrogen react over Cu-Zn-based catalysts to produce methanol, which is subsequently transformed into hydrocarbons with zeolite based catalysts [11,23]. In non-methanol-mediated synthesis, CO_2 is transformed to CO via the RWGS reaction and then CO further reacts with H_2 to form hydrocarbons.

In a methanol-mediated pathway, composite catalysts containing two different kinds of catalytic materials such as methanol synthesis catalysts (Cu/ZnO) and zeolites are utilized. They produce enhanced yields of C_2^+ hydrocarbons by joining methanol synthesis and the MTG (methanol-to-gasoline) reaction. Several interesting studies have appeared in the literature [68–72]. Fujiwara and Souma et al. [68,69] performed the hydrogenation of CO_2 reaction using a mixture of Fischer–Tropsch type-catalysts or alcohol synthesis catalysts with HY-type zeolite. The products obtained were light olefin-rich hydrocarbons. Zeolites with MFI and/or MFI-like structure such as H-Fe-silicate and H-Ga-silicate were found to produce a gasoline fraction [70]. Yoshie et al. [73] studied the hydrogenation of CO_2 using Cu-Fe-Na/zeolite composite catalysts and found that the ratio of SiO_2 to Al_2O_3 of zeolites affected the activity and selectivity of the composite catalysts. The authors further revealed that Na-rich Cu-Fe-Na, Na-type zeolite, and H-mordenite improved the activity of the catalysts and the hydrocarbon selectivity. Despite considerable efforts made in the development of composite catalysts, the process always produced light alkanes as major hydrocarbon products. Hence, it might serve to produce feedstock for the production of lighter alkenes, which are raw materials of the polymer industry.

FTS-based processes are an alternate for producing liquid fuels (diesel range) from carbon dioxide. Under FTS conditions, hydrogenation of CO_2 falls into two categories: (i) conversion of CO_2 to CO via the RWGS reaction followed by CO hydrogenation and (ii) direct hydrogenation (DH) of CO_2 to hydrocarbons by a mechanism distinct from CO hydrogenation, as illustrated below.

(i) RWGS:

$$CO_2 + H_2 \leftrightarrow CO + H_2O \qquad (4.2a)$$

FT:

$$CO + 2H_2 \rightarrow (-CH_2-) + H_2O \qquad (4.2b)$$

(ii) DH

$$CO_2 + 3H_2 \rightarrow (-CH_2-) + 2H_2O \qquad (4.3)$$

Iron and cobalt are the active metals used commercially in FTS. Among the two, iron possesses intrinsic water-gas shift activity, which is a prerequisite for converting CO_2 to hydrocarbons by the indirect route (i). Thus Fe-based FT catalysts are being extensively studied for CO_2 hydrogenation [11,12,74–80], and only a few studies have dealt with cobalt catalysts [75,76,81–84].

From a survey of the literature, it appears that the mechanism of CO_2 hydrogenation may be different on cobalt- and iron-based FT catalysts. In an early study, Russell and Miller [81] used copper-promoted cobalt catalysts for the

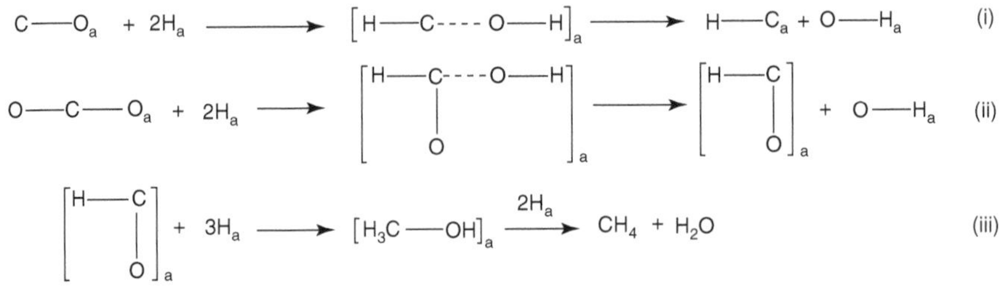

$$C{-}O_a + 2H_a \longrightarrow \left[H{-}C{-}{-}{-}{-}O{-}H\right]_a \longrightarrow H{-}C_a + O{-}H_a \quad (i)$$

$$O{-}C{-}O_a + 2H_a \longrightarrow \left[\begin{array}{c} H{-}C{-}{-}{-}{-}O{-}H \\ | \\ O \end{array}\right]_a \longrightarrow \left[\begin{array}{c} H{-}C \\ | \\ O \end{array}\right]_a + O{-}H_a \quad (ii)$$

$$\left[\begin{array}{c} H{-}C \\ | \\ O \end{array}\right]_a + 3H_a \longrightarrow \left[H_3C{-}OH\right]_a \xrightarrow{2H_a} CH_4 + H_2O \quad (iii)$$

Figure 4.5   Proposed mechanism for hydrocarbon formation from CO_2 using cobalt-based FT catalyst. Reproduced with permission from ref. 82.

synthesis of higher hydrocarbons from CO_2 and hydrogen at atmospheric pressure and found that catalysts containing no alkali produced no liquid products, or only traces, but did yield small amounts of liquid hydrocarbons after a suitable poisoning. Zhang et al. [82] proposed that the conversion of CO and CO_2 on cobalt FT catalyst occurs by different reaction pathways (Fig. 4.5). It is assumed that the hydrogenation and breaking of the two C–O bonds of the CO_2 provide the source of the different pathways.

The authors speculated that the breaking of the C–O bond (i), presumably by the addition of adsorbed H to form C–O–H, competes with, and probably leads, to the addition of adsorbed H to form the C–H bond. In the case of CO_2 (ii), there are two C–O bonds that must be broken prior to, or simultaneously with, the formation of the C–H bond. Based on the carbon mass balance, about 75% of the hydrogenation of CO_2 would proceed by reaction (iii) and the remainder would involve the breaking of the second C–O bond to continue along the normal FTS reaction pathway that is followed by CO hydrogenation.

Riedel and Schaub [79] used a Co/MnO/Aerosil catalyst and found that with increasing CO_2 content of the syngas the product distribution shifted from the FT process to exclusively methane. Dorner et al. [76] investigated the hydrogenation of CO_2 using a traditional Co-FT catalyst and claimed that cobalt selectively produces methane with some C_2, C_3, and C_4 hydrocarbon formation. Gnanamani et al. [75] conducted a very similar experiment using Pt-promoted $Co/\gamma - Al_2O_3$ catalyst. The authors showed that with cobalt catalysts CO_2 behaves like an inert gas when co-fed with CO and, moreover, by replacing CO with CO_2, the product composition for the cobalt catalyst changes from the FT type to mostly methanation (Table 4.1, Fig. 4.6a).

In contrast, Yao et al. [84] claimed that CO_2 and CO mixtures can be used as feed for the cobalt FT catalyst and that CO_2 is neither inert nor a diluent but rather can be converted to hydrocarbon products using syngas having high CO_2 content $[CO_2/(CO/CO_2) > 50\%]$. It appears from the literature as well as from our own studies that CO_2 hydrogenation over cobalt is more likely a methanation reaction. Riedel et al. [79] explained the difference in catalytic behavior between CO and CO_2 on cobalt by different modes of formation in the kinetic regime of FT synthesis. The authors proposed the selective inhibition of methane formation and the selective inhibition of product desorption as prerequisites for chain growth. The

TABLE 4.1   Effect of Feed Gas Composition on Product Selectivity for Cobalt FT Synthesis [75]

Feed Composition (vol %)				Selectivity (C, %)					H_2/CO Ratio		Exit P_{CO}
H_2	CO	N_2	CO_2	C_1	C_2	C_3	C_4	C_5^+	Inlet	Exit	(bar)
66.67	33.33	0	0	8.1	1.2	5.2	5.1	80.4	2.00	1.84	5.20
66.67	23.33	10.00	0	13.3	2.1	5.7	4.8	74.1	2.86	5.88	1.50
66.67	23.33	0	10.00	13.7	1.5	4.0	3.6	77.2	2.86	7.48	1.20
66.67	10.00	23.33	0	91.4	2.5	2.0	0.8	3.3	6.67	269	0.04
66.67	10.00	0	23.33	87.0	2.0	1.6	0.8	8.7	6.67	124	0.06
66.67	0	11.11	22.22	93.3	0.8	0.5	0.2	5.2	—	—	0.00

(Reaction conditions: T = 493 K, P = 1.99 bar, SV of feed gas = 5.0 slh^{-1} g^{-1} cat.)

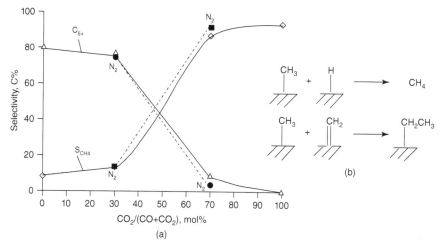

Figure 4.6   (a) Effect of feed gas composition on selectivity to CH_4 and C_{5+} for Co-FT synthesis. (reaction conditions: T = 493 K, P = 1.99 MPa, SV of feed gas = 5.0 slh^{-1} g^{-1} cat., dotted lines show the selectivity obtained for N_2 switch). (b) Selective inhibition of methane formation and product desorption as prerequisites for chain growth; adapted from ref. 79.

authors also speculated that with increasing CO coverage on the cobalt surface, the common bonding of CH_3 and H at the same cobalt atom becomes more and more improbable (Fig. 4.6b). Also, CO plays an important role in restructuring the cobalt metal surface [85], which may also affect the chain growth mechanism.

With iron, CO_2 first reacts with H_2 to form CO and H_2O in the RWGS reaction, followed by the FT reaction. The promoting effects of copper and potassium were found to increase RWGS activity and thereby increase the reactivity of CO_2. Various structural promoters, such as oxides of Si, Al, V, Cr, Mn, and Zn, offered higher selectivity to C_2^+ hydrocarbons.

In the past, many research publications have been devoted to investigating the nature of iron phases that exist during activation and FT synthesis [86,87]. However, the exact nature of active phase(s) remains controversial. For example, carburized iron is an active catalyst for FTS. Alkali metals promote catalyst carburization, apparently by facilitating the dissociation of CO at the iron catalyst surface [88]. Magnetite (Fe_3O_4) is often suggested to be the phase responsible for the water-gas shift reaction, with the FTS reaction taking place on different types of active sites on the catalyst. However, our recent study suggests that iron carbide is responsible for both FT and water-gas shift activity [89].

In the case of CO_2 hydrogenation, it is highly desirable that the iron catalyst have optimized RWGS and FT activity in order to promote higher hydrocarbon formation. Thus there is an increasing interest to better understand the catalyst structure-function relationships to achieve this goal, so that better formulations may be developed [88].

Contradictory results are often observed for product distribution for CO and CO_2 hydrogenation. In an early study, Riedel et al. [79] observed no excessive methane formation and the composition of the C_2^+ hydrocarbons for H_2/CO_2 was the same as that obtained with H_2/CO synthesis gas. The authors suggested that the main route of hydrocarbon formation from CO_2 includes first the conversion of CO_2 to CO, which then reacts to hydrocarbons. Alkalized iron catalysts are very active for the WGS reaction in both directions and even at the low CO partial pressures obtained, the FT selectivity remains apparently the same. Prasad et al. [11] concluded that Fe/Cu/Al/K is a suitable catalyst for FT synthesis from bio-syngas, giving high hydrocarbon yields and olefin selectivities. Fe/Cu/Si/K displayed lower activity and selectivity for bio-syngas hydrogenation.

In contrast, CAER researchers [74,75] found a considerable deviation in the product distribution obtained for CO and CO_2 hydrogenation using Fe catalysts (Table 4.2). An appreciable deviation was found in the low-alkali-content catalyst (low α -Fe) in which the methane selectivity for CO and CO_2 hydrogenation was found to be 5.8 and 37.8 C-%, respectively. The ASF distribution for CO and CO_2 hydrogenation using low α-Fe catalyst was 0.63 and 0.32, respectively. However, selectivity to methane appears to decrease with increasing CO_2 conversion by reducing the space velocity of H_2/CO_2 syngas as shown below (Table 4.2).

It appears that changes in the product distribution are a consequence of changes in the iron phases occurring during the reaction. In situ characterization studies of used catalysts indicate the essential formation of iron carbide phases [90,91]. Mössbauer studies on iron catalysts used for CO_2 hydrogenation revealed the presence of χ-Fe_5C_2, θ-Fe_3C and Fe_3O_4 phases [67]. Fiato et al. [92] observed a strong interrelation between iron carbide formation and CO_2 FT activity. A recent study by Riedel et al. [78] correlates several episodes of their kinetic regimes with compositional and structural changes of the catalyst. In the beginning of FT synthesis, the catalyst was not active, either for the WGS or for the FT synthesis, and the catalyst consisted of mainly α-Fe, Fe_3O_4, a small amount of Fe_2O_3, and only a small fraction of iron carbide. With the H_2/CO_2 syngas, only the Fe_5C_2 carbide was detected in the used catalyst, even after a long time at steady state. Remarkably, with H_2/CO_2 and the H_2/CO feeds at steady

TABLE 4.2 Hydrogenation of CO$_2$ over Fe-Based Catalysts [74,75]

Catal.	Feed	Feed Gas Composition, vol%			CO or CO$_2$ conv., %	Selectivity, C%						α (C$_8$-C$_{20}$)
		H$_2$	CO or CO$_2$	N$_2$		C$_1$	C$_2$	C$_3$	C$_4$	C$_5^+$	CO$_2$ * or CO	
lowα-Fe	H$_2$-CO[a]	17.5	25.0	57.5	32.8	5.8	4.0	7.2	4.3	46.9	31.8	0.63
	H$_2$-CO[b]	66.7	33.3	0	68.0	14.5	4.0	6.4	4.7	35.4	35.0	0.60
	H$_2$-CO$_2$[c]	75.0	25.0	0	20.9	37.8	14.5	12.5	5.7	5.4	24.1	0.32
	H$_2$-CO$_2$[c]	75.0	25.0	0	40.7[#]	28.1	9.4	6.1	2.5	43.6	10.3	0.60
high α-Fe	H$_2$-CO[a]	17.5	25.0	57.5	20.8	2.6	1.7	2.3	1.8	58.5	33.1	0.87
	H$_2$-CO$_2$[c]	75.0	25.0	0	19.7	16.5	7.6	4.8	2.3	53.8	15.0	0.75

Reaction condition: Low α-Fe catalyst: 543 K, 1.33 MPa, SV = 2 slh $^{-1}$ g $^{-1}$ cat; High α-Fe catalyst: 503 K, 1.33 MPa, SV = 2 slh $^{-1}$ g $^{-1}$ cat. ^{a}H$_2$/CO = 0.7 (e.g., coal-derived syngas, Texaco gasifier); ^{b}H$_2$/CO = 2.0 (e.g., stoichiometric FT, GTL); ^{c}H$_2$/CO$_2$ (stoichiometric ratio used).

$^\#$ Higher CO$_2$ conversion was achieved with a decrease of SV from 2.0 to 1.0 slh^{-1} g^{-1} cat.

* CO$_2$ when H$_2$-CO used as a feed and vice versa.

α: chain growth probability factor.

low α-Fe catalyst (100Fe:4.4Si:2.0Cu:0.5K); high α-Fe catalyst (100Fe:4.4Si:2.0Cu:5.0K) state, the product compositions were quite similar. The authors concluded that the surface of iron carbide is the true FTS catalyst for CO_2 hydrogenation.

In our recent Mössbauer study for hydrogenation of carbon oxides, we observed a phase change from χ-Fe_5C_2 to Fe_3O_4 (Fig. 4.7a) while switching from H_2/CO syngas to H_2/CO_2 in a 1L CSTR at 523 K and 2.03 MPa. The distribution

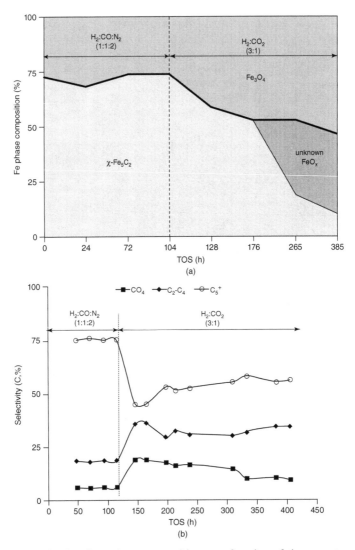

Figure 4.7 (a) Catalyst iron-phase composition as a function of time on stream during FTS, determined by Mössbauer spectroscopy; FTS performed at 523 K, 2.09 MPa, and a syngas space velocity of 3.0 sl/h/g Fe. (b) Comparison of hydrocarbon selectivity during the switch of CO to CO_2. FTS tested at 523 K, 2.09 MPa, and a syngas space velocity of 3.0 sl/h/g Fe.

of hydrocarbon products changes significantly with CO_2, but then it slowly transformed to regular FTS products as shown in Figure 4.7b. A correlation was obtained between the rates of FTS with % Fe carbide present in the catalyst.

From the above study, it appears that the abrupt change in product selectivity after switching from H_2/CO to H_2/CO_2 is not solely caused by changes in the iron phase as observed by Mössbauer analysis. It is reasonable to argue at this point that the surface carbon on iron would probably act as a buffer for product selectivity (Fig. 4.8). In the case of CO hydrogenation, CO dissociates on the iron surface and forms surface carbon and oxygen. The carbon diffuses into iron and retains a thin layer of iron carbide on the surface of iron [93]. The resulting iron carbide catalyzes the surface polymerization of carbon-containing intermediates to produce various hydrocarbons. Thus the activity and selectivity of iron catalysts for FTS depend on the relative percentages of iron carbide and Fe_3O_4 present in the catalyst samples. Also, we should keep in mind that the partial pressures of CO and water, as well as the coverages of intermediates, are key factors that influence the percentage composition of various carbides and oxides of iron present during FT synthesis.

With H_2/CO syngas, the relative percentages of carbides/oxides of iron should be maintained as much as possible under pseudo-steady-state conditions in order to promote FT chain growth. After switching from H_2/CO to H_2/CO_2 feeds, the relative fractions of carbides/oxides decrease substantially (Fig. 4.7a). This is due to the fact that carburization of iron occurs only in the presence of CO under FTS conditions. Thus a depletion of CO and a higher oxidizing atmosphere exist after switching from CO to CO_2. This in turn causes the iron catalyst system to change from carbide to oxide phase (χ-$Fe_5C_2 \rightarrow Fe_3O_4$). At the same time, the H_2/CO ratio in the exit gas stream was too high for CO_2 hydrogenation as compared to CO. As such, higher methane selectivity was obtained during the initial stage of FTS using H_2/CO_2 in correlation with the existing H_2/CO ratio; thus the partial pressure of CO during FT synthesis is a key factor that controls product selectivity.

From the kinetic point of view, hydrogen plays an important role in activating carbon dioxide over Fe- and Co-based FT catalysts. Our recent study [74] using iron

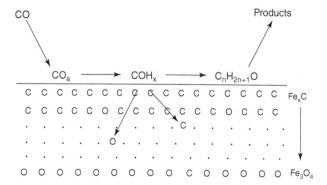

Figure 4.8 Schematic of the active low-temperature catalyst during FT synthesis (carbide outer layer with Fe_3O_4 inner core). Reproduced with permission from ref. 94.

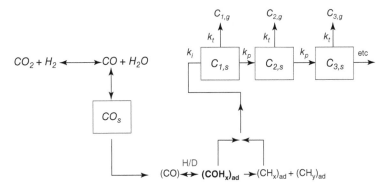

Figure 4.9   Plausible mechanism for hydrogenation of CO_2 using conventional Fe and Co catalysts (k^i, rate constant for initiation; k^p, rate constant for propagation; k^t rate constant for termination). Reproduced with permission from ref. 74.

and cobalt FT catalysts suggests that an inverse isotope effect exists for both CO and CO_2 hydrogenation. It was proposed that hydrogenation of CO is involved in the rate-limiting step, and, moreover, the inverse isotope effect appears to increase with increasing carbon number in the products (Fig. 4.9).

4.5   CONCLUSIONS

A reduction in CO_2 emitted to the atmosphere is a desirable and global goal. As a renewable feedstock, the transformation of CO_2 into fuels and chemicals is particularly promising and deserves worldwide research efforts. Hydrogenation of CO_2 is one of the interesting processes that utilizes a high-energy substance like H_2 to convert the low-cost CO_2 into fuels and other valuable products. The key issue of this process is the cost of H_2 in the overall scheme. Currently, the main sources of hydrogen are hydrocarbon feedstocks such as natural gas, coal, and petroleum; however, these feedstocks also produce CO_2. However, hydrogen can also be produced from cellulosic biomass, through a process much like coal gasification, from which the hydrogen can be removed and purified, or it can be produced from the reforming of bio-ethanol. On a net basis, the H_2 produced from bio-renewable resources may not be contributing to overall CO_2 emissions. Potential CO_2 -free sources for hydrogen production include nuclear power-driven electrolysis of water and solar processes.

As discussed in this review, heterogeneous catalytic hydrogenation of CO_2 may produce a wide variety of products including methane, methanol, dimethyl ether, higher hydrocarbons, and alcohols. CO_2 methanation has been extensively studied in the literature. Ni- and Ru-based catalysts were often used, and Co was used occasionally. Many suggest that methanation of CO_2 involves dissociative adsorption of CO_2 to form CO and atomic oxygen and the hydrogenation of CO is involved in the rate-limiting step.

With DRIFTS-IR, formates and carbon monoxide were identified as reaction intermediates for CO_2 hydrogenation over Ru/TiO_2 and it was proposed that CO_{ads} is the key intermediate for methanation.

Cu/ZnO-based catalysts were extensively studied for the hydrogenation of CO_2 to methanol. Cu is postulated to be the main active catalyst component, with ZnO being used to disperse Cu in a uniform manner. Significant progress has been made in shedding light on the chemistry of methanol formation by using modern surface characterization techniques. In general, two possible pathways are proposed: (i) direct hydrogenation of CO_2 via formate or (ii) the reduction of carbon dioxide to CO with subsequent hydrogenation to methanol. There are studies suggesting that, depending upon the partial pressure of CO and CO_2, either the hydrogenation of CH_3O species or the formation of CH_3O can be rate-limiting for methanol formation.

Compared to methanol, CO_2 to higher alcohols is a subject that has been less studied. Combinations of Cu-Zn-based methanol synthesis catalysts and zeolite catalysts were found to synthesize higher alcohols from CO_2. The mechanism of formation of higher alcohols synthesis may proceed through the reaction of CO insertion with hydrocarbon intermediates (RCH_n-). However, a direct nondissociative hydrogenation of CO_2 route has also been proposed.

Finally, we have discussed current understanding on catalytic reactivity and reaction mechanisms based on the literature as well as from our own studies of hydrogenation of CO_2 through a modified FTS process. Both cobalt- and iron-based catalysts were tested for hydrogenation of CO_2 to produce hydrocarbons. In general, cobalt catalysts produce exclusively methane under regular FTS conditions when CO_2 is used in the place of CO. On cobalt, CO_2 behaves like an inert gas in the presence of CO. Furthermore, by replacing CO with CO_2, the product composition changes from an FT-type distribution to mostly methanation. With iron, CO_2 first reacts with H_2 to form CO and H_2O (RWGS), followed by the FTS reaction. A considerable deviation in the product distribution was observed for CO and CO_2 hydrogenation over doubly promoted Fe-based catalysts. Finally, future research should strive to find novel catalytic systems that convert CO_2 to fuels and chemicals in a cost-effective way.

REFERENCES

1. S.N. Riduan and Y.G. Zhang (2010). *Dalton Trans.*, 39: 3347–3357.

2. G.A. Olah, A. Goeppert and G.K.S. Prakash (2009). *J. Org. Chem.*, 74: 487–498.

3. J. Tollefson (2009). *Nature*, 462: 966–967.

4. X.D. Xu and J.A. Moulijn (1996). *Energy Fuels*, 10: 305–325.

5. M. Mikkelsen, M. Jorgensen and F.C. Krebs (2010). *Energy Environ. Sci.*, 3: 43–81.

6. H. Yang, Z. Xu, M. Fan, R. Gupta, R.B. Slimane, A.E. Bland and I. Wright (2008). *J. Environ. Sci.*, 20: 14–27.

7. A.J. Hunt, E.H. K. Sin, R. Marriott and J.H. Clark (2010). *ChemSusChem*, 3: 306–322.

8. G. Ferey, C. Serre, T. Devic, G. Maurin, H. Jobic, P.L. Llewellyn, G.De Weireld, A. Vimont, M. Daturi and J.S. Chang (2011). *Chem. Soc. Rev.*, 40: 550–562.

9. C.S. Song (2006). *Catal. Today*, 115: 2–32.

10. J. Ma, N.N. Sun, X.L. Zhang, N. Zhao, F.K. Mao, W. Wei and Y.H. Sun (2009). *Catal. Today*, 148: 221–231.

11. P.S. Sai Prasad, J.W. Bae, K.-W. Jun, K.-W. Lee (2008). *Catal. Surv. Asia*, 12: 170–183.

12. R.W. Dorner, D.R. Hardy, F.W. Williams and H.D. Willauer (2010). *Energy Environ. Sci.*, 3: 884–890.

13. P.G. Jessop, T. Ikariya and R. Noyori (1995). *Chem. Rev.*, 95: 259–272.

14. H. Arakawa, M. Aresta, N.N. Armor, M.A. Barteau, E.J. Beckman, A.T. Bell, J.E. Bercaw, C. Creutz, E. Dinjus, D.A. Dixon, K. Domen, D.L. Dubois, J. Eckert, E. Fujita, D.H. Gibson, W.A. Goddard, D.W. Goodman, J. Keller, G.J. Kubas, H.H. Kung, J.E. Lyons, L.E. Manzer, T.J. Marks, K. Morokuma, K.M. Nicholas, R. Periana, L. Que, J.R.-Nielson, W.M.H. Sachtler, L.D. Schmidt, A. Sen, G.A. Somorjai, P.C. Stair, B.R. Stults and W. Tumas (2001). *Chem. Rev.*, 101: 953–996.

15. D.B. Dell'Amico, F. Calderazzo, L. Labella, F. Marchetti, and G. Pampaloni (2003). *Chem. Rev.*, 103: 3857–3897.

16. P. Sabatier, J.B. Senderens (1902). *Hebd. Seances Acad. Sci.*, 134: 514–516.

17. Z. Zhang, A. Kladi, and X.E. Verykios (1994). *J. Catal.*, 148: 737–747

18. V.M. Vlasenko, and G.E. Yuzefovich (1969). *Russ. Chem. Rev.*, 38: 728–739.

19. G.A. Mills, and F.W. Steffgen (1973). *Catal. Rev.*, 8: 159–210.

20. G. D. Weatherbee, and C.H. Bartholomew (1982). *J. Catal.*, 77: 460–472.

21. M. Greyson (1956). *Catalysis*, P.H. Emmett, ed., Reinhold, *New York*, Vol. 4, 473–511.

22. T.V. Herwijnen, H.V. Doesburg, and W.A. de Jong (1973). *J. Catal.*, 28: 391–402.

23. W. Wang, S. Wang, X. Ma, J. Gong (2011). *Chem. Soc. Rev.*, 40: 3703–3727.

24. M.H. Halmann, and M. Steinberg (1999). *Greenhouse Gas Carbon Dioxide Mitigation*, Lewis Publishers, Boca Raton, 334–343.

25. K.R. Thampi, J. Kiwi, M. Gratzel (1987). *Nature*, 327: 506–508.

26. T.M. Gur, R.A. Huggins (1983). *Science*, 219: 967–969.

27. D.E. Peebles, D.W. Goodman, J.M. White (1983), *J. Phys. Chem.*, 87: 4378–4387.

28. H. Baussart, R. Delobel, M. Le Bras, J.-M. Leroy (1987). *J.C.S. Faraday Trans. I*, 83: 1711–1718.

29. A. Trovarelli, C. De Leitenburg, G. Dolcetti (1991). *J.C.S. Chem. Commun.*, 472–473.

30. R. Dziembay, W. Makowski, H. Papp (1992). *J. Mol. Catal.*, 75, 81–99.

31. T. Tada, H. Habazaki, E. Akiyama, A. Kawashima, K. Asami, K. Hashimoto (1994). *Mater. Sci. Eng.*, 182: 1133–1136.

32. T. Inui, M. Funabiki, M. Suehiro, T. Sezume (1979). *J.C.S. Faraday I*, 75: 787–802.

33. J.A. Dalmon, and G.A. Martin (1979). *J.C.S. Faraday Trans. I*, 75: 1011–1015.

34. G.D. Weatherbee, and C.H. Bartholomew (1981). *J. Catal.*, 68: 67–76.

35. R. Barget and Y. Trambouze (1980). *C. R. Acad. Sci.*, 290: 153.

36. S. Fujita, H. Terunuma, H. Kobayashi, and N. Takezawa (1987). *React. Kinet. Catal. Lett.*, 33: 179.

37. M. Araki, and V. Ponec (1976). *J. Catal.*, 44: 439–448.

38. M. Marwood, R. Doepper and A. Renken (1997). *Appl. Catal. A*, 151: 223–246.

39. J. Lahtinen, T. Anraku, G.A. Somorjai (1994). *Catal. Lett.*, 25: 241–255.

40. K. Okabe, H. Yamada, T. Hanaoka, T. Matsuzaki, H. Arakawa, Y. Abe (2001). *Chem. Lett.*, 904–905.

41. M. Saito, T. Fujitani, M. Takeuchi, T. Watanabe (1996). *Appl. Catal. A: Gen.*, 138: 311–318.

42. J. Sloczynski, R. Grabowski, P. Olszewski, A. Kozlowska, J. Stoch, M. Lachowska, J. Skrzpek (2006). *Appl. Catal. A: Gen.*, 310: 127–137.

43. T. Inui, T. Takeguchi, A. Kohama, K. Tanida (1992). *Energy Convers. Mgmt.*, 33: 513–520.

44. K.C. Waugh (1999). *Catal. Lett.*, 58: 163–165.

45. C.V. Ovesen, B.S. Clausen, J. Schiatz, P. Stoltze, H. Topsoe, J.K. Norskov (1997). *J. Catal.*, 168: 133–142.

46. T. Fujitani, J. Nakamura (2000). *Appl. Catal. A: Gen.*, 191: 111–129.

47. P.L. Hansen, J.B. Wagner, S. Helveg, J.R.R.-Nielsen, B.S. Clausen, H. Topsoe (2002). *Science*, 295: 2053–2055.

48. R. Naumann d'Alnoncourt, X. Xia, J. Strunk, E. Loffler, O. Hinrichsen, M. Muhler (2006). *Phys. Chem. Chem. Phys.*, 8: 1525–1538.

49. F. Arena, G. Italiano, K. Barbera, S. Bordiga, G. Bonura, L. Spadaro, F. Frusteri (2008). *Appl. Catal. A: Gen.*, 350: 16–23.

50. Q. Sun, C.-W. Liu, W. Pan, Q.-M. Zhu, J.-F. Deng (1998). *Appl. Catal. A: Gen.*, 171: 301–308.

51. L.C. Grabow and M. Mavrikakis (2011). *ACS Catal.*, 1: 365–384.

52. H. Wei, X.K.-Chang, B.Y.-Bin, Y.L.-Hua (1999). *J. Natural Gas Chem.*, 8: 196–202.

53. G. Centi, S. Perathoner (2009). *Catal. Today*, 148: 191–205.

54. M.H. Halmann, M. Steinberg (1999). *Greenhouse Gas Carbon Dioxide Mitigation*, Lewis Publishers, Boca Raton, 357–358.

55. A. Okamoto, H. Arakawa (1994). *Chem. Ind. Chem.*, 47: 314.

56. T. Inui (1996). *Catal. Today*, 29: 329–335.

57. H. Kusama, K. Sayama, K. Okada, H. Arawaka (1994). *Preprints 74th Annual Meeting Catalysis Soc. Jpn.*, 430.

58. T. Inui, T. Yamamoto, M. Inoue, H. Hara, T. Takeguchi, J. Kim (1999). *Appl. Catal. A: Gen.*, 186: 395–406.

59. T. Inui (2002). *A.C.S Symposium Series 809*, CO_2 Conversion and Utilization, Eds. C. Song, A.M. Gaffney, K. *Fujimoto, Chapter 9*, 130–152.

60. H. Nagata, K. Yamada, M. Kishda, Y. Wada, K. Wakabayashi (1995). *Energy Convers Mgmt.*, 36: 657–660.

61. H. Kusama, K. Okabe, K. Sayama, H. Arawaka (1996). *Catal. Today*, 28: 261–266.

62. R.B. Anderson (1984). *The Fischer-Tropsch Synthesis*, Academic Press, Orlando, FL.

63. A.P. Steynberg, M.E. Dry (Eds.) (2004). Fischer-Tropsch Technology, *Stud. Surf. Sci. Tech.*, 152.

64. B.H. Davis, M.L. Occelli (Eds.) (2010). *Advances in Fischer-Tropsch Synthesis, Catalysts, and Catalysis, Chemical Industries*, 128.

65. M.E. Dry (1996). *Appl. Catal. A: Gen.*, 138: 319–344.

66. K. Fujimoto, T. Shikada (1987). *Appl. Catal.*, 31: 13–23.

67. J.F. Lee, W.S. Chern, M.D. Lee, T.Y. Dong (1992). *Can. J. Chem. Eng.*, 70: 511–515.

68. M. Fujiwara, R. Kieffer, H. Ando, Y. Souma (1995). *Appl. Catal. A: Gen.*, 121: 113–124.

69. M. Fujiwara, H. Ando, M. Tanaka, Y. Souma (1995). *Appl. Catal. A: Gen.*, 130: 105–116.

70. T. Inui, T. Makino, F. Okazumi, S. Nagano, A. Miyamoto (1987). *I.E.C. Res.*, 26: 647–652.

71. M. Fujiwara, H. Ando, M. Matsumoto, Y. Matsumura, M. Tanaka, Y. Souma (1995). *Chem. Lett.*, 839–840.

72. M. Fujiwara, R. Kieffer, H. Ando, Q. Xu, Y. Souma (1977). *Appl. Catal. A: Gen.*, 154: 87–101.

73. D. He, Q. Zhu, Q. Xu, F. Masahiro, S. Yoshie (1998). *J. Fuel Chem. Tech.*, 26: 201–202.

74. M.K. Gnanamani, G. Jacobs, W.D. Shafer, D.E. Sparks, B.H. Davis (2011). *Catal. Lett.*, 141: 1420–1428.

75. M.K. Gnanamani, W.D. Shafer, D.E. Sparks, B.H. Davis (2011). *Catal. Commun.*, 12: 936–939.

76. R.W. Dorner, D.R. Hardy, F.W. Williams, B.H. Davis, H.D. Willauer (2009). *Energy & Fuels*, 23: 4190–4195.

77. T. Herranz, S. Rojas, F.J.P. Alonso, M. Ojeda, P. Terreros, J.L.G. Fierro (2006). *Appl. Catal. A: Gen.*, 308: 19–30.

78. T. Riedel, H. Schulz, G. Schaub, K.-W. Jun, J.-S. Hwang, K.-W. Lee (2003). *Topics in Catal.*, 26: 41–54.

79. T. Riedel, M. Claeys, H. Schulz, G. Schaub, S.-S. Nam, K.-W. Jun, M.-J. Choi, G. Kishan, K.W. Lee (1999). *Appl. Catal. A: Gen.*, 186: 201–213.

80. S.-R. Yan, K.-W. Jun, J.-S. Hong, S.-B. Lee, M.-J. Choi, K.-W. Lee (1999). *Korean J. Chem. Eng.*, 16: 357–361.

81. W.W. Russell, G.H. Miller (1949). *J. Am. Chem. Soc.*, 72: 2446–2454.

82. Y. Zhang, G. Jacobs, D.E. Sparks, M.E. Dry, B.H. Davis (2002). *Catal. Today*, 71: 411–418.

83. C.G. Visconti, L. Lietti, E. Tronconi, P. Forzatti, R. Zennaro, E. Finocchio (2009). *Appl. Catal. A: Gen.*, 355: 61–68.

84. Y. Yao, D. Hildebrandt, D. Glasser, X. Liu (2010). *Ind. Eng. Chem. Res.*, 49: 11061–11066.

85. G.A. Somorjai (1994). *Ann. Rev. Phys. Chem.*, 45: 721.

86. E. de Smit, B.M. Weckhuysen (2008). *Chem. Soc. Rev.*, 37: 2758–2781.

87. J. B. Butt (1991). *Catal. Lett.*, 7: 61–81.

88. M.C. Ribeiro, G. Jacobs, B.H. Davis, D.C. Cronauer, A.J. Kropf, C.L. Marshall (2010). *J. Phys. Chem. C*, 114: 7895–7903.

89. V.R.R. Pendyala, G. Jacobs, J.C. Mohandas, M. Luo, H.H. Hamdeh, Y. Ji, M.C. Ribeiro, B.H. Davis (2010). *Catal. Lett.*, 140: 98–105.

90. D.B. Buker, K. Okabe, MP. Rsynek, C. Li, D. Wang, K.R.P.M. Rao, G.P. Huffman (1995). *J. Catal.*, 155: 353–365.

91. H. Schulz, G. Schaub, M. Claeys, T. Ridel (1999). *Appl. Catal. A: Gen.*, 186: 215–227.

92. R.A. Fiato, E. Iglesia, G.W. Rice, S.L. Soled (1998). *Stud. Surf. Sci. Catal.*, 114: 339–344.

93. J. Cheng, P. Hu, P. Ellis, S. French, G. Kelly, C.M. Lok (2010). *J. Phys. Chem. C*, 114: 1085–1093.

94. B.H. Davis (2009). *Catal. Today*, 141: 25–33.

CHAPTER 5

Direct Synthesis of Organic Carbonates from CO_2 and Alcohols Using Heterogeneous Oxide Catalysts

YOSHINAO NAKAGAWA, MASAYOSHI HONDA, and KEIICHI TOMISHIGE*

*Indicates the corresponding author.

Green Carbon Dioxide: Advances in CO_2 Utilization, First Edition.
Edited by Gabriele Centi and Siglinda Perathoner.
© 2014 John Wiley & Sons, Inc. Published 2014 by John Wiley & Sons, Inc.

5.1 INTRODUCTION

Organic carbonates are very important compounds. Dimethyl carbonate (DMC), the simplest organic carbonate, is produced at about 10^5 t/y worldwide and is used as a solvent, a carbonylation and methylating reagent, a starting material for polycarbonate resin, an electrolyte in lithium-ion batteries, and so on (Fig. 5.1A) [1,2]. In addition, it has been proposed that DMC can be used as an octane booster in gasoline and as an additive to diesel fuel to decrease the emission of particulate matter [3]. Dialkyl carbonates with longer chains, diphenyl carbonate, and cyclic carbonates are also industrially produced and are widely used as solvents and synthetic intermediates [4]. Because of the versatility, low toxicity, and biodegradability of organic carbonates, further growth of the international demand is expected.

The production methods of organic carbonates are summarized in Figure 5.1B [5]. The traditional methods for DMC synthesis is the phosgenation of methanol in the presence of pyridine. Most organic carbonates can be also produced by the phosgenation method [6]. However, the high toxicity and corrosive nature of phosgene and the low atom efficiency of the phosgenation method have encouraged the development of other approaches to the production of organic carbonates. Cyclic carbonates such as ethylene carbonate (EC) and propylene carbonate have been commercially produced by the addition reactions of epoxides with CO_2 [7]. Transesterification of EC with alcohols can produce dialkyl carbonates such as DMC [8]. The production of DMC with the transesterification method has been commercialized, although this method requires many steps and coproduces ethylene glycol [9]. Processes based on oxidative carbonylation of methanol to produce DMC have been also industrially carried out [10,11].

The direct synthesis of organic carbonates from alcohols and CO_2 is a very attractive alternative because of the high atom efficiency, low toxicity and price of the reactants, and simplicity of the operations [2]. In view of the chemical transformation of CO_2, highly oxidized organic carbonates are ideal targets because the large energy input in the transformation of CO_2 can be avoided [12,13]. A problem in the reaction is low conversion because of the equilibrium limitation. Conditions of high-pressure or even supercritical CO_2 are usually applied. A possible side-reaction is the formation of ethers such as dimethyl ether (DME) in DMC synthesis. The ether formation not only consumes the alcohol reactant but also produces water, shifting the equilibrium of the carbonate formation further to the reactant side. Therefore, the ether formation must be strictly suppressed.

Several types of catalysts have been reported for the direct synthesis of organic carbonates, especially DMC [2]: homogeneous metal complexes, complexes anchored to supports, and metal oxide catalysts. Organotin-based complexes such as $Me_2Sn(OMe)_2$ have been the most studied catalysts in the homogeneous systems for direct DMC synthesis [14–16]. High selectivities to DMC have been reported in these catalysts. However, the activity is far from satisfactory; only a few turnovers were obtained when no additive was used. These alkoxide catalysts can be decomposed by coproduced water and can be deactivated [17]. The difficulty in separation of the catalyst from the reaction solution after use is another problem.

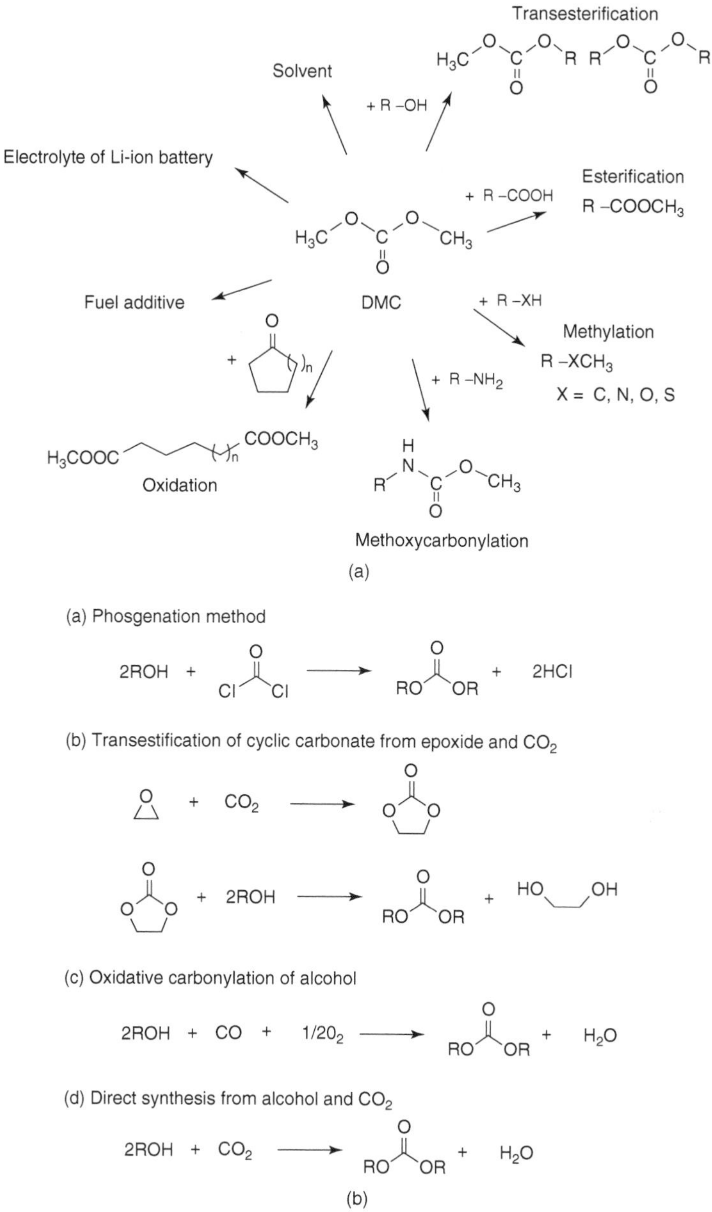

Transesterification

Solvent

+ R –OH

Electrolyte of Li-ion battery

Esterification

+ R –COOH

R –COOCH₃

DMC

Fuel additive

+ R –XH

Methylation

R –XCH₃

X = C, N, O, S

+ R –NH₂

COOCH₃

H₃COOC

Oxidation

Methoxycarbonylation

(a)

(a) Phosgenation method

$$2ROH + \underset{Cl \quad Cl}{\overset{O}{\parallel}} \longrightarrow \underset{RO \quad OR}{\overset{O}{\parallel}} + 2HCl$$

(b) Transestification of cyclic carbonate from epoxide and CO₂

$$\triangle\text{O} + CO_2 \longrightarrow \text{(cyclic carbonate)}$$

$$\text{(cyclic carbonate)} + 2ROH \longrightarrow \underset{RO \quad OR}{\overset{O}{\parallel}} + HO\_\_OH$$

(c) Oxidative carbonylation of alcohol

$$2ROH + CO + 1/2O_2 \longrightarrow \underset{RO \quad OR}{\overset{O}{\parallel}} + H_2O$$

(d) Direct synthesis from alcohol and CO₂

$$2ROH + CO_2 \longrightarrow \underset{RO \quad OR}{\overset{O}{\parallel}} + H_2O$$

(b)

Figure 5.1   (a) Utilization routes for DMC. (b) Synthesis of organic carbonates.

Immobilization of active complex catalysts on solid supports is one approach to overcome the separation problem of the homogeneous systems [18,19]. Fan et al. [20,21] prepared an organotin catalyst tethered on SBA-15 from the mixture of $(MeO)_2 ClSi(CH_2)_3 SnCl_3$ and tetraethyl orthosilicate. This material shows much higher activity than the sample prepared by the grafting method (Sn/SBA-15) in the direct synthesis of DMC from methanol and CO_2. Ballivet-Tkatchenko et al. [22] prepared n-$Bu_2Sn(OMe)_2$-based catalysts in which the active complex was anchored to the internal wall of SBA-15. These materials act as heterogeneous catalysts, and no leaching was observed. The problems with the immobilized catalysts include the complex preparation procedures, low reaction rates, and low turnover numbers (TONs). Supported or unsupported metal oxides are attractive materials for catalysts because of their low cost, easy handling, and mechanical and thermal stability. The application of metal oxide catalysts to direct synthesis of dialkyl carbonates is rather new: The first paper was published in 1999 in which ZrO_2 was used as a catalyst [23], while organotin catalysts were discovered in the 1970s [24]. In this chapter, the development of the metal oxide catalysts is reviewed. In addition to catalyst performance, improvement of the processes including the use of dehydrating agents is discussed.

5.2 CERIA-BASED CATALYSTS

Pure ceria and ceria-zirconia are frequently used in catalysis [25]. The redox properties have usually been used, and the best-known application is as additive to the three-way catalyst [26]. In the case of direct synthesis of organic carbonates, ceria is an effective catalyst, although the acid and base properties rather than the redox properties can be important.

5.2.1 Choice of Ceria Catalysts in Direct DMC Synthesis

Because of the importance of ceria in catalysis and material sciences, many kinds of "ceria" are available with various grades, morphology, surface area, and crystallinity from various suppliers. We prepared various ceria catalysts by calcination of five different precursors [Daiichi Kigenso CeO_2-HS, Daiichi Kigenso CeO_2-FN, Daiichi Kigenso CeO_2-FP, $Ce(OH)_4$ from Wako Pure Chemical, and $Ce(OH)_4$ from Soekawa Chemicals] at various temperatures (673, 873, 1073, and 1273 K) and compared the catalytic activities [27]. The reaction was carried out in a stainless steel autoclave reactor at 403 K. An equimolar mixture of methanol and CO_2 was used, and the total pressure at room temperature was about 5 MPa. The reaction time (2 h) was set to keep the DMC amount below the equilibrium level.

The selectivity to DMC was very high in all cases; no by-product such as DME was detected. Except for the catalysts calcined at 673 K, the activity of DMC formation was almost proportional to the BET surface area of the catalyst (Fig. 5.2a, dotted line). The activity of the catalysts calcined at 673 K was located below the line. Low activity per surface area over the catalysts calcined at lower temperature can be due to the catalyst structure. By means of FT-IR, the surface OH group (3499 cm^{-1}) of amorphous phase of CeO_2 [28] was significantly observed on the CeO_2 calcined at 673 K. Calcination at higher temperature decreased the band more significantly than those of the OH groups on crystalline CeO_2 (3724 and 3631 cm^{-1}). In addition, after methanol adsorption, both of the bands assigned to the OH groups

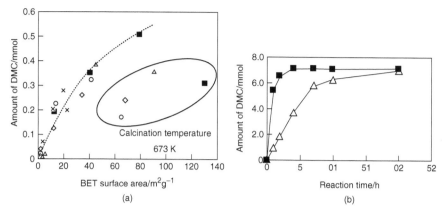

Figure 5.2 (a) Surface area dependence of DMC formation from methanol and CO_2 on various ceria catalysts [27]. Reaction conditions: methanol:CO_2 = 192 mmol:200 mmol, catalyst weight: 50 mg, reaction temperature: 403 K, reaction time: 2 h, Precursor: CeO_2-HS ($\diamond$), CeO_2-FN ($\blacksquare$), CeO_2 − FP ($\triangle$), $Ce(OH)_4$ from Wako Pure Chemicals ($\times$), and $Ce(OH)_4$ from Soekawa Chemicals ($\circ$). (b) Reaction time dependence of DMC formation from CH_3OH and CO_2 on CeO_2-HS calcined at 873 K [27]. Catalyst weight: 10 mg ($\triangle$), 100 mg ($\blacksquare$). Reaction conditions: CH_3OH :CO_2 = 192 mmol:200 mmol, reaction temperature: 403 K.

on crystalline CeO_2 disappeared on all the samples, and the band assigned to that on amorphous CeO_2 was unchanged.

The bulk structure observed by XRD was the fluorite structure, and no other crystalline phase was present regardless of the calcination temperature. These data suggested that stable crystallite surface is active for the reaction and the surface of amorphous phase is inactive. It has been reported that the most stable surface among the low index planes is a (111) surface [29–31], which may be the active site for this reaction. The catalytic activity of every sample decreased with increasing calcination temperature above 873 K, in accordance with the decrease of BET surface area. CeO_2-HS calcined at 873 K (BET surface area: 80 m^2/g) exhibited the highest activity. Aresta et al. [32] examined the catalytic activity and lifetime of commercial CeO_2 from Aldrich, CeO_2 synthesized by precipitation and calcination at 923 K, and Al/Ce (3–10% Al) or Fe/Ce (20% Fe) mixed oxides prepared by coprecipitation and calcination at 923 K. All catalysts showed good activity in the first run. However, the lifetime of commercial CeO_2 and Fe/Ce was quite short, as after the first cycle the activity decreases. The reduction of Ce^{IV} to Ce^{III} and the formation of formate were observed on commercial CeO_2 after the reaction by XPS and FT-IR, respectively. In contrast, synthesized CeO_2 and Al/Ce can be reused at least five times without loss of catalytic activity.

5.2.2 Performances of the Ceria Catalyst in DMC Synthesis

The catalytic performance of CeO_2-HS calcined at 873 K was investigated in detail (Table 5.1) [27]. Equilibrium limitation is critical in the direct DMC synthesis. The

TABLE 5.1 Syntheses of Organic Carbonates from Alcohols and CO_2 over CeO$_2$-HS Calcined at 873 K

Entry	Alcohol (amount, mmol)	CO_2 Amount, mmol	Catalyst Weight, mg	Temp., K	Time, h	Carbonate (amount, mmol)
1	Methanol (200)	200	10	403	2	DMC (0.24)
2	Methanol (200)	200	100	403	2	DMC (0.66)
3	Methanol (200)	200	100	423	2	DMC (0.65)
4	Methanol (200)	200	100	443	2	DMC (0.62)
5	Methanol (200)	200	10	383	2	DMC (0.05)
6	Methanol (200)	200	10	403	0.5	DMC (0.12)
7	Methanol (200)	50	10	403	0.5	DMC (0.11)
8	Methanol (200)	25	10	403	0.5	DMC (0.08)
9	Methanol (200)	3.1	10	403	0.5	DMC (0.04)
10	Ethanol (200)	200	10	403	2	DEC (0.09)
11	Ethanol (200)	200	100	403	2	DEC (0.41)
12	Methanol (100) + ethanol (100)	200	10	403	2	DMC (0.12), ethyl methyl carbonate (0.16), DEC (0.006)
13	Methanol (100) + ethanol (100)	200	100	403	2	DMC (0.17), ethyl methyl carbonate (0.28), DEC (0.04)
14	1-Propanol (192)	200	10	423	2	Di-*n*-propyl carbonate (0.27)
15	1-Propanol (192)	200	100	423	2	Di-*n*-propyl carbonate (0.49)
16	2-Propanol (192)	200	10	423	2	Diisopropyl carbonate (0.03)
17	2-Propanol (192)	200	100	423	2	Diisopropyl carbonate (0.15)
18	Phenol (100)	5 MPa	172	443	24	No reaction

equilibrium level of DMC yield was clearly demonstrated by the experiments in which a different amount of catalyst was used.

Figure 5.2b shows the reaction time dependence of DMC formation in the direct DMC synthesis. When 10 mg of catalyst (0.058 mmol Ce) was used, the DMC amount increased with the reaction time without an induction period (see also *entry 1* in Table 5.1) and reached about 0.71 mmol (0.74% yield based on methanol) after 20-h reaction. Considering that only part of Ce ions are located at the surface, the reaction of DMC formation proceeded catalytically.

In contrast, when 100 mg of catalyst was used, the formation rate became much higher and the DMC amount reached the same level even after 4-h reaction. This indicates that the plateau in Figure 5.2b is not due to the catalyst deactivation but to the reaction equilibrium.

At higher temperature, the amount of DMC produced over enough catalyst was slightly decreased (0.62 mmol at 443 K; *entries 2–4* in Table 5.1), indicating that the reaction is exothermic and a lower reaction temperature is more favorable for higher DMC yield. However, lower temperature decreased the catalytic activity: The rate at 383 K was about one-fifth as high as that at 403 K (*entry 5*). The CO_2 amount dependence of the DMC formation was also examined (*entries 6–9*). While the equilibrium level should be much affected by CO_2 pressure, the dependence of the rate on the CO_2 amount was rather small. Above 50 mmol of CO_2 (approximately 1.5 MPa), the rate of DMC formation was almost constant. This also demonstrates that DMC can be synthesized from methanol and CO_2 over ceria even at low CO_2 pressure.

5.2.3   Direct Synthesis of Various Organic Carbonates from Alcohols and CO_2 Without Additives

The system of this calcined CeO_2-HS catalyst was applied to the syntheses of other carbonates. The results are also shown in Table 5.1. Dialkyl carbonates with longer chains can be produced, although the yields were even lower than that of DMC (*entries 10–17*). Diphenyl carbonate was not produced (*entry 18*). The selectivities to the carbonates were always high, and ethers were not detected. As well as the equilibrium yield, the formation rate of dialkyl carbonates with longer chains was smaller than that of DMC. The reaction of the mixture of methanol and ethanol (mol/mol = 1/1) with CO_2 produced ethyl methyl carbonate (EMC), DMC, and diethyl carbonate (DEC) in the ratio of 1:1:0 at the initial stage (*entry 12*). When the reaction approached the equilibrium level, the selectivity to DEC increased and that to DMC decreased, approaching the values calculated by the binomial distribution (*entry 13*).

5.2.4   Reaction Mechanism

The mechanism of the homogeneous $Me_2Sn(OMe)_2$ catalyst for direct DMC synthesis has been investigated in detail [15,16]. The methoxide complex readily reacts with CO_2 at room temperature to give another methoxide, $Me_2Sn(OMe)(OCOOMe)$ (Fig. 5.3).

The structures of both methoxides have been confirmed by single-crystal X-ray analysis. The former methoxide is dinuclear and has both terminal and bridging methoxy ligands (t-OCH_3 and b-OCH_3, respectively). The latter methoxide is also dinuclear and has b-OCH_3 and monodentate monomethyl carbonate (t-OCOOMe) ligands. The thermolysis of $Me_2Sn(OMe)(OCOOMe)$ in CO_2 resulted in the formation of DMC in a reasonable yield. An intramolecular process for the reaction between the two ligands has been proposed (Fig. 5.3).

The adsorption of methanol and successive reaction with CO_2 on the calcined CeO_2-HS catalyst was investigated by FT-IR [27]. Figure 5.4a shows the spectra in the range of 1000–2000 cm^{-1}. The adsorption of methanol led to the appearance of bands at 1101 and 1052 cm^{-1}. These can be assigned to the C–O stretching modes of terminal (t-OCH_3) and bridged (b-OCH_3) methoxy species, respectively [33,34]. When 0.1 MPa of CO_2 was introduced to the CH_3OH-preadsorbed catalysts, the

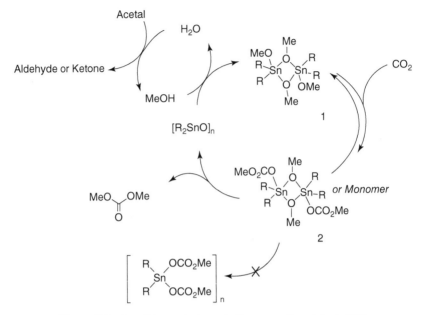

Figure 5.3   Possible catalytic cycle for organotin catalysts [15].

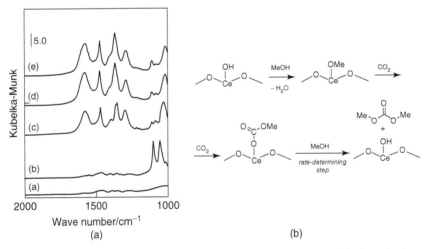

Figure 5.4   (a) DRIFT spectra of CH_3OH adsorption and successive CO_2 introduction at 403 K on CeO_2-HS calcined at 873 K [27]. (a) In N_2 flow, (b) after CH_3OH adsorption, (c) after CO_2 introduction at 0.1 MPa, (d) at 0.5 MPa, (e) at 5.0 MPa. Pretreatment: 773 K for 0.5 h in N_2 flow. (b) Reaction steps for the direct DMC synthesis over ceria catalyst.

band intensity at 1101 cm^{-1} decreased drastically and, at the same time, the bands at 1572, 1469, 1360, 1292, and 1109 cm^{-1} appeared. The bands at 1572, 1469, 1360, and 1109 cm^{-1} can be assigned to the monodentate monomethyl carbonate species [35]. The intensity of these bands was hardly affected by the increase of CO_2 pressure.

Aresta et al. [32] investigated the reaction of catalyst surface with methanol and CO_2 using an Al_2O_3-CeO_2 catalyst and ^{13}C CP-MAS NMR technique. The exposure of $Al_2O_3 - CeO_2$ catalyst to methanol at 408 K generated a new signal because of the $-OCH_3$ group at 49 ppm. After reaction of this sample with CO_2 at 408 K, the signal at 49 ppm disappeared and the signal of the methyl group of the monomethyl carbonate $-OCOOCH_3$ [36] very selectively appeared at 54 ppm. The signal of the OC(O) moiety was low in intensity and placed at 160 ppm. When the carboxylated sample was treated with gaseous methanol at 408 K in a closed tube, DMC was produced in the gas phase and the signals of monomethyl carbonate disappeared. These data suggest that the monomethyl carbonate species is easily formed on the catalyst surface and acts as the intermediate in the formation of organic carbonates (Fig. 5.4b).

The reaction of the monomethyl carbonate species with alcohol or alkoxide is the rate-determining step. The weak dependence of the reaction rate on CO_2 pressure supports the easy formation of the monomethyl carbonate intermediate. This reaction scheme can also explain the distribution of products in the reaction of methanol/ethanol mixture with CO_2 (*entries 12* and *13* in Table 5.1) [27]. First, both monomethyl carbonate and monoethyl carbonate are reversibly formed on the catalyst surface. Since the ethyl and methyl groups in surface monoalkyl carbonates are sufficiently far apart from the Ce−O bond, the coverage ratio of monomethyl carbonate to monoethyl carbonate can be equal to the molar ratio of methanol to ethanol. On the other hand, the reaction rate of monoalkyl carbonate with ethanol is lower than that with methanol, which was confirmed by the difference between the DMC and DEC formation rates in the reactions of methanol + CO_2 and ethanol +CO_2, respectively. Therefore, DMC and EMC are formed in the same ratio as that of methanol to ethanol at the initial stage by the reaction of monoalkyl carbonates with methanol. It is possible to form EMC via transesterification of DMC with ethanol. In order to evaluate the reaction rate of transesterification over the calcined CeO_2 catalyst, tests of the methanol + ethanol + CO_2 + DMC reaction were carried out [27]. At 403 K, the reaction rate of DMC with ethanol to EMC was much lower than the formation rate of EMC in the methanol + ethanol + CO_2 reaction. This indicates that the transesterification route is not the main route in the formation of EMC.

The state of the methanol in the reaction with surface monomethyl carbonate is not clarified. The $1020-1030 \text{ cm}^{-1}$ peak in the FT-IR spectrum of ceria treated with methanol and CO_2 (Fig. 5.4A) may be assigned to b-OCH_3 [27]. Considering the mechanism proposed for organotin catalysts, the reaction of b-OCH_3 with monomethyl carbonate may be the key step. In contrast, the ^{13}C NMR spectrum of $Al_2O_3 - CeO_2$ catalyst treated with methanol and CO_2 showed that no methoxy species was left [32]. On the basis of the NMR data and density functional theory

(DFT) calculations, Aresta et al. [32] proposed that the formation of DMC takes place via the interaction of surface monomethyl carbonate species with gas-phase methanol more than with the surface-bound methoxy species.

5.2.5 Ceria-Zirconia Catalysts

Ceria-zirconia mixed oxide is frequently used in the catalytic systems where pure ceria is effective [25]. One difference between ceria and ceria-zirconia is the higher thermal stability of ceria-zirconia. Ceria-zirconia mixed oxide catalyst was applied to the direct DMC synthesis [37,38]. The catalysts were prepared by calcining the hydroxides (available from Daiichi Kigenso) for 3 h under air atmosphere at various temperatures (573–1273 K). The molar ratio Ce/(Ce + Zr) was 0, 0.2, 0.33, and 0.5. All catalysts except ZrO_2 showed catalytic activity at 383 K, and the selectivity was always high (Table 5.2). With the Ce/(Ce + Zr) ratio of 0.2, the catalytic activity increased with increasing calcination temperature. On the other hand, the BET surface area decreased with increasing calcination temperature while the decrease [89 m^2/g (calcined at 673 K) → 20 m^2/g (1273 K)] was smaller than that observed in pure CeO_2 [131 m^2/g (673 K) → 13 m^2/g (1273 K)]. The same trends of activity-calcination temperature and surface area-calcination temperature were reported for ceria-zirconia with Ce/Zr = 1 by other researchers [39].

The XRD pattern of the catalysts showed that the solid solution with tetragonal structure was the only crystalline phase in ceria-zirconia [Ce/(Ce + Zr) = 0.2]. The crystal size becomes larger with higher calcination temperature, in accordance with the decrease of the BET surface area. One explanation for the high activity per surface area of the catalyst calcined at high temperature is that the active sites are the plain surface of the crystalline tetragonal phase with weaker acid–base properties than the rough surface [40]. Ceria-zirconia with higher Ce content [Ce/(Ce + Zr) = 0.5 showed lower activity than that with the Ce content of Ce/(Ce + Zr) = 0.2, probably because of the lower thermal stability of ceria than of ceria-zirconia. The

TABLE 5.2 Syntheses of DMC from Methanol and CO_2 over Ceria-Zirconia Catalysts[a]

Ce/ (Ce + Zr)	Calcination Temp., K	BET Surface Area, m^2 g^{-1}	DMC Amount, mmol
0	1273	10	0.00 (0.05[b])
0.2	1273	20	0.71
0.33	1273	19	0.73
0.5	1273	5	0.12
0.2	1073	35	0.48
0.2	773	65	0.06
0.2	573	120	0.00

[a] Reaction conditions: reaction temperature 383 K, methanol: CO_2 = 192 mmol: 200 mmol (ca. 5 MPa at room temperature), reaction time 2 h, catalyst weight 0.5 g.
[b] Reaction temperature, 443 K.

higher calcination temperature required to produce catalytically active surfaces in ceria-zirconia than in ceria may also be due to the difference of thermal stability. The catalytic activity of ceria-zirconia [Ce/(Ce + Zr) = 0.2] calcined at 1273 K was comparable to that of CeO_2-HS calcined at 873 K. The reaction temperature dependence of DMC formation showed that the ceria-zirconia catalyst can be used in a wide range of temperature, from even 343 K to 443 K. The catalytic performance of ceria-zirconia solid solution with a homogeneous tetragonal phase was also reported by other researchers [41].

The ceria-zirconia catalyst can be applied to the synthesis of cyclic carbonates such as EC and propylene carbonate from 1,2-diols and CO_2 [40,42]. The selectivity was very high below 403 K. At higher temperatures, a small amount of ethers was sometimes detected. Polycarbonates were not observed at all in any cases. The effect of solvents was investigated in the direct synthesis of propylene carbonate. The use of ethers and chloroform hardly changed the final yield. The use of dimethyl sulfoxide, N,N-dimethylformamide, and nitriles (acetonitrile and propionitrile) enhanced the carbonate yield by three- to fourfold. Acetonitrile was the most effective. It should be noted that the hydration of acetonitrile, which is an important reaction to shift the equilibrium and is discussed in Section 5.2.7, did not proceed in this case.

5.2.6 Modification of Ceria-Based Catalysts

Further modifications of ceria-zirconia catalysts have been reported. Zhang et al. [39] reported the modification of $Ce_{0.5}Zr_{0.5}O_2$ calcined at 1273 K with an ionic liquid. The catalytic activity was almost doubled when the ionic liquid 1-ethyl-3-methylimidazolium bromide ([EMIM]Br) was loaded on $Ce_{0.5}Zr_{0.5}O_2$ with a [EMIM]Br/mixed oxide ratio of 1/3. Lee et al. [43] supported Ga_2O_3 on $Ce_{0.6}Zr_{0.4}O_2$ catalyst calcined at 773 K. The $Ga_2O_3/Ce_{0.6}Zr_{0.4}O_2$ catalysts (1–15 wt%) showed a better catalytic performance than $Ce_{0.6}Zr_{0.4}O_2$ catalyst. In the XRD patterns of the catalysts, no characteristic peaks for Ga_2O_3 were found and a single cubic fluorite phase was present. The XRD peaks were slightly shifted to lower angles than in $Ce_{0.6}Zr_{0.4}O_2$. The catalyst with 5 wt% Ga_2O_3 showed the best catalytic performance. NH_3-TPD and CO_2-TPD results showed that the numbers of both acid and base sites decreased in the order of $Ga_2O_3/Ce_{0.6}Zr_{0.4}O_2$ (5 wt%) > $Ga_2O_3/Ce_{0.6}Zr_{0.4}O_2$ (1 wt%) > $Ga_2O_3/Ce_{0.6}Zr_{0.4}O_2$ (10 wt%) > $Ga_2O_3/Ce_{0.6}Zr_{0.4}O_2$ (15 wt%) > $Ce_{0.6}Zr_{0.4}O_2$, which was the same as the order of the catalytic activity. Acidity and/or basicity of the catalyst can play an important role in determining the catalytic performance.

5.2.7 Use of Acetonitrile as a Dehydrating Agent for DMC Synthesis

Since the carbonate yield is very low in the direct synthesis from alcohol and CO_2 because of the equilibrium limitations, dehydrating agent is frequently used to shift the reaction to the carbonate side [2]. However, the addition of inorganic dehydrating agents such as zeolite or magnesium sulfate does not improve the carbonate yield because the capability for dehydration is low at the high reaction temperatures. A reaction process separating the reaction part from the dehydration part at room temperature was very effective for the improvement of the DMC yield in the organotin-based homogeneous system [44]. Use of the hydration reaction of organic molecules such as orthoesters and 2,2-dimethoxypropane has been also attempted for the organotin-based systems, successfully enhancing the carbonate yields [15,17,45]. The disadvantages of this approach include the high cost of the dehydrating agents and the side-reactions involving the dehydrating agents.

The effect of the addition of acetonitrile to the DMC synthesis over ceria was investigated [46]. The hydration of acetonitrile produces acetamide (Eq. 5.1). A side reaction involving the acetonitrile hydration is the reaction of acetamide with methanol to methyl acetate and ammonia (Eq. 5.2). In addition, methyl carbamate can be formed by the reaction of DMC with NH_3 given by Eqs. 5.2 and 5.3.

$$CH_3CN + H_2O \rightarrow CH_3CONH_2 \qquad (5.1)$$

$$CH_3CONH_2 + CH_3OH \rightarrow CH_3COOCH_3 + NH_3 \qquad (5.2)$$

$$(CH_3O)_2CO + NH_3 \rightarrow CH_3O(CO)NH_2 + CH_3OH \qquad (5.3)$$

As a result, methyl carbamate can be derived from DMC and methyl acetate is derived from acetonitrile. Therefore, the sum of DMC + methyl carbamate can be almost the same as the sum of acetamide + methyl acetate, since almost all the water is removed by the hydration reaction of acetonitrile.

The results are summarized in Table 5.3. The reaction was conducted under conditions in which 0.04 mmol DMC (0.08% yield based on methanol) corresponds to the equilibrium (*entry 1*). The DMC amount increased with increasing acetonitrile amount in the range of 60–300 mmol, and it was saturated above 300 mmol (1.5 mmol DMC) (*entries 2–5*). The difference between DMC + methyl carbamate and acetamide + methyl acetate was small, and it can be explained by water as an impurity and by experimental errors. As a result, it was found that water formed with DMC from methanol and CO_2 reacted stoichiometrically with acetonitrile. The formation of dimethyl ether, which is an expected by-product in the DMC synthesis, was below the detection limit. At longer reaction time (*entries 6 and 7*), the methanol-based DMC yield reached 9%, which was more than 100 times as high as that at the equilibrium.

The formation rate of DMC increased with increasing CO_2 pressure in the case when no dehydrating agent was used; therefore a high CO_2 pressure is usually preferable for DMC synthesis. However, in this case the formation rate of DMC gradually decreased with increasing CO_2 pressure above 0.2 MPa (*entries 4, 8–11*). The decrease in DMC formation is interpreted as the hydration of acetonitrile being suppressed more significantly under higher CO_2 pressures. The hydration reaction of acetonitrile with a small amount of water catalyzed by ceria was carried out in the presence of methanol under various CO_2 pressures, and only acetamide and methyl acetate were detected and the amount of DMC was below the detection limit because of the presence of water in the reaction system (Fig. 5.5a). The amounts

TABLE 5.3 Syntheses of Organic Carbonates over CeO$_2$-HS Calcined at 873 K in Presence of Acetonitrile[a]

Entry	Methanol/ Acetonitrile, mmol/mmol	CO$_2$ Pressure, MPa[b]	Time, h	DMC	Methyl carbamate	Acetamide	Methyl acetate
1	100/0	0.5	2	0.04	—	—	—
2	100/60	0.5	2	0.50	0.05	0.42	0.11
3	100/180	0.5	2	1.15	0.06	1.24	0.11
4	100/300	0.5	2	1.47	0.06	1.72	0.09
5	100/420	0.5	2	1.52	0.08	2.03	0.10
6	200/600	0.5	2	2.01	0.00	2.28	0.06
7	200/600	0.5	48	8.94	1.08	8.58	1.31
8	100/300	0.08	2	0.70	0.03	0.98	0.06
9	100/300	0.2	2	1.73	0.06	1.74	0.08
10	100/300	1	2	1.14	0.05	1.26	0.09
11	100/300	5	2	0.52	0.00	0.16	0.05
12	200/600	0.2	2	1.14	0.00	1.59	0.00
13	200/600	0.2	24	6.07	0.33	6.60	0.66
14[c]	200/600	0.2	2	1.19[d]	0.11[e]	1.11	0.13[f]
15[c]	200/600	0.2	24	6.42[d]	0.79[e]	6.90	0.50[f]
16[g]	200/600	0.2	2	1.42[h]	0.21[i]	1.68	0.29[j]
17[g]	200/600	0.2	24	5.14[h]	0.41[i]	5.06	0.54[j]

[a] Reaction conditions: reaction temperature 423 K, catalyst weight 0.17 g.
[b] At room temperature.
[c] Ethanol was used.
[d] Amount of DEC.
[e] Amount of ethyl carbamate.
[f] Amount of ethyl acetate.
[g] 1-Propanol was used.
[h] Amount of di-n-propyl carbonate.
[i] Amount of n-propyl carbamate.
[j] Amount of n-propyl acetate.

of the products (acetamide and methyl acetate) gradually decreased with increasing CO$_2$ pressure, and it is clear that the dehydration of acetonitrile is strongly suppressed at high CO$_2$ pressure such as 5 MPa, suggesting that the formation of DMC in the presence of acetonitrile under low CO$_2$ pressure was controlled by the reaction rate of acetonitrile hydration. The coverage of adsorbed CO$_2$ on ceria can increase with increasing CO$_2$ pressure, and the active site of the acetonitrile hydration reaction can be covered with adsorbed CO$_2$ species.

The direct synthesis of organic carbonates has usually been carried out with high-pressure CO$_2$; therefore no attention has been paid to the CO$_2$-based yield of products. In contrast, in the present case, the reaction was evaluated under much lower CO$_2$ pressure, where the CO$_2$-based yield became much higher. The CO$_2$-based yield of DMC reached 40% when 0.2 MPa CO$_2$ was used (*entries 12* and *13*).

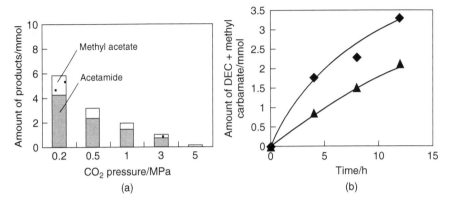

Figure 5.5 (a) Effect of CO_2 pressure on the amount of products in the mixture of methanol, acetonitrile, water, and CO_2 over ceria [47]. Reaction conditions: calcined CeO_2-HS 0.17 g, CH_3OH CH_3CN:H_2O = 100 mmol:300 mmol:10 mmol, reaction temperature 423 K, reaction time 2 h. (b) The addition effect of acetamide in DEC synthesis. Reaction conditions: CeO_2-HS (calcined at 873 K) 0.17 g, ethanol 100 mmol, acetonitrile 200 mmol, acetamide 0 (◆), or 5 mmol (▲), CO_2 0.5 MPa (at room temperature), 423 K.

The DMC formation rate at the initial stage was much higher than that at longer reaction times. An excessively long reaction time even lowered the DMC amount because of the formation of methyl carbamate (Eqs. 5.2 and 5.3), although most dehydrating agent remained in the reaction solution. The decrease of the DMC formation rate is probably due to the poisonous adsorption of products on the active sites. The lower reaction rate in the presence than in the absence of acetamide was observed in DEC synthesis (Fig. 5.5b) [47]. The amount of acetamide in this test (5 mol% of methanol) and the DMC formation tests (<5 mol%) is much larger than that of the active sites on ceria surface (total ceria/methanol = 1 mol%), suggesting the weak poisoning of acetamide by the reversible adsorption. The stability of the ceria catalyst was verified by the recyclability of the catalyst. The catalyst was recovered by filtration after the reaction and dried at 383 K for 2 h in the atmosphere, and then the catalyst was reused. The amount of products, BET surface area, and XRD patterns were almost unchanged in this stability test (Table 5.4). These results demonstrated that the ceria catalyst was stable under the reaction conditions and that the deactivation at a longer reaction time was not due to the changes of the ceria structure or the valence of cerium ions.

5.2.8 Use of Acetonitrile as Dehydrating Agent for Synthesis of Various Carbonates

The synthesis of other dialkyl carbonates with the use of acetonitrile as a dehydrating agent was investigated (*entries 14–17* in Table 5.3) [47]. Ethanol and 1-propanol were applied as reactants. Carbamates and esters were also formed by the side-reactions. The yield of DEC and di-*n*-propyl carbonate based on CO_2

TABLE 5.4   Recycle Test of CeO$_2$-HS Catalyst in DMC Synthesis[a]

| Usage Times | Amount of Products, mmol | | | | BET Surface Area After Each Test, m^2 g^{-1} |
	DMC	Methyl carbamate	Acetamide	Methyl acetate	
1	1.47	0.06	1.72	0.09	72
2	1.45	0.04	1.54	0.10	71
3	1.48	0.05	1.60	0.13	71

[a] Reaction conditions: catalyst 0.17 g, methanol 100 mmol, acetonitrile 300 mmol, CO$_2$ 0.5 MPa (at room temperature), 423 K, 2 h.

reached 42% and 33%, respectively, under low-pressure CO$_2$ (0.2 MPa). The yields of the three dialkyl carbonates were saturated at almost the same level, suggesting that the produced acetamide suppressed the formation of dialkyl carbonates similarly. The order of the dialkyl carbonate amount after short reaction time was di-*n*-propyl carbonate > DEC > DMC, while the formation rate of DEC was smaller than that of DMC in the absence of acetonitrile and it is more difficult to synthesize dialkyl carbonate with a longer alkyl chain. Therefore, it is expected that the order of the formation amount is caused by the difference in the activity of the acetonitrile hydration, not by the formation rate of dialkyl carbonates. When the performance in acetonitrile hydration in methanol, ethanol, and 1-propanol under pressurized CO$_2$ was compared, the order of the hydration rate was indeed 1-propanol > ethanol > methanol. One possible explanation of the order is based on the solubility of CO$_2$. It is known that the order in the solubility of CO$_2$ is methanol > ethanol > 1-propanol. Lower solubility of CO$_2$ can decrease the CO$_2$ coverage of ceria surface, which is connected to higher nitrile hydration activity. The acidity of the alcohols may also affect the reactivity order, although the difference in the acidity, that is, pK_a values, is rather small.

5.2.9   Use of Benzonitrile as Dehydrating Agent

Benzonitrile hydration was also applied to an in situ dehydration for DMC synthesis [48]. Similarly to the case of DMC synthesis in combination with acetonitrile hydration, a small amount of ester (methyl benzoate) and methyl carbamate was detected. The sum of DMC + methyl carbamate should be equal to the sum of benzamide + methyl benzoate. In fact, the former sum observed experimentally was slightly smaller than the latter. The difference was possibly due to the remaining water in reactants, catalysts, and reactors. A higher yield of DMC was obtained than that with acetonitrile hydration (*entry 1* in Table 5.5 and *entry 4* in Table 5.3). At optimum conditions, the maximum methanol-based DMC yield was 47% (*entry 5* in Table 5.5). Under low CO$_2$ pressure (0.1 MPa), the CO$_2$-based yield reached 70% (*entry 6*). The selectivities were also higher than those with acetonitrile hydration. The selectivity to DMC [DMC/(DMC + methyl carbamate)] and benzamide [benzamide/(benzamide + methylbenzoate)] was calculated to be 99% and 99%, respectively, at 9.4% methanol-based conversion at 423 K (*entry*

TABLE 5.5 Syntheses of Organic Carbonates over CeO_2-HS Calcined at 873 K in Presence of Benzonitrile[a]

Entry	Alcohol	Alcohol/ Benzonitrile, mmol/mmol	CO_2 Pressure, MPa^b	Time, h	Dialkyl carbonate	Alkyl carbamate	Benzamide	Alkyl benzoate
					Amount of Products, mmol			
1	Methanol	100/300	0.5	2	4.7	0.05	5.3	0.05
2	Methanol	100/200	1.0	2	4.5	0.07	4.7	0.06
3	Methanol	100/200	1.0	24	9.6	0.00	10.7	0.07
4	Methanol	100/500	1.0	2	2.8	0.00	2.9	0.02
5	Methanol	100/500	1.0	86	23.5	1.7	26.7	2.6
6	Methanol	200/400	0.1	24	5.3	0.00	7.5	0.55
7	Ethanol	100/200	1.0	2	5.1	0.02	5.5	0.04
8	Ethanol	100/200	1.0	24	10.5	0.44	11.4	0.48
9	1-Propanol	100/200	1.0	2	5.1	0.00	5.3	0.07
10	1-Propanol	100/200	1.0	24	8.6	0.12	8.5	0.38
11	2-Propanol	100/200	1.0	2	4.4	0.00	4.4	0.12
12	2-Propanol	100/200	1.0	24	6.5	0.00	10.1	0.03
13	Benzyl alcohol	100/200	1.0	2	2.6	0.06	3.4	0.10
14	Benzyl alcohol	100/200	1.0	24	2.4	0.10	3.5	0.17
15[c]	Methanol	100/200	0.1	1	0.0	0.0	19.9	1.2
16[c]	Methanol	100/200	1.0	1	0.0	0.0	2.9	0.12
17[c]	Methanol	100/200	5.0	1	0.0	0.0	1.0	0.05
18[d]	Methanol	100/200	1.0	2	1.6	0.0	1.3	0.03
19[d]	Methanol	100/200	1.0	24	5.1	0.0	5.1	0.02
20[e]	Methanol	100/200	1.0	2	0.85	0.0	3.5	0.04
21[f]	Methanol	100/200	1.0	2	0.0	0.0	11.5	0.05

[a] Reaction conditions: reaction temperature 423 K, catalyst weight 0.17 g.
[b] At room temperature.
[c] Water (20 mmol) was added.
[d] Catalyst weight 0.05 g.
[e] Benzamide (2 mmol) was added.
[f] Benzamide (10 mmol) was added.

1). In the case of acetonitrile hydration, the selectivity to DMC and acetamide was 85% and 85%, respectively, at 8.2% methanol-based conversion. The selectivity to benzamide is important, considering the recyclability of benzonitrile: High selectivity may enable the reproduction of benzonitrile by benzamide dehydration.

This system was applied to the direct synthesis of dialkyl carbonates from various alcohols and CO_2; ethanol and 1-propanol were chosen as the alcohols with longer alkyl chains (*entries 7–10*), 2-propanol as a secondary alcohol (*entries 11* and *12*), and benzyl alcohol as a sterically bulky alcohol (*entries 13* and *14*). In all cases, the dialkyl carbonates were synthesized from their corresponding alcohols with much higher yields than at the equilibrium point. Similar to the DMC synthesis from methanol and CO_2, benzamide, carbamate, and esters were also detected as side products. The orders of the formation rate and final yield of dialkyl carbonates (DEC $\geq$ DMC $\geq$ di-*n*-propyl carbonate > diisopropyl carbonate > dibenzyl carbonate) may be dependent on the activity of dialkyl carbonate formation, the acidity of alcohols, and the solubility of CO_2 in the corresponding alcohols.

To investigate the benzonitrile hydration, the reaction was conducted in the presence of water (*entries 15–17*). Notably, the hydration reaction did not proceed (<0.005% yield of benzamide) in the absence of the catalyst. DMC was not detected at all under these hydration conditions, and the main product was benzamide. A small amount of methyl benzoate was found as a by-product, and the selectivity was lower than that to methyl acetate when acetonitrile was used instead of benzonitrile (Fig. 5.5a).

Similar to the case of acetonitrile hydration, the amount of products decreased with increasing CO_2 pressure, indicating that the dehydration of benzonitrile is strongly suppressed by higher CO_2 pressures. The reaction rate of benzonitrile hydration was higher than that of acetonitrile hydration, especially in the case of low CO_2 pressure. The higher activity and selectivity of benzonitrile hydration over acetonitrile hydration have been reported for other catalyst systems, such as $[Cp'_2Mo(OH)(OH_2)]^+$ [49], $[(dippe)Ni(\eta^2\text{-}NCR)]$ [50], and $[\{Rh(OMe)(cod)\}_2]PCy_3$[51]. The higher selectivity for benzonitrile hydration was interpreted in terms of the higher stability to alcoholysis of benzamide than acetamide. The reaction of benzamide with alcohol produced intermediates [e.g., $PhC(OH)(OMe)NH_2$] that had no resonance between the substituent and phenyl ring. The energy difference between benzamide, which is stabilized by resonance, and the intermediate is larger than that between acetamide and the corresponding intermediate.

5.2.10   Deactivation of the Ceria Catalyst in the Presence of Benzonitrile

The initial rate of DMC formation was almost proportional to the catalyst weight (*entries 2* and *18* in Table 5.5). The yield of DMC was saturated after a long reaction time. The final yield of DMC was changed when a different amount of catalyst was used (*entries 3* and *19*), showing that the yield was not determined by the equilibrium. Considering the case with acetonitrile hydration, the poisoning by the product benzamide was a plausible explanation. The activity test in the presence of benzamide showed that the presence of benzamide from the initial stage decreased the formation rate of DMC significantly (*entries 20* and *21*). To characterize the adsorbed species over the used catalyst, the FT-IR spectra of the catalyst after the catalytic use were measured at 423 K (Fig. 5.6).

The bands at 3367, 3178, 3075, 1656, 1623, 1577, 1452, 1407, 1317, 1186, 1145, and 1124 cm^{-1} were observed (bands at higher wave number not shown), all of which can be assigned to benzamide [52]. The spectrum was almost unchanged even after the introduction of methanol or CO_2. On the other hand, adsorption of methanol and CO_2 was clearly observed on the sample after calcination. These results suggested that the benzamide formed in the reaction covers the surface of

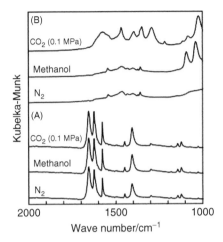

Figure 5.6   FT-IR spectra of the ceria catalyst after the catalytic use in direct DMC synthesis in the presence of benzonitrile. (A) After the catalytic use, (B) after the subsequent calcination at 873 K. Conditions for catalysis: CeO_2 0.17 g, $CH_3OH:C_6H_5CN = 100$ mmol:200 mmol, CO_2 1 MPa (at room temperature), 423 K, 2 h. Spectra were measured at 423 K.

ceria and inhibits the adsorption of methanol and CO_2. To estimate the amount of benzamide on the catalyst surface, the TG-DTA profile of the used catalyst was measured. Analysis of the weight loss data indicated that the benzamide weight loss contribution was about 0.31 mmol g_{cat}^{-1}. This value is larger than the adsorption amount of CO_2 (0.067 mmol CO_2 g_{cat}^{-1}) [40], and the coverage of the catalyst surface with benzamide explained the suppression of CO_2 adsorption. To reuse the catalyst, the adsorbed benzamide was removed by calcination treatment. The catalytic performance of the used catalyst after calcination at 873 K was almost the same as that of the fresh catalyst. In addition, the BET surface area of the catalyst was almost constant during the repeated reaction tests. The crystal size of the fresh and used catalyst, estimated by using XRD with the Sherrer equation, was also unchanged (about 10 nm).

5.2.11   Use of Other Dehydrating Agents

The hydration of 2,2-dimethoxypropane produces acetone (Eq. 5.4). A side reaction involving the 2,2-dimethoxypropane hydration is the decomposition of 2,2-dimethoxypropane to acetone and dimethyl ether (Eq. 5.5).

$$(CH_3)_2 C(OCH_3)_2 + H_2O \rightarrow CH_3COCH_3 + 2CH_3OH \qquad (5.4)$$

$$(CH_3)_2 C(OCH_3)_2 \rightarrow CH_3COCH_3 + CH_3OCH_3 \qquad (5.5)$$

The use of 2,2-dimethoxypropane as a dehydrating agent was applied to ceria-based catalysts [46]. However, this approach was less effective for DMC synthesis over ceria (CeO_2-HS calcined at 873 K) than the use of nitriles. A large amount of acetone and dimethyl ether was coproduced by the decomposition of 2,2-dimethoxypropane (Eq. 5.5).

In the case of ceria-zirconia catalysts, higher utilization efficiency of 2,2-dimethoxypropane was observed [53]. The yield of DMC reached 6.9 mmol (7% methanol conversion). This value was 8 times higher than the equilibrium yield under the conditions without 2,2-dimethoxypropane. The yield of dimethyl ether was only 0.2 mmol. However, the DMC formation rate decreased with the addition of 2,2-dimethoxypropane, and this level of conversion required about 100 h of reaction time. Too much 2,2-dimethoxypropane also induced the decomposition of 2,2-dimethoxypropane, lowering the utilization efficiency of the dehydrating agent. Zhang et al. [39] applied the use of trimethoxymethane (trimethyl orthoformate) as a dehydrating agent to the ceria-zirconia catalyst. Hydration of trimethoxymethane produces methanol and methyl formate (Eq. 5.6).

$$HC(OCH_3)_3 + H_2O \rightarrow HCOOCH_3 + 2CH_3OH \tag{5.6}$$

The methanol conversion increased as the amount of trimethoxymethane increased. The methanol conversion reached 10.4%, which is much higher (>5 times) than the equilibrium conversion, when 50 wt% trimethoxymethane was introduced at 12 MPa CO_2 for 34 h.

5.3 ZIRCONIA-BASED CATALYSTS

5.3.1 Structure and Catalytic Performance of Zirconia

Zirconia is also an active catalyst for the direct synthesis of organic carbonates. In contrast to ceria or ceria-zirconia, zirconia has multiple stable crystalline phases. Therefore, the relation between the phase and the catalytic performance is very important. Our group prepared various zirconia catalysts by calcination of a commercially available $ZrO_2 \cdot xH_2O$ (Nacalai tesque Inc.) at different temperatures (388–1073 K) [23,37]. From the XRD pattern (Fig. 5.7a), the zirconia catalysts prepared by the calcination below 623 K had amorphous structure. Metastable tetragonal structure was mainly formed by the calcination at 673–773 K. Monoclinic phase was predominant at higher calcination temperature above 773 K. Tetragonal phase gradually decreased with increasing calcination temperature and totally disappeared at 1073 K.

From the Raman spectra, which are sensitive to near-surface structures, the major phase of the sample calcined above 673 K was the monoclinic phase. The surface of the sample calcined at 673 K consisted of a major monoclinic phase and a minor tetragonal phase. The BET surface area was greatest in the catalysts with a calcination temperature of 473 K and decreased with increasing calcination temperature above 473 K. These catalysts were applied to direct DMC synthesis (Table 5.6).

The catalytic activity showed a maximum at the calcination temperature at 673 K, and then it decreased with the calcination temperature above 773 K. The zirconia calcined at 1073 K showed almost no activity. Formation of dimethyl ether was

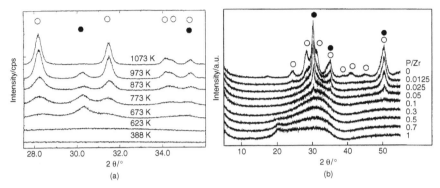

Figure 5.7 (a) XRD patterns of zirconium oxides prepared by calcination at various calcination temperatures [23]. Peaks marked with ∘ and • are assigned to the monoclinic and tetragonal phases, respectively. (b) XRD patterns of H_3PO_4/ZrO_2 with various H_3PO_4 loadings [55]. Peaks marked with ∘ and • are assigned to the monoclinic and tetragonal phases, respectively.

TABLE 5.6   Syntheses of DMC from Methanol and CO_2 over Zirconia Catalysts[a]

Catalyst	Calcination Temp., K	BET Surface Area, $m^2 g^{-1}$	Reaction Temp., K	DMC Amount, mmol
ZrO_2	388	307	443	0.18
ZrO_2	473	311	443	0.22
ZrO_2	573	201	443	0.26
ZrO_2	623	205	443	0.28
ZrO_2	673	118	443	0.30
ZrO_2	773	69	443	0.24
ZrO_2	873	32	443	0.12
ZrO_2	973	11	443	0.08
ZrO_2	1073	9	443	0.00
ZrO_2	673	118	403	0.08
H_3PO_4/ZrO_2 (P/Zr = 0.0125)	673	178	403	0.20
H_3PO_4/ZrO_2 (P/Zr = 0.025)	673	189	403	0.26
H_3PO_4/ZrO_2 (P/Zr = 0.05)	673	216	403	0.30
H_3PO_4/ZrO_2 (P/Zr = 0.1)	673	231	403	0.29
H_3PO_4/ZrO_2 (P/Zr = 0.3)	673	214	403	0.05
H_3PO_4/ZrO_2 (P/Zr = 0.5)	673	210	403	0.00
H_3PO_4/ZrO_2 (P/Zr = 1)	673	11	403	0.00

[a] Reaction conditions: methanol: CO_2 = 192 mmol:200 mmol (ca. 5 MPa at room temperature), reaction time 2 h, catalyst weight 0.5 g.

below the detection limit in all cases. On the basis of these data, metastable tetragonal ZrO_2 can be the most active phase for DMC formation. It should be noted that both the tetragonal ZrO_2 and catalytically active ceria-zirconia have the same distorted fluorite structure, and the ceria, which is also catalytically active, has a fluorite structure. The catalytic activity of the most active zirconia (calcined at 673 K) showed lower activity than ceria or ceria-zirconia. The reaction over zirconia catalysts proceeded at around 400 K or higher. The zirconia catalysts showed almost no activity at the reaction temperature of 383 K, where ceria and ceria-zirconia catalysts can be used. Even at 443 K, the catalytic activity of zirconia was less than half of that of ceria-zirconia.

5.3.2 Modification of Zirconia Catalysts

Surface modification of zirconia was performed in order to promote the reaction activity. The best-known method for the enhancement of surface acidity is the modification of zirconia with sulfate ions, known as a superacidic catalyst [54]. As a suitable acid-strength modification, phosphoric acid with medium strength acidity was chosen [55,56]. The catalysts were prepared by impregnating the $ZrO_2 \bullet xH_2O$ with aqueous H_3PO_4 solution. Water was removed by heating, and the sample was dried at 393 K for 10 h, followed by calcination at different temperatures (473–923 K) for 3 h. A dramatic additive effect on direct DMC synthesis was observed at all reaction temperatures (383–443 K). The reaction temperature could be lowered by 40 K by the addition of phosphoric acid (Table 5.6). The amounts of dimethyl ether and CO were smaller than the detection limit. The H_3PO_4/ZrO_2 catalyst with P/Zr ratio of 0.05 and a calcination temperature of 673 K showed the highest activity, although ceria-zirconia or CeO_2-HS calcined at 873 K showed even higher activity. DMC formation was not observed at P/Zr > 0.5. The catalyst prepared by direct loading of H_3PO_4 on precalcined ZrO_2 also showed no activity. The surface area reached a maximum at P/Zr = 0.1 and then decreased at P/Zr > 0.1, which did not correspond to the order of catalytic activity.

XRD patterns of H_3PO_4/ZrO_2 calcined at 673 K (Fig. 5.7b) showed that the monoclinic phase decreased with the loading of phosphoric acid. Metastable tetragonal phase was predominantly formed on the sample with P/Zr = 0.05. At P/Zr > 0.3, the H_3PO_4/ZrO_2 catalysts did not show the diffraction patterns of either the tetragonal or the monoclinic phase. With a higher phosphoric acid content, a new peak appeared around $2\theta = 20°$. This peak has not been identified, but it seems to be due to some salt compound formed between the zirconium and phosphate ions. The ^{31}P MAS NMR of the H_3PO_4/ZrO_2 catalysts showed that the catalysts with P/Zr = 0.0125, 0.05, and 0.1 contained monophosphates or the end-chain pyrophosphates and no middle chain of pyrophosphates. This indicates that the dispersion of phosphate species on H_3PO_4/ZrO_2 (P/Zr = 0.0125, 0.05 and 0.1) is very high. Laser Raman spectra of the H_3PO_4/ZrO_2 showed that the tetragonal phase was predominantly formed on the surface of the catalysts with P/Zr = 0.0125, 0.025 and 0.5, while the monoclinic phase was the major phase on the surface of unmodified zirconia. On the basis of these data, it is thought that the surface of tetragonal zirconia, which is stabilized by the addition of phosphoric acid, is very important for DMC synthesis. The NH_3-TPD profiles of unmodified zirconia and H_3PO_4/ZrO_2 (P/Zr = 0.05) showed that the modification of zirconia with phosphoric acid led to an increase in the number of weak acid sites (desorption temperature < 473 K) and did not change the number of stronger acid sites (desorption temperature > 473 K).

Jiang et al. [57] incorporated $H_3PW_{12}O_{40}$ into zirconia by using the sol–gel technique. The catalysts after calcination at 573 K contained tetragonal zirconia and WO_3 phases, while the characteristic bands of the Keggin unit of $H_3PW_{12}O_{40}$ were still observed in the IR spectra. The catalytic activity was much enhanced by the incorporation of $H_3PW_{12}O_{40}$, especially at low reaction temperatures. The formed DMC amount on $H_3PW_{12}O_{40}/ZrO_2$ (23 wt%) at 403 K became about 28 times larger than that on ZrO_2. The catalyst had some activity even at 343 K. DME formation was not detected at all.

5.3.3   Reaction Mechanism over Zirconia-Based Catalysts

The FT-IR spectra of unmodified zirconia and H_3PO_4/ZrO_2 (P/Zr = 0.05) were measured to investigate the adsorbed species on the catalyst during the reaction cycle (Fig. 5.8) [56]. On both catalysts, methanol adsorption led to the appearance of the bands at 1160 and 1054 cm^{-1}. These can be assigned to the C–O stretching modes of terminal (t-OCH_3) and bridging (b-OCH_3) methoxy species, respectively [35]. When CO_2 was introduced to the methanol-preadsorbed catalysts, the band intensity at 1160 cm^{-1} cm^{-1} decreased and the bands at 1600, 1474, 1370, and 1130 cm^{-1} appeared. The intensity of these bands increased with the increase in

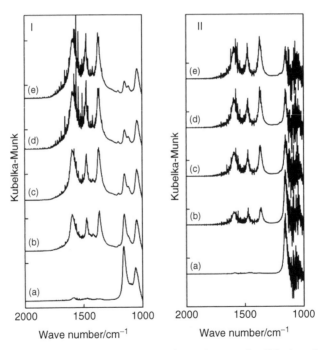

Figure 5.8   FT-IR spectra of methanol adsorption and successive CO_2 introduction at 443 K on ZrO_2 (I) and H_3PO_4/ZrO_2 (P/Zr = 0.05) (II): (a) after CH_3OH adsorption, (b) after CO_2 introduction under 0.1 MPa, (c) at 1.0 MPa, (d) at 3.0 MPa, and (e) at 5.0 MPa (subtracted spectra) [56]. Pretreatment: 673 K for 0.5 h under N_2 flow.

CO_2 pressure. These bands can be assigned to the monomethyl carbonate-like species [35].

CO_2 introduction hardly affected the band at 1054 cm^{-1} assigned to b-OCH$_3$ (slightly shifted to the lower wave number). In the spectra of high-wave number region, the bands at 3750 and 3680 cm^{-1} were observed on both of the catalysts before methanol adsorption. These bands can be assigned to the O–H stretching modes of terminal (t-OH) and bridging (b-OH) hydroxy groups, respectively [35]. After methanol adsorption, neither t-OH nor b-OH on unmodified zirconia was observed. On the other hand, a small amount of b-OH was present on H_3PO_4/ZrO_2 (P/Zr = 0.05), while t-OH totally disappeared. It appears that the remaining b-OH on H_3PO_4/ZrO_2 (P/Zr = 0.05) corresponds to the weak Brønsted acid sites. Methanol adsorption also led to the appearance of the bands at 2924 and 2820 cm^{-1} on both catalysts. These bands can be assigned to the C–H stretching modes of t- and b-OCH$_3$ species, respectively [35]. The reaction scheme of direct DMC synthesis on zirconia-based catalysts can be postulated as follows:

1. For methanol activation on base sites, t- and b-OH are converted to t- and b-OCH$_3$ when methanol is adsorbed.

2. For CO_2 activation on base sites, t-OCH$_3$ reacts with CO_2 to form monomethyl carbonate.

3. For methanol activation on acid sites, DMC is formed from the reaction between monomethyl carbonate and the methyl cation formed on Lewis acid sites on zirconia and/or the Brønsted acid sites on H_3PO_4/ZrO_2.

This scheme is essentially the same for that proposed for ceria catalysts (Fig. 5.4b).

Jung and Bell [58,59] investigated in detail the reactions over monoclinic zirconia by FT-IR. Pure monoclinic zirconia was prepared by boiling a solution of zirconyl chloride for 240 h at pH 1.5. The precipitated material was recovered, washed, and dried at 383 K. Calcination at 573 K for 3 h gave the catalyst. XRD showed that only the monoclinic phase of zirconia was present in the calcined material. The BET surface area was 110 m^2/g. Adsorption of methanol on zirconia at 298 K decreased the bands at 3768, 3745 (sh), and 3672 cm^{-1} associated with terminal, bibridged, and tribridged OH groups, respectively. At the same time, the bands at 2923 and 2817 cm^{-1} associated with C–H stretching vibration of monodentate (terminal) and bidentate (bridging) methoxy groups appeared. The growing intensity of the bands at 1157 and 1032 cm^{-1} associated with bending vibrations for both types of methoxy groups was also observed in the lower wave number range. The rates of hydroxyl group consumption and methoxide group formation were essentially equivalent. It was also evident that terminal and tribridged groups reacted at equivalent rates and that the rate of the formation of t-OCH$_3$ groups was slightly faster than the rate of the formation of b-OCH$_3$ groups. The exposure of monoclinic zirconia containing methoxy groups to CO_2 at 298 K resulted in the decrease of the bands for methoxy groups (2923, 2817, 1157 and 1032 cm^{-1}) and the growth of bands at 2965, 2883, 1600, 1497 (sh), 1474, 1389 (sh), 1370, 1200, and 1113 cm^{-1}, which could be assigned to adsorbed monodentate monomethyl carbonate species (Fig. 5.9a).

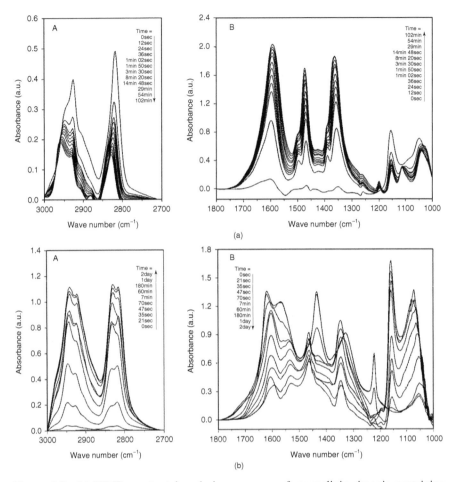

Figure 5.9 (a) FT-IR spectra taken during exposure of monoclinic zirconia containing preadsorbed methanol to CO_2 at 298 K [58]. (b) FT-IR spectra taken during exposure of monoclinic zirconia containing preadsorbed CO_2 to methanol at 298 K [58].

The dynamics of t-OCH_3 consumption and monodentate monomethyl carbonate formation were nearly identical. The exposures in the order of $CO_2 \rightarrow$ methanol were also investigated. The exposure of CO_2 to monoclinic zirconia at 298 K resulted in the rapid growth of the bands at 1625, 1437, and 1225 cm^{-1} that can be assigned to bidentate hydrogencarbonate species and a band at 1325 cm^{-1} that can be assigned to bidentate carbonate species. The adsorption of CO_2 was very rapid, and the apparent first-order rate coefficient was 1 order higher than that of methanol. When the monoclinic zirconia containing preadsorbed CO_2 was exposed to methanol vapor at 298 K (Fig. 5.9b), the bands associated to bidentate hydrogencarbonate species first disappeared and new bands appeared that are characteristic of monodentate monomethyl carbonate species (1608, 1464, and 1360 cm^{-1}).

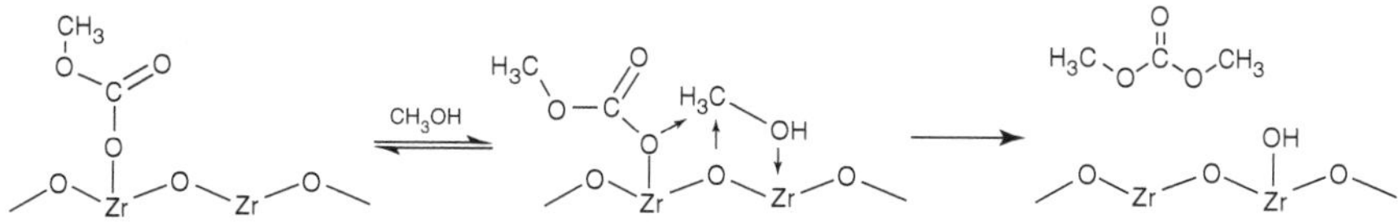

Figure 5.10   Acid-base pairs of the zirconia surface in the DMC formation.

These spectrum changes were slower than those when methanol-preadsorbed zirconia was exposed to CO_2. The bands for t- and b-OCH_3 (1157 and 1032 cm^{-1}, respectively) were grown later. The FT-IR spectra were also obtained on exposure of monoclinic zirconia to DMC vapor at 298 K. After enough time to remove the effect of adsorbed trace water, the spectrum became characteristic of that for a mixture of monodentate monomethyl carbonate species and methoxy species (t-OCH_3 and b-OCH_3). Based on data and the concept of microreversibility, which requires that the decomposition of DMC occurs via the same elementary steps by which it forms, the authors concluded that DMC is formed by the transfer of methyl group of methanol to the terminal O atom of monodentate monomethyl carbonate species.

The mechanism is essentially the same as that proposed for ceria and H_3PO_4/ZrO_2 as discussed above (Fig. 5.4B), although interpretations of the change of the band for b-OCH_3 species are different. The authors also proposed that the formations of monodentate monomethyl carbonate species and DMC are facilitated by the Lewis acid–base pairs of surface zirconia sites (Fig. 5.10).

The same authors (Jung and Bell) investigated the elementary processes of direct DMC synthesis in the case of tetragonal zirconia [59]. The pure tetragonal zirconia was prepared by the precipitation from zirconyl chloride solution with ammonia hydroxide at pH 10. The precipitated material was washed, dried at 373 K, and then calcined in pure O_2 at 973 K. The BET area of the material was 187 m^2/g. The presence of tetragonal zirconia as the only phase was confirmed by XRD. From the FT-IR study, similar species were observed for both tetragonal and monoclinic catalysts. The reaction rates were different. The dissociative adsorption of methanol to form methoxy species was approximately twice as fast on monoclinic zirconia as on tetragonal zirconia. CO_2 insertion to form monodentate monomethyl carbonate species occurred more than an order of magnitude more rapidly over monoclinic zirconia. The transfer of methyl group from methanol to monodentate monomethyl carbonate species and the resulting formation of DMC proceeded roughly twice as fast over monoclinic zirconia. DMC adsorbed molecularly on tetragonal zirconia but decomposed to form monomethyl carbonate and methoxy species on monoclinic zirconia. From these spectroscopic data, the authors concluded that the monoclinic phase is catalytically more active than the tetragonal phase, although direct comparison of the DMC formation rate between these phases was not conducted. The inconsistency between these data and the data shown in Section 5.3.1 may be due to the difference in crystal faces between the samples with different preparation procedures.

5.3.4 Combination of Dehydrating Agents with Zirconia-Based Catalysts

The use of a dehydrating agent to improve carbonate yield has been less studied in the case of zirconia-based catalysts. The use of acetonitrile was tested in DMC synthesis over zirconia calcined at 673 K. In the conditions of 1 mmol catalyst, methanol:acetonitrile = 200:600 mmol, 0.5 MPa CO_2, and 423 K, the amount of DMC reached 0.32 mmol after 24 h while methyl carbamate (0.19 mmol), acetamide (0.37 mmol) and methyl acetate (0.51 mmol) were also produced [46]. The yield was four times higher than what can be obtained at equilibrium in the absence of acetonitrile. However, the yield and selectivity [=DMC/(DMC + methyl carbamate)] were much lower than those obtained over CeO_2-HS catalyst calcined at 873 K in the same conditions (7.3 mmol DMC with 93% selectivity). This indicates that the consecutive reaction of acetamide (Eq. 5.2) can proceed more easily than that the hydration of acetonitrile on zirconia.

Eta et al. [60,61] investigated direct DMC synthesis using potassium-doped zirconia catalysts and various dehydrating agents. Potassium-doped zirconia was prepared by impregnation of KCl on ZrO_2 and calcination at 1073 K. The DMC yield of 4.9% based on methanol was obtained in the conditions of 463 mmol methanol, 14.5 mmol butylene oxide as a dehydrating agent, 1 g catalyst, 9.5 MPa CO_2, and 423 K. The yield at equilibrium in the same conditions without dehydrating agent was 0.50%. Addition of magnesium turnings (21 mmol) as a promoter under similar reaction conditions resulted in a 1.5-fold increase in the amount of DMC (7.2% yield), although the effect of magnesium alone was not catalytic since the amount of DMC formed was lower than the amount of magnesium used. Butylene oxide showed better water-absorbing properties under the employed reaction conditions with a better yield of DMC than trimethoxymethane and dimethoxymethane. The selectivity from methanol to DMC was 53%, and that from methylene oxide to butylene glycol was 54%. A mechanism in which DMC and butylene glycol were formed via the reaction of adsorbed monomethyl carbonate intermediate and methoxybutanol was proposed [61]. The same research group has very recently reported the effect of ionic liquids with methoxide anion on water trapping and on the yield and selectivity to DMC for zirconia-magnesia-catalyzed direct DMC synthesis [62]. The maximum yield of 11% based on methanol was obtained under the conditions of 247 mmol methanol, 1 g ZrO_2-MgO (Zr/Mg = 15 mol/mol), 4 g 1-butyl-3-methylimidazolium methoxide ([C4MIM][OMe], 23.5 mmol), 8 MPa CO_2, and 393 K. The maximum yield of DMC without ionic liquid or dehydrating agent in the same conditions was 0.3%. The turnover number (TON), defined as the number of moles of DMC formed per mole of ionic liquid used, was 0.44. After the reaction, the counter-anion of the ionic liquid was converted to hydroxide (Eq. 5.7), as confirmed by NMR. The recovered ionic liquid reacted with a solution of sodium methoxide in methanol resulted in the regeneration of methoxide ionic liquid. The performances of the fresh and regenerated ionic liquids were similar in the conversion of methanol and the yield of DMC. Therefore, the ionic liquid with methoxide anion can act a dehydrating agent and enhance the DMC yield beyond equilibrium.

$$[CnMIM][OMe] + H_2O \rightarrow [CnMIM][OH] + CH_3OH \qquad (5.7)$$

5.4 OTHER METAL OXIDE CATALYSTS

We examined the activity of various metal oxide catalysts in direct DMC synthesis [63]. Only ceria, zirconia, and stannic oxide showed activity to DMC formation among monometallic oxides. Silica, alumina, titania, magnesia, yttria, hafnia, lanthana, praseodymium oxide, molybdenum trioxide, zinc oxide, gallium oxide, germanium oxide, indium oxide, antimony (III) oxide, and bismuth oxide were inactive. In the case of alumina and titania, a small amount of dimethyl ether was detected. The activity of stannic oxide, which has a tetragonal rutile structure, was lower than that of ceria or zirconia. The low activity of stannic oxide was also reported by Aymes et al. [64]. Acidic zeolites such as H-ZSM-5 as catalysts gave DME as a main product, and DMC formation was not detected at all. Wu et al. [65] investigated the catalytic activity of unmodified and modified V_2O_5 with phosphoric acid in the flowing system. Unmodified V_2O_5 showed small activity. Modification of V_2O_5 with phosphoric acid greatly enhanced the activity [8 times higher conversion on H_3PO_4/V_2O_5 (P/V = 0.20]. The selectivity to DMC was around 90% over both unmodified and modified V_2O_5 catalysts. XRD patterns of the catalysts showed that the catalysts with $P/V \leq 0.05$ had the orthorhombic V_2O_5 structure. When increasing P/V ratio to 0.15, the orthorhombic phase gradually decreased with further increasing H_3PO_4 content, while the tetragonal phase increased with increasing H_3PO_4 content. Both orthorhombic and tetragonal phases were predominantly formed for H_3PO_4/V_2O_5 catalysts when P/V ratio > 0.15. The tetragonal phase appears to be catalytically active, although the orthorhombic phase may have some activity.

5.5 CONCLUSIONS AND OUTLOOK

Very few metal oxides are catalytically active for the direct carbonate synthesis from alcohol and CO_2. Oxides of cerium, zirconium, tin, and vanadium are examples. In conditions without any dehydrating agents, ceria and ceria-zirconia are very active, selective, and stable catalysts. However, the yields are inevitably very low because of the equilibrium limitation. The use of organic dehydrating agents such as nitriles, acetals, and orthoesters is a promising approach to overcome the limitation. In this approach, side-reactions involving the dehydrating agents and the hydrated products frequently occur depending on the types of dehydrating agents and catalysts. Deactivation of the catalyst by the dehydrating agents or the derivatives is also a frequently encountered problem. In most cases, the DMC amount stops rising and even starts to drop before total consumption of the dehydrating agents. The combination of ceria catalyst and nitriles as dehydrating agents gave the highest ratio of (obtained DMC yield)/(equilibrium yield in the absence of dehydrating agent) to date. The choice of the dehydrating agents and the design of the catalysts are important to suppress the side reactions and the deactivation. The price and recyclability of the dehydrating agents are also important in view of practical use. Despite the difficulties, this research field is new and rapidly growing in these few years. Without the side-reactions, very high yield and selectivity can be theoretically obtained when organic dehydrating agents are used.

REFERENCES

1. Y. Ono (1997). *Appl. Catal. A: Gen.*, 155: 133–166.
2. T. Sakakura, K. Kohno (2009). *Chem. Commun.*, 1312–1330.
3. M. A. Pacheco, C. L. Marshall (1997). *Energy Fuels*, 11: 2–29.
4. A.-A. G. Shaikh, S. Sivaram (1996). *Chem. Rev.*, 96: 951–976.
5. D. Delledonne, F. Rivetti, U. Romano (2001). *Appl. Catal. A: Gen.*, 221: 241–251.
6. H. Babad, A. G. Zeller (1973). *Chem. Rev.*, 73: 75–91.
7. W. J. Peppel (1958). *Ind. Eng. Chem.*, 50: 767–770.
8. J. F. Knifton, R. G. Duranleau (1991). *J. Mol. Catal.*, 67: 389–399.
9. S. Fukuoka, M. Kawamura, K. Komiya, M. Tojo, H. Hachiya, K. Hasegawa, M. Aminaka, H. Okamoto, I. Fukawa and S. Konno (2003). *Green Chem.*, 5: 497–507.
10. T. Matsuzaki, A. Nakamura (1997). *Catal. Surv. Jpn.*, 1: 77–88.
11. U. Romano, R. Tesei, M. M. Mauri, P. Rebora (1980). *Ind. Eng. Chem. Prod. Res. Dev.*, 19: 396–403.
12. K. M. K. Yu, I. Curcic, J. Gabriel, S. C. E. Tsang (2008). *ChemSusChem*, 1: 893–899.
13. T. Sakakura, J.-C. Choi, H. Yasuda (2007). *Chem. Rev.*, 107: 2365–2387.
14. J. Kizlink (1993). *Collect. Czech. Chem. Commun.*, 58: 1399–1402.
15. T. Sakakura, J.-C. Choi, Y. Saito, T. Sako (2000). *Polyhedron*, 19: 573–576.
16. J.-C. Choi, T. Sakakura, T. Sako (1999). *J. Am. Chem. Soc.*, 121: 3793–3794.
17. T. Sakakura, Y. Saito, M. Okano, J.-C. Choi, T. Sako (1998). *J. Org. Chem.*, 63: 7095–7096.
18. A. Corma, H. Garcia (2006). *Adv. Synth. Catal.*, 348: 1391–1412.
19. M. H. Valkenberg, W. F. Holderich (2002). *Catal. Rev. – Sci. Eng.*, 44: 321–374.
20. B. Fan, H. Li, W. Fan, J. Zhang, R. Li (2010). *Appl. Catal. A: Gen.*, 372: 94–102.
21. B. Fan, J. Zhang, R. Li, W. Fan (2008). *Catal. Lett.*, 121: 297–302.
22. D. Ballivet-Tkatchenko, F. Bernard, F. Demoisson, L. Plasseraud, S. R. Sanapureddy (2011). *ChemSusChem*, 4: 1316–1322.
23. K. Tomishige, T. Sakaihori, Y. Ikeda, K. Fujimoto (1999). *Catal. Lett.*, 58: 225–229.
24. S. Sakai, T. Fujinami, T. Yamada and S. Furusawa (1975). *Nippon Kagaku Kaishi*, 1789–1794 (in Japanese).
25. A. Trovarelli (1996). *Catal. Rev. – Sci. Eng.*, 38: 439–520.
26. H. C. Yao, Y. F. Yu Yao, *J. Catal.* 1984, 86, 254–265.

27. Y. Yoshida, Y. Arai, S. Kado, K. Kunimori, K. Tomishige (2006). *Catal. Today*, 115: 95–101.

28. A. Sahibed-Dine, A. Aboulayt, M. Bensitel, A. B. Mohammed Saad, M. Daturi, J. C. Lavalley (2000). *J. Mol. Catal. A: Chem.*, 162: 125–134.

29. Y. Namai, K. Fukui, Y. Iwasawa (2003). *Catal. Today*, 85: 79–91.

30. J. C. Conesa (1995). *Surf. Sci.*, 339: 337–352.

31. T. X. T. Sayle, S. C. Parker, C. R. A. Catlow (1994). *Surf. Sci.*, 316: 329–336.

32. M. Aresta, A. Dibenedetto, C. Pastore, A. Angelini, B. Aresta, I. Pápai (2010). *J. Catal.*, 269: 44–52.

33. M. Daturi, C. Binet, J. C. Lavalley, A. Galtayries, R. Sporken, *Phys. Chem. Chem. Phys.* 1999, 1, 5717–5724.

34. E. Finocchio, M. Daturi, C. Binet, J. C. Lavalley, G. Blanchard (1999). *Catal. Today*, 52: 53–63.

35. M. Bensitel, V. Moravek, J. Lamotte, O. Saur, J. C. Lavalley (1987). Spectrochim. *Acta*, 43A: 1487–1491.

36. A. Dibenedetto, C. Pastore, M. Aresta (2006). *Catal. Today*, 115: 88–94.

37. K. Tomishige, Y. Ikeda, T. Sakaihori, K. Fujimoto (2000). *J. Catal.*, 192: 355–362.

38. K. Tomishige, Y. Furusawa, Y. Ikeda, M. Asadullah, K. Fujimoto (2001). *Catal. Lett.*, 76: 71–74.

39. Z.-F. Zhang, Z.-W. Liu, J. Lu, Z.-T. Liu (2011). *Ind. Eng. Chem. Res.*, 50: 1981–1988.

40. K. Tomishige, H. Yasuda, Y. Yoshida, M. Nurunnabi, B. Li, K. Kunimori (2004). *Green Chem.*, 6: 206–214.

41. Z.-F. Zhang, Z.-T. Liu, Z.-W. Liu, J. Lu (2009). *Catal. Lett.*, 129: 428–436.

42. K. Tomishige, H. Yasuda, Y. Yoshida, M. Nurunnabi, B. Li, K. Kunimori (2004). *Catal. Lett.*, 95: 45–49.

43. H. J. Lee, S. Park, I. K. Song, J. C. Jung (2011). *Catal. Lett.*, 141: 531–537.

44. J.-C. Choi, L.-N. He, H. Yasuda, T. Sakakura (2002). *Green Chem.*, 4: 230–234.

45. T. Sakakura, J.-C. Choi, Y. Saito, T. Masuda, T. Sako, T. Oriyama (1999). *J. Org. Chem.*, 64: 4506–4508.

46. M. Honda, A. Suzuki, N. Begum, K. Fujimoto, K. Suzuki, K. Tomishige (2009). *Chem. Commun.*, 4596–4598.

47. M. Honda, S. Kuno, N. Begum, K. Fujimoto, K. Suzuki, Y. Nakagawa, K. Tomishige (2010). *Appl. Catal. A: Gen.*, 384: 165–170.

48. M. Honda, S. Kuno, S. Sonehara, K. Fujimoto, K. Suzuki, Y. Nakagawa, K. Tomishige (2011). *ChemCatChem*, 3: 365–370.

49. K. L. Breno, M. D. Pluth, D. R. Tyler (2003). *Organometallics*, 22: 1203–1211.

50. M. G. Crestani, A. Arévalo, J. J. García (2006). *Adv. Synth. Catal.*, 348: 732–742.

51. A. Goto K. Endo, S. Saito (2008). *Angew. Chem. Int. Ed.* 2008, 47: 3607–3609.

52. H. G. Brittain (2009). *Cryst. Growth Des.*, 9: 2492–2499.

53. K. Tomishige, K. Kunimori (2002). *Appl. Catal. A: Gen.*, 237: 103–109.

54. K. Arata (1990). *Adv. Catal.*, 37: 165–211.

55. Y. Ikeda, T. Sakaihori, K. Tomishige, K. Fujimoto (2000). *Catal. Lett.*, 66: 59–62.

56. Y. Ikeda, M. Asadullah, K. Fujimoto, K. Tomishige (2001). *J. Phys. Chem. B*, 105: 10653–10658.

57. C. Jiang, Y. Guo, C. Wang, C. Hu, Y. Wu, E. Wang (2003). *Appl. Catal. A: Gen.*, 26: 203–212.

58. K. T. Jung, A. T. Bell (2001). *J. Catal.*, 204: 339–347.

59. K. T. Jung, A. T. Bell (2002). *Top. Catal.*, 20: 97–105.

60. V. Eta, P. Mäki-Arvela, A.-R. Leino, K. Kordás, T. Salmi, D. Y. Murzin, J.-P. Mikkola (2010). *Ind. Eng. Chem. Res.*, 49: 9609–9617.

61. V. Eta, P. Mäki-Arvela, J. Wärnå, T. Salmi, J.-P. Mikkola, D. Y. Murzin (2011). *Appl. Catal. A: Gen.*, 404: 39–46.

62. V. Eta, P. Mäki-Arvela, E. Salminen, T. Salmi, D. Y. Murzin, J.-P. Mikkola (2011). *Catal. Lett.* 141: 1254–1261.

63. K. Tomishige (2002). *Curr. Top. Catal.*, 3: 81–101.

64. D. Aymes, D. Ballivet-Tkatchenko, K. Jeyalakshmi, L. Saviot, S. Vasireddy (2009). *Catal. Today*, 147: 62–67.

65. X. L. Wu, M. Xiao, Y. Z. Meng, Y. X. Lu (2002). *J. Mol. Catal. A: Chem.*, 238: 158–162.

CHAPTER 6

High-Solar-Efficiency Utilization of CO_2: the STEP (Solar Thermal Electrochemical Production) of Energetic Molecules

STUART LICHT

6.1 INTRODUCTION

Anthropogenic release of carbon dioxide (CO_2) and atmospheric CO_2 have reached record levels. One path toward CO_2 reduction is to utilize renewable energy to

Green Carbon Dioxide: Advances in CO_2 Utilization, First Edition.
Edited by Gabriele Centi and Siglinda Perathoner.
© 2014 John Wiley & Sons, Inc. Published 2014 by John Wiley & Sons, Inc.

produce electricity. Another, less explored, path is to utilize renewable energy to directly produce societal staples such as metals, bleach, and fuels, including carbonaceous fuels. Whereas solar-driven water splitting to generate hydrogen fuels has been extensively studied [1,2], there have been few studies of solar-driven carbon dioxide splitting, although recently we introduced a global process for the solar thermal electrochemical production (STEP) of energetic molecules, including CO_2 splitting [3–8]. "CO_2 is a highly stable, noncombustible molecule, and its thermodynamic stability makes its activation energy demanding and challenging." [9] In search of a solution for climate change associated with increasing levels of atmospheric CO_2, the field of carbon dioxide splitting (solar or otherwise), while young, is growing rapidly and, as with water splitting, includes the study of photo-electrochemical, biomimetic, electrolytic, and thermal pathways of carbon dioxide splitting [10,11].

The direct thermal splitting of CO_2 requires excessive temperatures to drive any significant dissociation. As a result, lower-temperature thermochemical processes using coupled reactions have recently been studied [12–16]. The coupling of multiple reactions steps decreases the system efficiency. To date, such challenges, and the associated efficiency losses, have been an impediment to the implementation of the related, extensively studied field of thermochemical splitting of water [2]. Photoelectrochemistry probes the energetics of illuminated semiconductors in an electrolyte, and provides an alternative path to solar fuel formation. Photoelectro-chemical solar cells (PECs) can convert solar energy to electricity [17–21] and, with inclusion of an electrochemical storage couple, have the capability for internal energy storage, to provide a level output despite variations in sunlight [22,23]. Solar to photoelectrochemical energy can also be stored externally in chemical form, when it is used to drive the formation of energetically rich chemicals. Photochemical, and photoelectrochemical, splitting of carbon dioxide [24–29] have demonstrated selective production of specific fuel products. Such systems function at low current density and efficiencies of ~1% and, as with photoelectrochemical water splitting, face stability and bandgap challenges related to effective operation with visible light [21,30,31].

The electrically driven (nonsolar) electrolysis of dissolved CO_2 is under investigation at or near room temperature in aqueous, nonaqueous, and PEM media [32–41]. These are constrained by the thermodynamic and kinetic challenges associated with ambient temperature, endothermic processes, high electrolysis potential, large overpotential, low rate, and low electrolysis efficiency. High-temperature, solid oxide electrolysis of carbon dioxide dates back to the 1960s, with suggestions to use such cells to renew air for a space habitat [42–44], and the sustainable rate of the solid oxide reduction of carbon dioxide is improving rapidly [45–51]. Molten carbonate, rather than solid oxide, fuel cells running in the reverse mode were also studied to renew air in 2002 [52]. In a manner analogous to our 2002 high-temperature solar water splitting studies (described below) [53–55], we showed in 2009 that molten carbonate cells are particularly effective for the solar-driven electrolysis of CO_2 [3,4,8] and also CO_2-free iron metal production [5,6].

Light-driven water splitting was originally demonstrated with TiO_2 (a semiconductor with a bandgap, $E_g > 3.0$ eV) [56]. However, only a small fraction of sunlight has sufficient energy to drive TiO_2 photoexcitation. Studies had sought to tune (lower) the semiconductor bandgap to provide a better match to the electrolysis potential [57]. In 2000, we used external multiple bandgap photovoltaics (PVs) to generate H_2 by splitting water at 18% solar energy conversion efficiency [58]. However, that room temperature process does not take advantage of additional available thermal energy.

An alternative to tuning a semiconductor bandgap to provide a better match to the solar spectrum is an approach to tune (lower) the electrolysis potential [55]. In 2002 we introduced a photo electrochemical thermal water splitting theory [53], which was verified by experiment in 2003, for H_2 generation at over 30% solar energy conversion efficiency, providing the first experimental demonstration that a semiconductor, such as Si ($E_g = 1.1$ eV), with bandgap lower than the standard water splitting potential [$E°_{H_2O}(25°C) = 1.23$ V] can directly drive hydrogen formation [55]. With increasing temperature, the quantitative decrease in the electrochemical potential to split water to hydrogen and oxygen had been well known by the 1950s [59a,59b]. In 1976 Wentworth and Chen wrote on "simple thermal decomposition reactions for storage of solar energy," with the limitation that the products of the reaction must be separated to prevent back reaction (and without any electrochemical component) [60], and as early as 1980 it was noted that thermal energy could decrease the necessary energy for the generation of H_2 by electrolysis [61]. However, the process combines elements of solid-state physics, insolation, and electrochemical theory, complicating rigorous theoretical support of the process. Our photo electrochemical thermal water splitting model for solar/H_2 by this process was the first derivation of bandgap-restricted, thermal-enhanced, high solar water splitting efficiencies. The model, predicting solar energy conversion efficiencies that exceed those of conventional PVs, was initially derived for Air Mass (AM)1.5 and terrestrial insolation and later expanded to include sunlight above the atmosphere (AM0 insolation) [53,54]. Experimental confirmation followed and established that the water splitting potential can be specifically tuned to match efficient photo-absorbers [55], eliminating the challenge of tuning (varying) the semiconductor bandgap, which can lead to over 30% solar-to-chemical energy conversion efficiencies. Our early process was specific to H_2 and did not incorporate the additional temperature enhancement of excess super-bandgap energy and concentration enhancement of excess reactant to further decrease the electrolysis potential in our contemporary STEP process.

6.2   SOLAR THERMAL ELECTROCHEMICAL PRODUCTION OF ENERGETIC MOLECULES: AN OVERVIEW

6.2.1   STEP Theoretical Background

A single, small bandgap junction, such as in a silicon PV, cannot generate the minimum photopotential required to drive many room temperature electrolysis

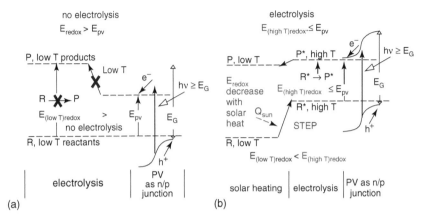

Figure 6.1 Comparison of PV and STEP solar-driven electrolysis energy diagrams. STEP uses sunlight to drive otherwise energetically forbidden pathways of charge transfer. The energy of photo-driven charge transfer is insufficient (a) to drive (unheated) electrolysis but is sufficient (b) to drive endothermic electrolysis in the solar-heated synergistic process. The process uses both visible and thermal solar energy for higher efficiency; thermal energy decreases the electrolysis potential, forming an energetically allowed pathway to drive electrochemical charge transfer.

reactions, as shown on the left side of Figure 6.1. The advancement of such studies had focused on tuning semiconductor bandgaps [57] to provide a better match to the electrochemical potential (specifically, the water splitting potential) or utilizing more complex, multiple bandgap structures using multiple photon excitation [58]. These structures are not capable of excitation beyond the band edge and cannot make use of longer-wavelength sunlight. PVs are limited to super-bandgap sunlight, $h\nu > E_g$, precluding use of long-wavelength radiation, $h\nu < E_g$. STEP instead directs this IR sunlight to heat electrochemical reactions, and uses visible sunlight to generate electronic charge to drive these electrolyses.

Rather than tuning the bandgap to provide a better energetic match to the electrolysis potential, the STEP process instead tunes the redox potential to match the bandgap. The right side of Figure 6.1 presents the energy diagram of a STEP process. The high-temperature pathway decreases the free energy requirements for processes whose electrolysis potential decreases with increasing temperature. STEP uses solar energy to drive otherwise energetically forbidden pathways of charge transfer. The process combines elements of solid-state physics, insolation (solar illumination), and high-temperature electrochemical energy conversion. Kinetics improve and endothermic thermodynamic potentials decrease with increasing temperature. The result is a synergy, making use of the full spectrum of sunlight and capturing more solar energy. STEP is intrinsically more efficient than other solar energy conversion processes, as it utilizes not only the visible sunlight used to drive PVs but also the previously detrimental (due to PV thermal degradation) thermal component of sunlight for the electrolytic formation of chemicals.

   The two bases for improved efficiencies with the STEP process are the fact that (i) excess heat, such as unused heat in solar cells, can be used to increase the temperature of an electrolysis cell, such as for electrolytic CO_2 splitting while (ii) the product-to-reactant ratio can be increased to favor the kinetic and energetic formation of reactants. With increasing temperature, the quantitative decrease in the electrochemical potential to drive a variety of electrochemical syntheses is well known, substantially decreasing the electronic energy (the electrolysis potential) required to form energetic products. The extent of the decrease in the electrolysis potential, E_{redox}, may be tuned by choosing the constituents and temperature of the electrolysis. The process distinguishes radiation that is intrinsically energy sufficient to drive PV charge transfer and applies all excess solar thermal energy to heat the electrolysis reaction chamber.

   Figure 6.2 summarizes the charge, heat, and molecular flow for the STEP process; the high-temperature pathway decreases the potential required to drive endothermic electrolyses, and also facilitates the kinetics of charge transfer (i.e., decreases overpotential losses), which arise during electrolysis. This process consists of

1. Sunlight harvesting and concentration
2. Photovoltaic charge transfer driven by super-bandgap energy
3. Transfer of sub-bandgap and excess super-bandgap radiation to heat the electrolysis chamber
4. High-temperature, low-energy electrolysis forming energy-rich products
5. Cycle completion by preheating of the electrolysis reactant through heat exchange with the energetic electrolysis products

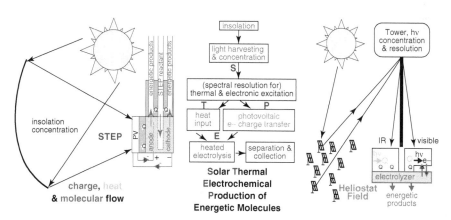

Figure 6.2   Global use of sunlight to drive the formation of energy-rich molecules. *Left*: Charge and heat flow in STEP: heat flow (arrows containing the letter Q), electron flow (arrows containing the letter e), and reagent flow (arrows near the anode, STEP reactant, and cathode). *Right*: Beam splitters redirect sub-bandgap sunlight away from the PV onto the electrolyzer. Modified with permission from ref. 3.

As indicated on the right side of Figure 6.2, the light harvesting can use various optical configurations, for example, in lieu of parabolic, or Fresnel, concentrators, a heliostat/solar tower with secondary optics can achieve higher process temperatures (>1000°C) with concentrations of ~ 2000 suns. Beam splitters can redirect sub-bandgap radiation away from the PV (minimizing PV heating) for a direct heat exchange with the electrolyzer.

Solar heating can decrease the energy to drive a range of electrolyses. Such processes can be determined using available entropy, S, enthalpy, H, and free-energy, G, data [59b], and are identified by their negative isothermal temperature coefficient of the cell potential [59a]. This coefficient, $(dE/dT)_{isoth}$, is the derivative of the electromotive force of the isothermal cell:

$$(dE/dT)_{isoth} = \Delta S/nF = (\Delta H - \Delta G)/nFT \qquad (6.1)$$

The starting process of modeling any STEP process is the conventional expression of a generalized electrochemical process, in a cell that drives an n electron charge transfer electrolysis reaction, comprising "x" reactants, R_i, with stoichiometric coefficients r_i, and yielding "y" products, C_i, with stoichiometric coefficients c_i.

$$\text{Electrode 1| Electrolyte |Electrode 2}$$

Using the convention of $E = E_{cathode} - E_{anode}$ to describe the positive potential necessary to drive a nonspontaneous process, by transfer of n electrons in the electrolysis reaction:

$$n \text{ electron transfer electrolysis reaction} : \sum_{i=1\text{ to }x} r_i R_i \rightarrow \sum_{i=1\text{ to }y} c_i C_i \qquad (6.2)$$

At any electrolysis temperature, T_{STEP}, and at unit activity, the reaction has electrochemical potential, E_T°. This may be calculated from consistent, compiled unit activity thermochemical data sets, such as the NIST condensed phase and fluid properties data sets [59b], as:

$$E_T^\circ = -\Delta G^\circ (T = T_{STEP})/nF; E_{ambient}^\circ \equiv E_T^\circ (T_{ambient}); \text{ here}$$
$$T_{ambient} = 298.15K = 25°C,$$

and:

$$\Delta G^\circ (T = T_{STEP}) = \sum_{i=1\text{ to }y} c_i (H^\circ(C_i, T) - TS^\circ(C_i, T))$$
$$- \sum_{i=1\text{ to }x} r_i (H^\circ(R_i, T) - TS^\circ(R_i, T)) \qquad (6.3)$$

Compiled thermochemical data are often based on different reference states, while a consistent reference state is needed to understand electrolysis-limiting processes, including water [62,63]. This challenge is overcome by modification of the unit activity ($a = 1$) consistent calculated electrolysis potential to determine the potential at other reagent and product relative activities via the Nernst equation [64,65]. Electrolysis provides control of the relative amounts of reactant and generated product in a system. A substantial activity differential can also drive STEP improvement at elevated temperature, and will be derived. The potential variation with activity, a, of the reaction $\sum_{i=1 \text{ to } x} r_i R_i \rightarrow \sum_{i=1 \text{ to } y} c_i C_i$, is given by:

$$E_{T,a} = E_T^\circ - (RT/nF) \cdot \ln(\Pi_{i=1 \text{ to } x} a(R_i)^{r_i} / \Pi_{i=1 \text{ to } y} a(C_i)^{c_i}) \qquad (6.4)$$

Electrolysis systems with a negative isothermal temperature coefficient tend to cool as the electrolysis products are generated. Specifically, in endothermic electrolytic processes, the Eq. 6.4 free-energy electrolysis potential, E_T, is less than the enthalpy-based potential. This latter value is the potential at which the system temperature would remain constant during electrolysis. This thermoneutral potential, E_{tn}, is given by:

$$E_{tn}(T_{STEP}) = -\Delta H(T)/nF; \Delta H(T_{STEP})$$
$$= \sum_{i=1 \text{ to } b} c_i H(C_i, T_{STEP}) - \sum_{i=1 \text{ to } a} r_i H(R_i, T_{STEP}) \qquad (6.5)$$

Two general STEP implementations are being explored. Both can provide the thermoneutral energy to sustain a variety of electrolyses. The thermoneutral potential, determined from the enthalpy of a reaction, describes the energy required to sustain an electrochemical process without cooling. For example, the thermoneutral potential we have calculated and reported for CO_2 splitting to CO and O_2 at unit activities, from Eq. 6.5, is $1.46(\pm0.01)$ V over the temperature range of $25-1400°C$. As represented in Figure 6.3 on the left, the standard electrolysis potential at room temperature, E°, can comprise a significant fraction of the thermoneutral potential. The first STEP mode, energetically represented next to the room temperature process in the scheme, separates sunlight into thermal and visible radiation. The solar visible component generates electronic charge that drives electrolysis charge transfer. The solar thermal component heats the electrolysis and decreases both the E° at this higher T and the overpotential. The second mode, termed Hy-STEP (on the right) from "hybrid-STEP," does not separate sunlight, and instead directs all sunlight to heating the electrolysis, generating the highest T and smallest E, while the electrical energy for electrolysis is generated by a separate source (such as by photovoltaic, solar thermal electric, wind turbine, hydro, nuclear, or fossil fuel generated electronic charge). As shown on the right side of Figure 6.3, high relative concentrations of the electrolysis reactant (such as CO_2 or iron oxide) will further decrease the electrolysis potential.

6.2.2   STEP Solar-to-Chemical Energy Conversion Efficiency

The Hy-STEP mode is being studied outdoors with wind or solar concentrator photovoltaic (CPV)-generated electricity to drive $E_{electrolysis}$. The STEP mode is

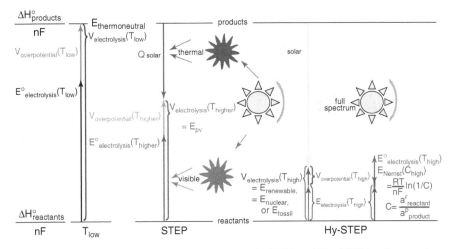

Figure 6.3 Comparison of solar energy utilization in STEP and Hy-STEP implementations of the solar thermal electrochemical production of energetic molecules.

experimentally more complex and is presently studied indoors under solar simulator illumination. Determination of the efficiency of Hy-STEP with solar electric is straightforward in the domain in which $E_{electrolysis} < E_{thermoneutral}$ and the coulombic efficiency is high. Solar thermal energy is collected at an efficiency of $\eta_{thermal}$ to decrease the energy from $E_{thermoneutral}$ to $E_{electrolysis}$, and then electrolysis is driven at a solar electric energy efficiency of $\eta_{solar-electric}$:

$$\eta_{Hy\text{-}STEP\ solar} = (\eta_{thermal} \cdot (E_{thermoneutral} - E_{electrolysis})$$
$$+ \eta_{solar-electric} \cdot E_{electrolysis}) / E_{thermoneutral} \qquad (6.6)$$

$\eta_{thermal}$ is higher than $\eta_{solar-electric}$, and gains in efficiency occur in Eq. 6.6 in the limit as $E_{electrolysis}$ approaches 0. $E_{electrolysis} = 0$ is equivalent to thermochemical, rather than electrolytic, production. As seen in Figure 6.4, at unit activity $E°_{CO2/CO}$ does not approach 0 until 3000°C. Material constraints inhibit approach to this higher temperature, while electrolysis also provides the advantage of spontaneous product separation. At lower temperature, small values of $E_{electrolysis}$ can occur at higher reactant and lower product activities, as described in Eq. 6.4. In the present configuration sunlight is concentrated at 75% solar-to-thermal efficiency, heating the electrolysis to 950°C, which decreases the high current density CO_2 splitting potential to 0.9 V, and the electrolysis charge is provided by CPV at 37% solar to electric efficiency. The solar-to-chemical energy conversion efficiency is in accordance with Eq 6.6:

$$\eta_{Hy\text{-}STEP\ solar} = (75\% \cdot (1.46\,V - 0.90\,V) + 37\% \cdot 0.90\,V)/1.46\,V = 52\% \qquad (6.7)$$

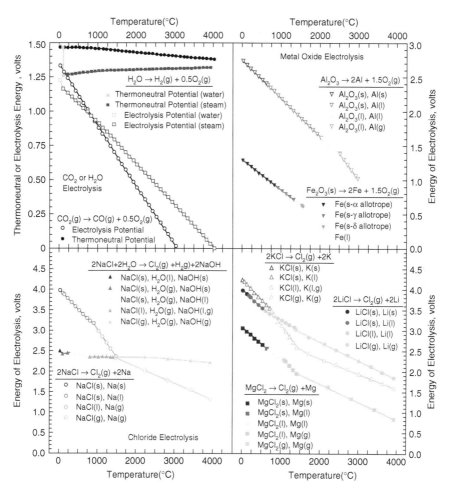

Figure 6.4 The calculated potential to electrolyze selected oxides (*top*) and chlorides (*bottom*). The indicated decrease in electrolysis energy with increase in temperature provides energy savings in the STEP process in which high temperature is provided by excess solar heat. Energies of electrolysis are calculated from Eq. 6.3, with consistent thermochemical values at unit activity using NIST gas and condensed-phase Shomate equations [59b]. Note that with water excluded, the chloride electrolysis decreases (*lower left*). All other indicated electrolysis potentials, including that of water or carbon dioxide, decrease with increasing temperature. Thermoneutral potentials are calculated with Eq. 6.5. Modified with permission from ref. 3.

A relatively high concentration of reactants lowers the voltage of electrolysis via the Nernst term in Eq 6.4. With the appropriate choice of high-temperature electrolyte, this effect can be dramatic, for example, both in STEP iron and in comparing the benefits of the molten carbonate to solid oxide (gas phase) reactants for STEP CO_2 electrolytic reduction, sequestration, and fuel formation. Fe(III) (as found in the common iron ore hematite) is nearly insoluble in sodium carbonate, while it is soluble to over 10 m (molar) in lithium carbonate [6], and as discussed in Section 6.2.3, molten carbonate electrolyzer provides 10^3 to 10^6 times higher concentration of reactant at the cathode surface than a solid oxide electrolyzer.

In practice, for STEP iron or carbon capture, we simultaneously drive lithium carbonate electrolysis cells together in series, at the CPV maximum power point (Fig. 6.5). Specifically, a Spectrolab CDO-100-C1MJ concentrator solar cell is used to generate 2.7 V at maximum power point, with solar-to-electrical energy efficiencies of 37% under 500 suns illumination. As seen in Figure 6.5, at maximum power, the 0.99-cm^2 cell generates 1.3 A at 100 suns, and when masked to 0.2 cm^2 area generates 1.4 A at 500 suns. Electrolysis electrode surface areas were chosen to match the solar cell-generated power. At 950°C at 0.9 V, the electrolysis cells generate carbon monoxide at 1.3 to 1.5 A (the electrolysis current stability is shown at the bottom right of Fig. 6.5).

In accord with Eq. 6.6 and Figure 6.3, Hy-STEP efficiency improves with temperature increase to decrease overpotential and $E_{electrolysis}$, and with increase in the relative reactant activity. Higher solar efficiencies will be expected, both with more effective carbonate electrocatalysts (as morphologies with higher effective surface area and lower overpotential are developed) and also as PV efficiencies increase. Increases in solar to electric (PV, CPV, and solar thermal-electric) efficiencies continue to be reported and will improve Eq. 6.7 efficiency. For example, multijunction CPVs improved to $\eta_{PV} = 40.7\%$ have been reported [69].

Engineering refinements will improve some aspects, and decrease other aspects, of the system efficiency. Preheating the CO_2, by circulating it as a coolant under the CPV (as we currently do in the indoor STEP experiments but not outdoor Hy-STEP experiments) will improve the system efficiency. In the present configuration outgoing CO and O_2 gases at the cathode and anode heat the incoming CO_2. Isolation of the electrolysis products will require heat exchangers with accompanying radiative heat losses, and for electrolyses in which there are side reactions or product recombination losses, $\eta_{Hy-STEP\,solar}$ will decrease proportional to the decrease in coulombic efficiency. At present, wind turbine-generated electricity is more cost effective than solar-electric, and we have demonstrated a Hy-STEP process with wind-electric for CO_2 free production of iron (delineated in Section 6.3.3). Addition of long-term (overnight) molten salt-insulated storage will permit continuous operation of the STEP process. Both STEP implementations provide a basis for practical, high solar efficiencies.

Components for STEP CO_2 capture and conversion to solid carbon are represented on the left side of Figure 6.5, and are detailed in refs. 4–7. A 2.7-V CPV photopotential drives three in series electrolyses at 950°C. Fundamental details of the heat balance are provided in ref. 4. The CPV has an experimental solar efficiency of 37%, and the 63% of insolation not converted to electricity comprises a significant heat source. The challenge is to direct a substantial fraction of this heat

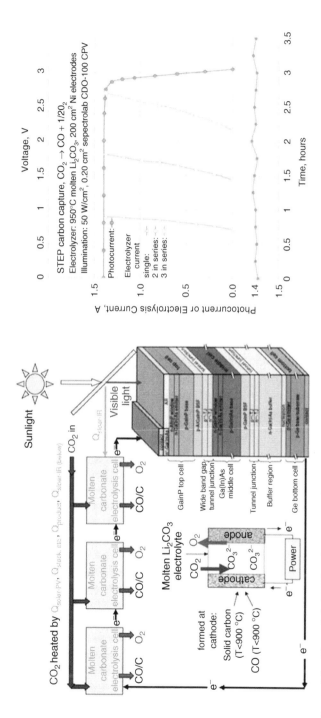

Figure 6.5 *Upper left*: STEP carbon capture in which three molten carbonate electrolysis cells in series are driven by a concentrator photovoltaic (CPV). Sunlight is split into two spectral regions; visible radiation drives the CPV and thermal radiation heats the electrolysis cell. In Hy-STEP (not shown), sunlight is not split and the full spectrum heats the electrolysis cell, and electronic charge is generated separately by solar, wind, or other source. *Right*: The maximum power point photovoltage of one Spectrolab CPV is sufficient to drive three in series CO_2-splitting 950°C molten Li_2CO_3 electrolysis cells. Photocurrent at 500 suns [masked (0.20 cm²) Spectrolab CDO-100 CPV], or electrolysis current, versus voltage; electrolysis current is shown for one, two, or three 950°C Li_2CO_3 electrolysis cells in series with 200-cm² Ni electrodes. Three electrolysis cells in series provide a power match at the 2.7 V maximum power point of the CPV at 950°C; similarly (not shown), two 750°C Li_2CO_3 electrolysis cells in series provide a power match at 2.7 V to the CPV. *Lower left*: Stable carbon capture (with 200 cm² "aged" Ni electrodes at 750°C); fresh electrodes (not shown) exhibit an initial fluctuation as carbon forms at the cathode and Ni oxide layer forms on the anode. The rate of solid carbon deposition gradually increases as the cathode surface area slowly increases in time. Modified with permission from ref. 4.

159

to the electrolysis. An example of this challenge is in the first stage of heating, in which higher temperature increases CO_2 preheat but diminishes the CPV power. Heating of the reactant CO_2 is a three-tier process in the current configuration:

1. Preheating of room temperature CO_2 consists of
   (a) Flow-through a heat exchange fixed to the back of the concentrator solar cell and/or
   (b) Preheating to simulate CO_2 extracted from an available heat source such as a hot smoke (flue) stack.
2. Secondary heating consists of passing this CO_2 through a heat exchange with the outgoing products.
3. Tertiary heat is applied through concentrated, split solar thermal energy (Fig. 6.5).

An upper limit to the energy required to maintain a constant system temperature is given in the case in which neither solar IR, excess solar visible, nor heat exchange from the environment or products would be applied to the system. When an 0.90-V electrolysis occurs, an additional 0.56 V, over $E_{tn} = 1.46V$, is required to maintain a constant system temperature. Hence, in the case of three electrolyses in series, as in Figure 6.5, an additional $3 \times 0.56V = 1.68V$ will maintain constant temperature. This is less than the 63% of the solar energy (equivalent to 4.6 V) not used in generating the 2.7 V of maximum power point voltage of electronic charge from the CPV in this experiment. Heating requirements are even less when the reactant activity is maintained at a level that is higher than the product activity. For example, this is accomplished when products are continuously removed to ensure that the partial pressure of the products is lower than that of the CO_2. This lowers the total heat required for temperature neutrality to below that of the unit activity thermoneutral potential 1.46 V.

The STEP effective solar energy conversion efficiency, η_{STEP}, is constrained by both photovoltaic and thermal boost conversion efficiencies, η_{PV} and $\eta_{thermal-boost}$ [8]. Here, the CPV sustains a conversion efficiency of $\eta_{PV} = 37.0\%$. In the system, passage of electrolysis current requires an additional, combined (ohmic and anodic + cathodic over−) potential above the thermodynamic potential. However, mobility and kinetics improve at higher temperature to decrease this overpotential. The generated CO contains an increase in oxidation potential compared to CO_2 at room temperature $[E_{CO_2/CO}(25°C) = 1.33V$ for $CO_2 \rightarrow CO + 1/2O_2$ in Fig. 6.4], an increase of 0.43 V compared to the 0.90 V used to generate the CO. The electrolysis efficiency compares the stored potential to the applied potential [4]:

$$\eta_{thermal-boost} = E°_{electrolysis}(25°C)/V_{electrolysis}(T)$$

Given a stable temperature electrolysis environment, the experimental STEP solar to CO carbon capture and conversion efficiency is the product of this relative gain in energy and the electronic solar efficiency:

$$\eta_{STEP} = \eta_{PV} \cdot \eta_{thermal-boost} = 37.0\% \cdot (1.33V/0.90V) = 54.7\% \qquad (6.8)$$

Ohmic and overpotential losses are already included in the measured electrolysis potential. This 54.7% STEP solar conversion efficiency is an upper limit of the present experiment, and, as with the Hy-STEP mode, improvements are expected in electrocatalysis and CPV efficiency. Additional losses will occur when beam splitter and secondary concentrator optics losses and thermal systems matching are incorporated, but this serves to demonstrate that the synergy of this solar/photo/electrochemical/thermal process leads to energy efficiency higher than that for solar-generated electricity [69] or for photochemical [70], photoelectrochemical [21,27], solar thermal [71], or other CO_2 reduction processes [72].

The CPV does not need, or function with, sunlight of energy less than that of the 0.67 eV bandgap of the multijunction Ge bottom layer. From our previous calculations, this thermal energy comprises 10% of AM1.5 insolation, which will be further diminished by the solar thermal absorption efficiency and heat exchange to the electrolysis efficiency [54], and under 0.5 MW m^{-2} of incident sunlight (500 suns illumination), yields $\sim 50\,\mathrm{kW\,m^{-2}}$, which may be split off as thermal energy toward heating the electrolysis cell without decreasing the CPV electronic power. The CPV, while efficient, utilizes less than half of the super-bandgap ($h\nu > 0.67\,\mathrm{eV}$) sunlight. A portion of this $>\sim250$ kWm^{-2} available energy, is extracted through heat exchange at the back side of the CPV. Another useful source for consideration as supplemental heat is industrial exhaust. The temperature of industrial flue stacks varies widely with fossil fuel source and application and ranges up to 650°C for an open-circuit gas turbine. The efficiency of thermal energy transfer will limit use of this available heat.

A lower limit to the STEP efficiency is determined when no heat is recovered, either from the CPV or remaining solar IR, when heat is not recovered via heat exchange from the electrolysis products, and when an external heat source is used to maintain a constant electrolysis temperature. In this case, the difference between the electrolysis potential and the thermoneutral potential represents the enthalpy required to keep the system from cooling. In this case, our 0.9-V electrolysis occurs at an efficiency of (0.90 V/1.46 V) · 54.7% = 34%. While the STEP energy analysis, detailed in Section 6.4.2, for example, for CO_2 to CO splitting, is more complex than that of the Hy-STEP mode, more solar thermal energy is available including a PV's unused or waste heat to drive the process and to improve the solar-to-chemical energy conversion efficiency. We determine the STEP solar efficiency over the range from inclusion of no solar thermal heat (based on the enthalpy, rather than free energy, of reaction) to the case in which the solar thermal heat is sufficient to sustain the reaction (based on the free energy of reaction). This determines the efficiency range, as chemical flow out to the solar flow in (as measured by the increase in chemical energy of the products compared to the reactants), from 34% to over 50%.

6.2.3   Identification of STEP Consistent Endothermic Processes

The electrochemical driving force for a variety of chemicals of widespread use by society will be shown to significantly decrease with increasing temperature.

As calculated and summarized in the top left of Figure 6.4, the electrochemical driving force for electrolysis of either CO_2 or water significantly decreases with increasing temperature. The ability to remove CO_2 from exhaust stacks or atmospheric sources provides a response to linked environmental impacts, including global warming due to anthropogenic CO_2 emission. From the known thermochemical data for CO_2, CO, and O_2, and in accord with Eq. 6.1, CO_2 splitting can be described by:

$$CO_2(g) \rightarrow CO(g) + 1/2O_2(g);$$
$$E°CO_2 \text{split} = (G°CO + 0.5G°O_2 - G°CO_2)/2F; E°(25°C) = 1.333 \text{ V} \quad (6.9)$$

As an example of the solar energy efficiency gains, this progress report focuses on CO_2 splitting potentials and provides examples of other useful STEP processes. As seen in Figure 6.4, CO_2 splitting potentials decrease more rapidly with temperature than those for water splitting, signifying that the STEP process may be readily applied to CO_2 electrolysis. Efficient, renewable, non-fossil fuel energy-rich carbon sources are needed, and the product of Eq. 6.9, carbon monoxide, is a significant industrial gas with a myriad of uses, including the bulk manufacturing of hydrocarbon fuels, acetic acid, and aldehydes (and detergent precursors) and for use in industrial nickel purification [66]. To alleviate challenges of fossil-fuel resource depletion, CO is an important syngas component and methanol is formed through the reaction with H_2. The ability to remove CO_2 from exhaust stacks or atmospheric sources also limits CO_2 emission. Based on our original analogous experimental photo-thermal electrochemical water electrolysis design [55], the first CO_2 STEP process consists of solar-driven and solar thermal-assisted CO_2 electrolysis. In particular, in a molten carbonate bath electrolysis cell, fed by CO_2:

$$\text{cathode :} \quad 2CO_2(g) + 2e^- \rightarrow CO_3 = (\text{molten}) + CO(g)$$
$$\text{anode :} \quad CO_3^=(\text{molten}) \rightarrow CO_2(g) + 1/2O_2(g) + 2e^-$$
$$\text{cell :} \quad CO_2(g) \rightarrow CO(g) + 1/2O_2(g) \quad (6.10)$$

Molten alkali carbonate electrolyte fuel cells typically operate at 650°C. Li, Na, or K cation variation can affect charge mobility and operational temperatures. Sintered nickel often serves as the anode and porous lithium-doped nickel oxide often as the cathode, while the electrolyte is suspended in a porous, insulating, chemically inert $LiAlO_2$ ceramic matrix [67].

Solar thermal energy can be used to favor the formation of products for electrolysis characterized by a negative isothermal temperature coefficient but will not improve the efficiency of heat-neutral or exothermic reactions. An example of this restriction occurs for the electrolysis reaction currently used by industry to generate chlorine. During 2008, the generation of chlorine gas (principally for use as bleach and in the chlor-alkali industry) consumed approximately 1% of the world's electricity [68], prepared in accord with the industrial electrolytic process:

$$2NaCl + 2H_2O \rightarrow Cl_2 + H_2 + 2NaOH; E°(25°C) = 2.502\,V \qquad (6.11)$$

In the lower left portion of Figure 6.4, the calculated electrolysis potential for this industrial chlor-alkali reaction exhibits little variation with temperature, and hence the conventional generation of chlorine by electrolysis would not benefit from the inclusion of solar heating. This potential is relatively invariant, despite a number of phase changes of the components (indicated on the figure and which include the melting of NaOH or NaCl). However, as seen in Figure 6.4, the calculated potential for the anhydrous electrolysis of chloride salts is endothermic, including the electrolyses to generate not only chlorine, but also metallic lithium, sodium, and magnesium, and can be greatly improved through the STEP process:

$$MCl_n \rightarrow n/2Cl_2 + M;$$

$$E°\,MCl_n\,split\,(25°C) = 3.98\,V\text{-M} = Na, 4.24\,V\text{-K}, 3.98\,V\text{-Li}, 3.07\,V\text{-Mg} \quad (6.12)$$

The calculated decreases for the anhydrous chloride electrolysis potentials are on the order of volts per 1000°C temperature change. For example, from 25°C up to the $MgCl_2$ boiling point of 1412°C, the $MgCl_2$ electrolysis potential decreases from 3.07 V to 1.86 V. This decrease provides a theoretical basis for significant, non-CO_2-emitting, non-fossil fuel-consuming processes for the generation of chlorine and magnesium, delineated in Section 6.3.4 and occurring at high solar efficiency analogous to the similar CO_2 STEP process.

In Section 6.3.2 the STEP process is derived for the efficient solar removal/recycling of CO_2. In addition, thermodynamic calculation of metal and chloride electrolysis rest potentials identifies electrolytic processes that are consistent with endothermic processes for the formation of iron, chlorine, aluminum, lithium, sodium, and magnesium, via CO_2-free pathways. As shown, the conversion and replacement of the conventional, aqueous, industrial alkali-chlor process, with an anhydrous electrosynthesis, results in a redox potential with a calculated decrease of 1.1 V from 25°C to 1000°C.

As seen in the top right of Figure 6.4, the calculated electrochemical reduction of metal oxides can exhibit a sharp, smooth decrease in redox potential over a wide range of phase changes. These endothermic processes provide an opportunity for the replacement of conventional industrial processes by the STEP formation of these metals. In 2008, industrial electrolytic processes consumed ~5% of the world's electricity, including for aluminum (3%), chlorine (1%), and lithium, magnesium, and sodium production. This 5% of the global 19×10^{12} kWh of electrical production is equivalent to the emission of 6×10^8 metric tons of CO_2 [68]. The iron and steel industry accounts for a quarter of industrial direct CO_2 emissions. Currently, iron is predominantly formed through the reduction of hematite with carbon, emitting CO_2:

$$Fe_2O_3 + 3C + 3/2O_2 \rightarrow 2Fe + 3CO_2 \qquad (6.13)$$

A non-CO_2-emitting alternative is provided by the STEP-driven electrolysis of Fe_2O_3:

$$Fe_2O_3 \rightarrow 2Fe + 3/2O_2 E° = 1.28 \text{ V} \qquad (6.14)$$

As seen in the top right of Figure 6.4, the calculated iron generating electrolysis potential drops 0.5 V (a 38% drop) from 25°C to 1000°C and, as with the CO_2 analog, will be expected to decrease more rapidly with non-unit activity conditions, as will be delineated in a future study. Conventional industrial processes for these metals and chlorine, along with CO_2 emitted from power and transportation, are responsible for the majority of anthropogenic CO_2 release. The STEP process, to efficiently recover CO_2 and in lieu of these industrial processes, can provide a transition beyond the fossil fuel-electric grid economy.

The top left of Figure 6.4 includes calculated thermoneutral potentials for CO_2 and water splitting reactions. At ambient temperature, the difference between E_{th} and E_T does not indicate an additional heat requirement for electrolysis, as this heat is available via heat exchange with the ambient environment. At ambient temperature, $E_{tn} - E_T$ for CO_2 or water is 0.13 and 0.25 V, respectively, and is calculated (not shown) as 0.15 ± 0.1 V for Al_2O_3 and Fe_2O_3 and 0.28 ± 0.3 V for each of the chlorides. We find that molten electrolytes present several fundamental advantages compared to solid oxides for CO_2 electrolysis.

1. The molten carbonate electrolyzer provides 10^3 to 10^6 times higher concentration of reactant at the cathode surface than a solid oxide electrolyzer. Solid oxides utilize gas-phase reactants, whereas carbonates utilize molten-phase reactants. Molten carbonate contains 2×10^{-2} mol reducible tetravalent carbon/cm^3. The density of reducible tetravalent carbon sites in the gas phase is considerably lower. Air contains 0.03% CO_2, equivalent to only 1×10^{-8} mol of tetravalent carbon/cm^3, and flue gas (typically) contains 10–15% CO_2, equivalent to 2×10^{-5} mol reducible C(IV)/cm^3. Carbonate's higher concentration of active, reducible tetravalent carbon sites logarithmically decreases the electrolysis potential, and can facilitate charge transfer at low electrolysis potentials.

2. Molten carbonates can directly absorb atmospheric CO_2, whereas solid oxides require an energy-consuming preconcentration process.

3. Molten carbonates electrolyses are compatible with both solid- and gas-phase products.

4. Molten processes have an intrinsic thermal buffer not found in gas-phase systems. Sunlight intensity varies over a 24-h cycle and more frequently with variations in cloud cover. This disruption to other solar energy conversion processes is not necessary in molten salt processes. For example as discussed in Section 6.4.3, the thermal buffer capacity of molten salts has been effective for solar-to-electric power towers to operate 24/7. These towers concentrate solar thermal energy to heat molten salts, which circulate and via heat exchange boil water to drive conventional mechanical turbines.

6.3 DEMONSTRATED STEP PROCESSES

6.3.1 STEP Hydrogen

STEP occurs at both higher electrolysis and higher solar conversion efficiencies than conventional room temperature photovoltaic (PV) generation of H_2. Experimentally, we demonstrated a sharp decrease in the water splitting potential in an unusual molten sodium hydroxide medium (Fig. 6.6) and, as shown in Figure 6.7, three series-connected Si CPVs efficiently driving two series-connected molten hydroxide water-splitting cells at 500°C to generate hydrogen [55].

Recently we have considered the economic viability of solar hydrogen fuel production. That study provided evidence that the STEP system is an economically viable solution for the production of hydrogen [55].

6.3.2 STEP Carbon Capture

In this process carbon dioxide is captured directly, without the need to preconcentrate dilute CO_2, with a high-temperature electrolysis cell powered by sunlight in a single step. Solar thermal energy decreases the energy required for the endothermic conversion of CO_2 and kinetically facilitates electrochemical reduction, while solar visible generates electronic charge to drive the electrolysis. CO_2 can be captured as solid carbon and stored or used as carbon monoxide to feed chemical or synthetic fuel production. Thermodynamic calculations are used to determine, and

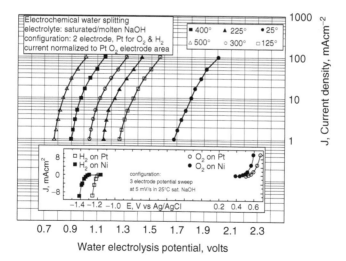

Figure 6.6 V_{H_2O}, measured in aq. saturated or molten NaOH, at 1 atm. Steam is injected in the molten electrolyte. O_2 anode is 0.6 cm² Pt foil. IR and polarization losses are minimized by sandwiching 5 mm from each side of the anode, oversized Pt gauze cathode. *Inset*: At 25°C, 3 electrode values comparing Ni and Pt working electrodes and with a Pt gauze counterelectrode at 5 mV/s.

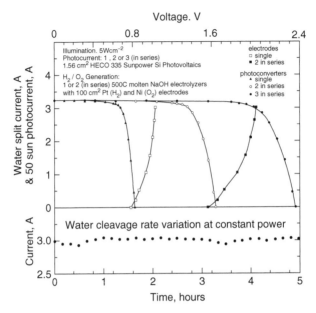

Figure 6.7   Photovoltaic and electrolysis charge transfer of STEP hydrogen, using Si CPVs driving molten NaOH water electrolysis. Photocurrent is shown for one, two, or three 1.561-cm^2 HECO 335 Sunpower Si photovoltaics in series at 50 suns. The CPVs drive 500°C molten NaOH steam electrolysis using Pt gauze electrodes. *Lower inset*: electrolysis current stability.

then demonstrate, a specific low-energy molten carbonate salt pathway for carbon capture.

Prior investigations of the electrochemistry of carbonates in molten salts tended to focus on reactions of interest to fuel cells [67], rather than the (reverse) electrolysis reactions of relevance to the STEP reduction of CO$_2$, typically in alkali carbonate mixtures. Such mixtures substantially lower the melting point compared to the pure salts, and would provide the thermodynamic maximum voltage for fuel cells. However, the electrolysis process is maximized in the opposite temperature domain of fuel cells, that is, at elevated temperatures that decrease the energy of electrolysis, as schematically delineated in Figure 6.1. These conditions provide a new opportunity for effective CO$_2$ capture.

CO$_2$ electrolysis splitting potentials are calculated from the thermodynamic free energy components of the reactants and products [3,4,59b] as $E = -\Delta G(\text{reaction})/nF$, where $n = 4$ or 2 for the respective conversion of CO$_2$ to the solid carbon or carbon monoxide products. As calculated using the available thermochemical enthalpy and entropy of the starting components, and as summarized in the left side of Figure 6.8, molten Li$_2$CO$_3$, via a Li$_2$O intermediate, provides a preferred, low-energy route compared to Na$_2$CO$_3$ or K$_2$CO$_3$ (via Na$_2$O or K$_2$O) for the conversion of CO$_2$. High temperature is advantageous as it decreases the free energy necessary to drive the STEP endothermic

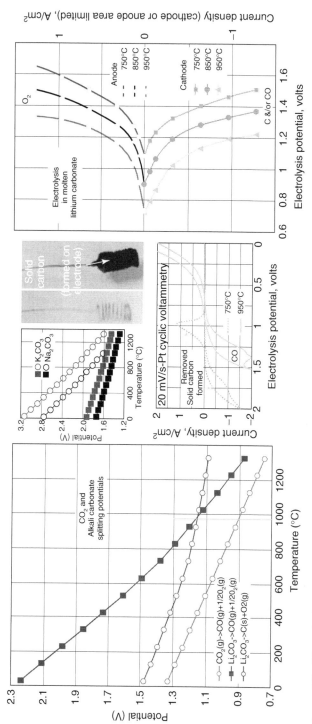

Figure 6.8 The calculated (*left*) and measured (*right*) electrolysis of CO² in molten carbonate. *Left*: The calculated thermodynamic electrolysis potential for carbon capture and conversion in Li_2CO_3 (*main figure*), or Na_2CO_3 or K_2CO_3 (*middle upper left*): squares refer to $M_2CO_3 \rightarrow C + M_2O + O_2$ and circles to $M_2CO_3 \rightarrow CO + M_2O + 1/2O_2$. To the left of the vertical brown line (*left*), solid carbon is the thermodynamically preferred (lower energy) product. To the right of the vertical line, CO is preferred. Carbon dioxide fed into the electrolysis chamber is converted to solid carbon in a single step. *Photographs* (*middle upper right*): coiled Pt cathode before (*left*), and after (*right*), CO_2 splitting to solid carbon at 750°C in molten carbonate with a Ni anode. *Right*: The electrolysis full cell potential is measured, under anode or cathode limiting conditions, at a Pt electrode for a range of stable anodic and cathodic current densities in molten. Li_2CO_3 *Lower middle*: cathode size-restricted full cell cyclic voltammetry (CV) of Pt electrodes in molten Li_2CO_3. Modified with permission from ref. 4.

167

process. The carbonates, Li_2CO_3, Na_2CO_3, and K_2CO_3, have respective melting points of 723°C, 851°C, and 891°C. Molten Li_2CO_3 not only requires lower thermodynamic electrolysis energy but in addition has higher conductivity (6 S cm^{-1}) than Na_2CO_3 (3 S cm^{-1}) or K_2CO_3 (2 S cm^{-1}) near the melting point [73]. Higher conductivity is desired as it leads to lower electrolysis ohmic losses. Low carbonate melting points are achieved by a eutectic mix of alkali carbonates ($T_{mp}Li_{1.07}Na_{0.93}CO_3$: 499°C; $Li_{0.85}Na_{0.61}K_{0.54}CO_3$:393°C). Mass transport is also improved at higher temperature; the conductivity increases from 0.9 to 2.1 S cm^{-1} with temperature increase from 650°C to 875°C for a 1:1:1 by mass mixture of the three alkali carbonates [74].

In 2009 we showed that molten carbonate electrolyzers can provide an effective medium for solar splitting of CO_2 at high conversion efficiency. In 2010 Lubormirsky et al. and our group separately reported that molten lithiated carbonates provide a particularly effective medium for the electrolytic reduction of CO_2 [4,75]. As we show in the photograph in Figure 6.8, at 750°C CO_2 is captured in molten lithium carbonate electrolyte as solid carbon by reduction at the cathode at low electrolysis potential. It is seen in the cyclic voltammetry (CV) that a solid carbon peak that is observed at 750°C is not evident at 950°C. At temperatures less than ~ 900°C in the molten electrolyte, solid carbon is the preferred CO_2 splitting product, while carbon monoxide is the preferred product at higher temperature. As seen on the left side of Figure 6.8, the electrolysis potential is <1.2 V at either 0.1 or 0.5 A/cm^2, respectively, at 750 or 850°C. Hence, the electrolysis energy required at these elevated, molten temperatures is less than the minimum energy required to split CO_2 to CO at 25°C:

$$CO_2 \rightarrow CO + 1/2O_2 \qquad E°(T = 25°C) = 1.33 \text{ V} \qquad (6.15)$$

The observed experimental carbon capture correlates with:

$$Li_2CO_3 \text{ (molten)} \rightarrow C \text{ (solid)} + Li_2O\text{(dissolved)} + O_2 \text{ (gas)} \qquad (6.16A)$$

$$Li_2CO_3 \text{ (molten)} \rightarrow CO\text{(gas)} + Li_2O\text{(dissolved)} + 1/2O_2 \text{ (gas)} \qquad (6.16B)$$

When CO_2 is bubbled in, a rapid reaction back to the original lithium carbonate is strongly favored:

$$Li_2O\text{ (dissolved)} + CO_2 \text{ (gas)} \rightarrow Li_2CO_3 \text{ (molten)} \qquad (6.17A)$$

$$Li_2CO_3 \leftrightarrow Li_2O + CO_2 \qquad (6.17B)$$

In the presence of CO_2, reaction 6.17A is strongly favored (exothermic), and the rapid reaction back to the original lithium carbonate occurs while CO_2 is bubbled into molten lithium carbonate containing the lithium oxide. The carbon capture reaction in molten carbonate, combines Eqs. 6.16 and 6.17:

$$CO_2 \text{ (gas)} \rightarrow C\text{(solid)} + O_2 \text{ (gas)} \, T \leq 900°C \qquad (6.18A)$$

$$CO_2 \text{ (gas)} \rightarrow CO\text{(gas)} + 1/2O_2 \text{ (gas)} \, T \geq 950°C \qquad (6.18B)$$

The electrolysis of carbon capture in molten carbonates can occur at lower experimental electrolysis potentials than the unit activity potentials calculated in Figure 6.8. A constant influx of CO_2 to the cell maintains a low concentration of Li_2O, in accord with reaction 23. The activity ratio, Θ, of the carbonate reactant to the oxide product in the electrolysis chamber, when high, decreases the cell potentials with the Nernst concentration variation of the potential in accord with Eq. 6.16, as:

$$E_{CO2/X}(T) = E°_{CO2/X}(T) - 0.0592V \cdot T(K)/(n \cdot 298K) \cdot \log(\Theta);$$

$$N = 4 \text{ or } 2, \text{ for } X = C_{solid} \text{ or CO product} \tag{6.19}$$

For example, from Eq. 6.19, the expected cell potential at 950°C for the reduction to the CO product is $E_{CO2/CO} = 1.17V - (0.243V/2) \cdot 4 = 0.68V$, with a high $\Theta = 10,000$ carbonate/oxide ratio in the electrolysis chamber. As seen in the Figure 6.8 photo, CO_2 is captured in 750°C Li_2CO_3 as solid carbon by reduction at the cathode at low electrolysis potential. The carbon formed in the electrolysis in molten Li_2CO_3 at 750°C is in quantitative accord with the 4 e⁻ reduction of Eq. 6.16A, as determined by

1. Mass, at constant 1.25 A for both 0.05 and 0.5 A/cm² (large and small electrode) electrolyses (the carbon is washed in a sonicator and dried at 90°C)
2. Ignition (furnace combustion at 950°C)
3. Volumetric analysis in which KIO_3 is added to the carbon and converted to CO_2 and I_2 in hot phosphoric acid ($5C + 4KIO_3 + 4H_3PO_4 \rightarrow 5CO_2 + 2I_2 + 2H_2O + 4KH_2PO_4$), and the liberated I_2 is dissolved in 0.05 M KI and titrated with thiosulfate using a starch indicator

We also observe the transition to the carbon monoxide product with increasing temperature. Specifically, while at 750°C the molar ratio of solid carbon to CO gas formed is 20:1, at 850° in molten Li_2CO_3 the product ratio is 2:1, at 900°C the ratio is 0.5:1, and at 950°C the gas is the sole product. Hence, in accord with Figure 6.5, switching between the C or CO product is temperature programmable.

We have replaced Pt with Ni, nickel alloys (inconel and monel), Ti, and carbon, and each is an effective carbon capture cathode material. Solid carbon deposits on each of these cathodes at similar overpotential in 750°C molten Li_2CO_3. For the anode, both platinum and nickel are effective, while titanium corrodes under anodic bias in molten Li_2CO_3. As seen in the right side of Figure 6.8, electrolysis anodic overpotentials in Li_2CO_3 electrolysis are comparable but larger than cathodic overpotentials, and current densities of >1 A cm⁻² can be sustained. Unlike other fuel cells, carbonate fuel cells are resistant to poisoning effects [67] and are effective with a wide range of fuels, and this appears to be the same for the case in the reverse mode (to capture carbon, rather than to generate electricity). Molten Li_2CO_3 remains transparent and sustains stable electrolysis currents after extended (hours/days) carbon capture over a wide range of electrolysis current densities and temperatures.

As delineated in Section 6.2.3, in practice, either STEP or Hy-STEP modes are useful for efficient solar carbon capture. CO_2 added to the cell is split at 50% solar-to-chemical energy conversion efficiency by series-coupled lithium carbonate electrolysis cells driven at maximum power point by an efficient CPC. Experimentally, we observe the facile reaction of CO_2 and Li_2O in molten Li_2CO_3. We can also calculate the thermodynamic equilibrium conditions between the species in the system, Eq. 6.3B. Using the known thermochemistry of Li_2O, CO_2 and Li_2CO_3 [59b], we calculate the reaction free energy of Eq. 6.1, and from this calculate the thermodynamic equilibrium constant as a function of temperature. From this equilibrium constant, the area above the curve on the left side of Figure 6.9 presents the wide domain (above the curve) in which Li_2CO_3 dominates, that is, where excess CO_2 reacts with Li_2O such that $p_{CO2} \cdot a_{Li_2O} < aLi_2CO_3$. This is experimentally verified when we dissolve Li_2O in molten Li_2CO_3 and inject CO_2 (gas). Through the measured mass gain, we observe the rapid reaction to Li_2CO_3. Hence, CO_2 is flowed into a solution of 5% by weight Li_2O in molten Li_2CO_3 at 750°C; the rate of mass gain is only limited by the flow rate of CO_2 into the cell (using an Omega FMA 5508 mass flow controller) to react one equivalent of CO_2 per dissolved Li_2O. As seen in the measured thermogravimetric analysis on the right side of Figure 6.9, the mass loss in time is low in lithium carbonate heated in an open atmosphere ($\sim 0.03\%$ CO_2) up to 850°C but accelerates when heated to 950°C. However, the 950°C mass loss falls to nearly zero when heated under pure (1 atm) CO_2. Also in accord with Eq. 6.1, added Li_2O shifts the equilibrium to the left. As seen in Figure 6.9 in an open atmosphere, there is no mass loss in a 10% Li_2O, 90% Li_2CO_3 mixture at 850°C, and the Li_2O-containing electrolyte absorbs CO_2 (gains mass) at 750°C to provide for the direct carbon capture of atmospheric CO_2, without a CO_2 preconcentration stage. This consists of the absorption of atmospheric CO_2 (in molten Li_2CO_3 containing Li_2O, to form Li_2CO_3), combined with a facile rate of CO_2 splitting due to the high carbonate concentration, compared to the atmospheric concentration of CO_2, and the continuity of the steady-state concentration of Li_2O, as Li_2CO_3 is electrolyzed in Eq. 6.16.

6.3.3 STEP Iron

A fundamental change in the understanding of iron oxide thermochemistry can open a facile, new CO_2-free route to iron production. Along with control of fire, iron production is one of the founding technological pillars of civilization, but it is a major source of CO_2 emission. In industry, iron is still produced by the carbothermal greenhouse gas-intensive reduction of iron oxide by carbon-coke, and a CO_2-free process to form this staple is needed.

The earliest attempt at electrowinning iron (the formation of iron by electrolysis) from carbonate appears to have been in 1944 in the unsuccessful attempt to electrodeposit iron from a sodium carbonate, peroxide, metaborate mix at 450–500°C, which deposited sodium and magnetite (iron oxide) rather than iron [76,77]. Other attempts [77] have focused on iron electrodeposition from molten mixed halide electrolytes, which has not provided a successful route to form iron [78,79], or

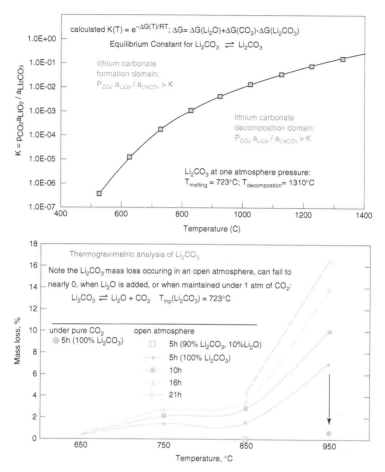

Figure 6.9 *Top*: Species stability in the lithium carbonate, lithium oxide, carbon dioxide system, as calculated from Li_2CO_3, Li_2O, and CO_2 thermochemical data. *Bottom*: Thermogravimetric analysis of lithium carbonate. The measured mass loss in time of Li_2CO_3. *Not shown*: The Li_2CO_3 mass loss rate also decreases with an increasing ratio of Li_2CO_3 mass to the surface area of the molten salt exposed to the atmosphere. This increased ratio may increase the released partial pressure of CO_2 above the surface, increase the rate of the back reaction ($Li_2O + CO_2 \rightarrow Li_2CO_3$), and therefore result in the observed decreased mass loss. Hence, under an open atmosphere at 950°C, the mass loss after 5 h falls from 7% to 4.7%, when the starting mass of pure Li_2CO_3 in the crucible is increased from 20 to 50 g. Under these latter conditions (open atmosphere, 950°C, 50 g total electrolyte), but using a 95% Li_2CO_3, 5% Li_2O mix, the rate of mass loss is only 2.3%. Modified with permission from ref. 8.

aqueous iron electrowinning [80–83], which is hindered by the high thermody-namic potential ($E° = 1.28\,V$) and diminished kinetics at low temperature.

We present a novel route to generate iron metal by the electrolysis of dissolved iron oxide salts in molten carbonate electrolytes, unexpected because of the reported insolubility of iron oxide in carbonates. We report high solubility of lithiated iron oxides and facile charge transfer that produces the staple iron at a high rate and low electrolysis energy and can be driven by conventional electrical sources but is also demonstrated with STEP processes that decrease or eliminate a major global source of greenhouse gas emission [3,4].

As recently as 1999, the solubility of ferric oxide, Fe_2O_3, in 650°C molten carbonate was reported as very low, a $10^{-4.4}$ mole fraction in lithium/potassium carbonate mixtures, and was reported as invariant of the fraction of Li_2CO_3 and K_2CO_3 [84]. Low solubility, of interest to the optimization of molten carbonate fuel cells, had likely discouraged research into the electrowinning of iron metal from ferric oxide in molten lithium carbonate. Rather than the prior part per million reported solubility, we find higher Fe(III) solubilities, on the order of 50% in carbonates at 950°C. The CV of a molten Fe_2O_3 Li_2CO_3 mixture presented in Figure 6.10 exhibits a reduction peak at −0.8 V on Pt (gold curve), which is more pronounced at an iron electrode (light gold curve). At constant current, iron is clearly deposited. The cooled deposited product contains pure iron metal and trapped salt and changes to a rust color with exposure to water (Fig. 6.10 photo). The net electrolysis is the redox reaction of ferric oxide to iron metal and O_2, Eq. 6.14. The deposit is washed and dried, is observed to be reflective gray metallic, responds to an external magnetic field, and consists of dendritic iron crystals.

The two principal natural ores of iron are hematite (Fe_2O_3) and the mixed-valence $Fe^{2+/3+}$ magnetite (Fe_3O_4). We observe that Fe_3O_4 is also highly soluble in molten Li_2CO_3 and may also be reduced to iron with the net electrolysis reaction:

$$Fe_3O_4 \rightarrow 3Fe + 2O_2 \qquad E° = 1.32\,V, E_{thermoneutral} = 1.45\,V \qquad (6.20)$$

Fe_3O_4 electrolysis potentials run parallel, but ∼ 0.06 V higher, than those of Fe_2O_3 in Figure 6.4. The processes are each endothermic; the required electrolysis poten-tial decreases with increasing temperature. For Fe_3O_4 in Figure 6.10, unlike the single peak evident for Fe_2O_3, two reduction peaks appear in the CV at 800°C. After the initial cathodic sweep (indicated by the left arrow), the CV exhibits two reduction peaks, again more pronounced at an iron electrode (gray curve), that appear to be consistent with the respective reductions of Fe^{2+} and Fe^{3+}. In either Fe_2O_3 or Fe_3O_4, the reduction occurs at a potential before we observe any reduc-tion of the molten Li_2CO_3 electrolyte, and at constant current iron is deposited. After 1 h of electrolysis at either 200 or 20 mA/cm^2 of iron deposition, as seen in the Figure 6.10 photographs and as with the Fe_2O_3 case, the extracted cooled elec-trode, after extended electrolysis and iron formation, contains trapped electrolyte. After washing, the product weight is consistent with the eight electrons per Fe_3O_4 coulombic reduction to iron.

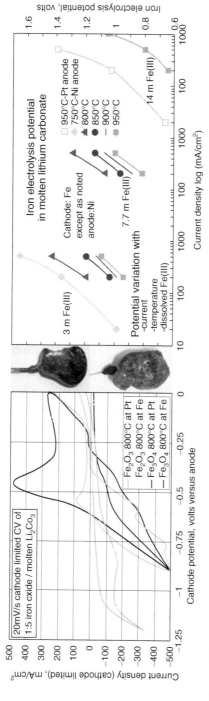

Figure 6.10 *Middle*: Photographs of electrolysis products from 20% Fe₂O₃ or Fe₃O₄ by mass in 800°C Li₂CO₃; after extended 0.5 A electrolysis at a coiled wire (Pt or Fe) cathode with a Ni anode. *Left*: cathode-restricted CV in Li₂CO₃, containing 1:5 by weight of either Fe₂O₃ or Fe₃O₄. *Right*: The measured iron electrolysis potentials in molten Li₂CO₃, as a function of the temperature, current density, and concentration of dissolved Fe(III). Modified with permission from ref. 5.

173

The solid products of the solid reaction of Fe$_2$O$_3$ and Li$_2$CO$_3$ have been characterized [85,86]. We prepare and probe the solubility of lithiated iron oxide salts in molten carbonates and report that high Fe(III) solubilities, on the order of 50% in molten carbonates, are achieved via the reaction of Li$_2$O with Fe$_2$O$_3$, yielding an effective method for CO$_2$-free iron production.

Lithium oxide, as well as Fe$_2$O$_3$ and Fe$_3$O$_4$, each have melting points above 1460°C. Li$_2$O dissolves in 400–1000°C molten carbonates. We find that the solubility of Li$_2$O in molten Li$_2$CO$_3$ increases from 9 to 14 m from 750° to 950°C. After preparation of specific iron oxide salts, we add them to molten alkali carbonate. The resultant Fe(III) solubility is similar when either LiFeO$_2$ or LiFeO$_2$, as Fe$_2$O$_3$ + Li$_2$O, is added to the Li$_2$CO$_3$. As seen in the left side of Figure 6.11, the solubility of LiFeO$_2$ is over 12 m above 900C° in Li$_2$CO$_3$.

Solid reaction of Fe$_2$O$_3$ and Na$_2$CO$_3$ produces both NaFeO$_2$ and NaFe$_5$O$_8$ products [87]. As seen in Figure 6.11, unlike the Li$_2$CO$_3$ electrolyte, our measurements in either molten Na$_2$CO$_3$ or K$_2$CO$_3$ exhibit <<1 wt % iron oxide solubility, even at 950°C. However, the solubility of (Li$_2$O + Fe$_2$O$_3$) is high in the alkali carbonate eutectic mix, Li$_{0.87}$Na$_{0.63}$K$_{0.50}$CO$_3$, and is approximately proportional to the Li fraction in the pure Li$_2$CO$_3$ electrolyte. Solubility of this lithiated ferric oxide in the Li$_x$Na$_y$K$_z$CO$_3$ mixes provides an alternative molten medium for iron production, which, compared to pure lithium carbonate, has the disadvantage of lower conductivity [5] but the advantage of even greater availability and a wider operating temperature domain range (extending several hundred degrees lower than the pure lithium system).

Fe$_2$O$_3$ or LiFe$_5$O$_8$ dissolves rapidly in molten Li$_2$CO$_3$ but reacts with the molten carbonate as evident in a mass loss, which evolves one equivalent of CO$_2$ per Fe$_2$O$_3$, to form a steady-state concentration of LiFeO$_2$ in accord with the reaction of Eq. 6.21 (but occurring in molten carbonate) [6]. However, 1 equivalent of Li$_2$O and 1 equivalent of Fe$_2$O$_3$ or LiFeO$_2$ dissolves without the reactive formation of CO$_2$. This is significant for the electrolysis of Fe$_2$O$_3$ in molten carbonate. As LiFeO$_2$ is reduced Li$_2$O is released, Eq. 6.22, facilitating the continued dissolution of Fe$_2$O$_3$ without CO$_2$ release or change in the electrolyte. More concisely, iron production via hematite in Li$_2$CO$_3$ is given by I and II:

I Dissolution in molten carbonate: $\text{Fe}_2\text{O}_3 + \text{Li}_2\text{O} \rightarrow 2\text{LiFeO}_2$ (6.21)

II Electrolysis, Li$_2$O regeneration: $2\text{LiFeO}_2 \rightarrow 2\text{Fe} + \text{Li}_2\text{O} + 3/2\text{O}_2$ (6.22)

Iron production, Li$_2$O unchanged (I + II): $\text{Fe}_2\text{O}_3 \rightarrow 2\text{Fe} + 3/2\text{O}_2$ (6.23)

As indicated in Figure 6.9, a molar excess of greater than 1:1 of Li$_2$O to Fe$_2$O$_3$ in molten Li$_2$CO$_3$ will further inhibit the eq 1 disproportionation of lithium carbonate. The right side of Figure 6.11 summarizes the thermochemical calculated potentials constraining iron production in molten carbonate. Thermodynamically, it is seen that at higher potential steel (iron containing carbon) may be directly formed via the concurrent reduction of CO$_2$, which we observe in the Li$_2$CO$_3$ at higher electrolysis

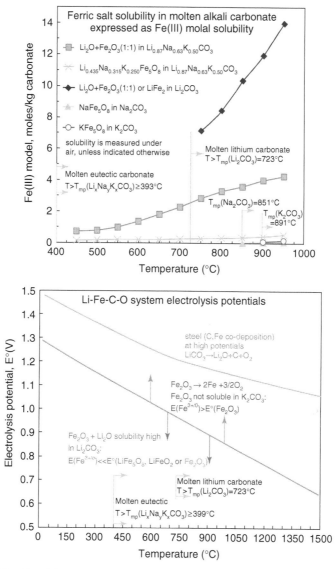

Figure 6.11 *Top*: Measured ferric oxides solubilities in alkali molten carbonates. *Bottom*: Calculated unit activity electrolysis potentials of $LiFe_5O_8$, Fe_2O_3, or Li_2CO_3. Vertical arrows indicate Nernstian shifts at high or low Fe(III). Modified with permission from [5].

potential, as $Li_2CO_3 \rightarrow C + Li_2O + O_2$, followed by carbonate regeneration via Eq. 6.3, to yield by electrolysis in molten carbonate:

Steel production:

$$Fe_2O_3 + 2xCO_2 \rightarrow 2FeC_x + (3/2 + 2x)O_2 \qquad (6.24)$$

From the kinetic perspective, a higher concentration of dissolved iron oxide improves mass transport, decreases the cathode overpotential, and permits higher steady-state current densities of iron production, and will also substantially decrease the thermodynamic energy needed for the reduction to iron metal. In the electrolyte Fe(III) originates from dissolved ferric oxides, such as LiFeO$_2$ or LiFe$_5$O$_8$. The potential for the 3e$^-$ reduction to iron varies in accord with the general Nernstian expression, for a concentration [Fe(III)], at activity coefficient α:

$$E_{Fe(III/0)} = E^°_{Fe(III/0)} + (RT/nF) \log (α_{Fe(III)} [Fe(III)])^{1/3} \qquad (6.25)$$

This decrease in electrolysis potential is accentuated by high temperature and is a ~ 0.1 V per decade 10 increase in Fe(III) concentration at 950°C. A higher activity coefficient, $α_{Fe(III)} > 1$, would further decrease the thermodynamic potential to produce iron. The measured electrolysis potential is presented on the right of Figure 6.10 for dissolved Fe(III) in molten Li$_2$CO$_3$, and is low. For example, 0.8V sustains a current density of 500 mA·cm^{-2} in 14 m Fe(III) in Li$_2$CO$_3$ at 950°C. Higher temperature, and higher concentration, lowers the electrolysis voltage, which can be considerably less than the room potential required to convert Fe$_2$O$_3$ to iron and oxygen. When an external source of heat, such as solar thermal, is available, the energy savings over room temperature iron electrolysis are considerable.

Electrolyte stability is regulated through control of CO$_2$ pressure and/or by dissolution of excess Li$_2$O. Electrolyte mass change was measured in 7 m LiFeO$_2$ and 3.5 m Li$_2$O in molten Li$_2$CO$_3$ after 5 h. Under argon there is a 1, 5, or 7 wt% loss at 750°C, 850°C or 950°C, respectively, through CO$_2$ evolution. Little loss occurs under air (0.03% CO$_2$), while under pure CO$_2$ the electrolyte gains 2–3 wt% (external CO$_2$ reacts with dissolved Li$_2$O to form Li$_2$CO$_3$).

The endothermic nature of the new synthesis route, that is, the decrease in iron electrolysis potential with increasing temperature, provides a low-free energy opportunity for the STEP process. In this process, solar thermal provides heat to decrease the iron electrolysis potential, Figure 6.10, and solar visible generates electronic charge to drive the electrolysis. A low-energy route for the CO$_2$-free formation of iron metal from iron ores is accomplished by the synergistic use of both visible and infrared sunlight. This provides high solar energy conversion efficiencies, Figure 6.5, when applied to Eqs. 6.14 and 6.20 in a molten carbonate electrolyte. We again use a 37% solar energy conversion efficient concentrator photovoltaic (CPV) as a convenient power source to drive the low-electrolysis energy iron deposition without CO$_2$ formation in Li$_2$CO$_3$ [3], as schematically represented in Figure 6.12.

A solar/wind hybrid solar thermal electrochemical production (Hy-STEP) iron electrolysis process is also demonstrated [6]. In lieu of solar electric, electronic energy can be provided by alternative renewables, such as wind. As shown on the right side of Figure 6.12, in this Hy-STEP example the electronic energy is driven by a wind turbine and concentrated sunlight is only used to provide heat to decrease the energy required for iron splitting. In this process, sunlight is concentrated to provide effective heating but is not split into separate spectral regions as in our alternative implementation. Hy-STEP iron production is measured with a 31.5″ × 44.5″

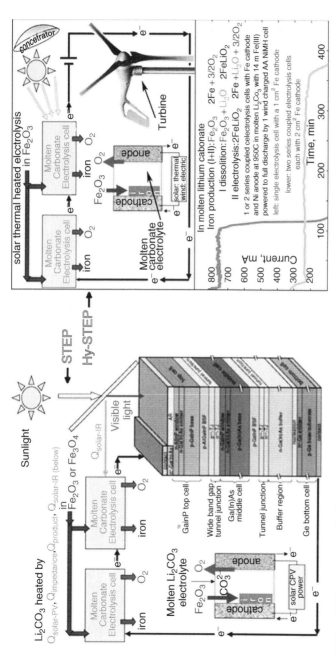

Figure 6.12 STEP and (wind) Hy-STEP iron. *Left:* STEP iron production in which two molten carbonate electrolysis in series are driven by a concentrator photovoltaic (CPV). The 2.7 V maximum power of the CPV can drive either two 1.35 V iron electrolyses at 800°C (schematically represented) or three 0.9 V iron electrolyses at 950°C. At 0.9V, rather than at E° (25°C) = 1.28 V, there is a considerable energy savings, achieved through the application of external heat, including solar thermal, to the system. *Upper right:* The Hy-STEP solar thermal/wind production of CO₂-free iron. Concentrated sunlight heats and wind energy drives electronic transfer into the electrolysis chamber. The required wind-powered electrolysis energy is diminished by the high temperature and the high solubility of iron oxide. *Lower right:* Iron is produced at high current density and low energy at an iron cathode and with a Ni anode in 14 m Fe₂O₃ + 14 m Li₂O dissolved in molten Li₂CO₃. Modified with permission from ref. 6.

177

Fresnel lens (Edmund Optics), which concentrates sunlight to provide temperatures of over 950°C, and a Sunforce-44444 400 W wind turbine provides electronic charge, charging series nickel metal hydride (MH) cells at 1.5V. Each MH cell provides a constant discharge potential of 1.0–1.3 V, each of which are used to drive one or two series-connected iron electrolysis cells as indicated in the right side of Figure 6.12, containing 14 m Fe(III) molten Li_2CO_3 electrolysis cells. Electrolysis current is included in the lower right of Figure 6.12. Iron metal is produced. Steel (iron containing carbon) may be directly formed via the concurrent reduction of CO_2, as will be delineated in an expanded study.

6.3.4   STEP Chlorine and Magnesium Production (Chloride Electrolysis)

The predominant salts in seawater (global average $3.5 \pm 0.4\%$ dissolved salt by mass) are NaCl (0.5 M) and $MgCl_2$ (0.05 M). The electrolysis potential for the industrial chlor-alkali reaction exhibits little variation with temperature, and hence the conventional generation of chlorine by electrolysis, Eq. 6.11, would not benefit from the inclusion of solar heating [3]. However, when confined to anhydrous chloride splitting, as exemplified in the lower portion of Figure 6.4, the calculated potential for the anhydrous electrolysis of chloride salts is endothermic for the electrolyses, which generate a chlorine and metal product. The application of excess heat, as through the STEP process, decreases the energy of electrolysis and can improve the kinetics of charge transfer for the Eq. 6.12 range of chloride splitting processes. The thermodynamic electrolysis potential for the conversion of NaCl to sodium and chlorine decreases, from 3.24V at the 801°C melting point to 2.99 V at 1027°C [3]. Experimentally, at 850°C in molten NaCl, we observe the expected, sustained generation of yellow-green chlorine gas at a platinum anode and of liquid sodium (mp 98°C) at the cathode. Electrolysis of a second chloride salt, $MgCl_2$, is also of particular interest. The magnesium, as well as the chlorine, electrolysis products are significant societal commodities. Magnesium metal, the third most commonly used metal, is generally produced by the reduction of calcium magnesium carbonates by ferrosilicons at high temperature [88], which releases substantial levels of CO_2 contributing to the anthropogenic greenhouse effect. However, traditionally, magnesium has also been produced by the electrolysis of magnesium chloride, using steel cathodes and graphite anodes, and alternative materials have been investigated [89].

Of significance here to the STEP process, is the highly endothermic nature of anhydrous chloride electrolysis, such as for $MgCl_2$ electrolysis, in which solar heat will also decrease the energy (voltage) needed for the electrolysis. The rest potential for electrolysis of magnesium chloride decreases from 3.1 V at room temperature to 2.5 V at the 714°C melting point. As seen in Figure 6.13, the calculated thermodynamic potential for the electrolysis of magnesium chloride continues to decrease with increasing temperature, to ~ 2.3 V at 1000°C. The 3.1 V energy stored in the magnesium and chlorine room temperature products, when formed

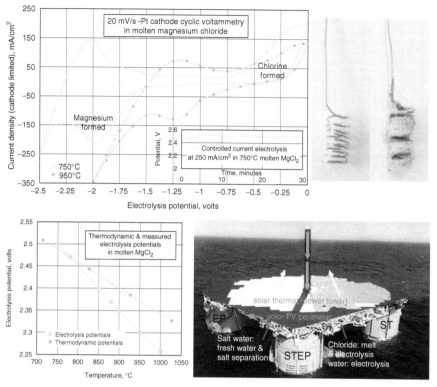

Figure 6.13 *Photographs, upper right*: coiled platinum before (*left*), and after (*right*), $MgCl_2$ electrolysis forming Mg metal on the cathode (shown) and evolving chlorine gas on the anode. *Main figure, upper left*: cathode size-restricted cyclic voltammetry of Pt electrodes in molten $MgCl_2$. *Inset*: The measured full cell potential during constant-current electrolysis at 750°C in molten $MgCl_2$. *Lower left*: Thermodynamic and measured electrolysis potentials in molten $MgCl_2$ as a function of temperature. Electrolysis potentials are calculated from the thermodynamic free energy components of the reactants and products as $E = -\Delta G$ (reaction)/2F. Measured electrolysis potentials are stable values on Pt at 0.250 A/cm^2 cathode [8]. *Lower right:* A schematic representation of a separate (i) solar thermal and (ii) photovoltaic field to drive water purification, hydrogen generation, and the endothermic electrolysis of the separated salts to useful products. Modified with permission from ref. 8.

at 2.3 V, provides an energy savings of 35%, if sufficient heat applied to the process can sustain this lower formation potential. Figure 6.13 also demonstrates the experimental decrease in the $MgCl_2$ electrolysis potential with increasing temperature in the lower left portion. In the top portion of Figure 6.13, the concurrent shift in the cyclic voltammogram is evident, decreasing the potential peak of magnesium formation, with increasing temperature from 750°C to 950°C. Sustained electrolysis and generation of chlorine at the anode and magnesium at the cathode (Fig. 6.13, photo inset) is evident at platinum electrodes. The measured potential during constant-current electrolysis at 750°C in molten $MgCl_2$ at the electrodes is included in Figure 6.13.

In the magnesium chloride electrolysis cell, nickel electrodes yield results similar to platinum and can readily be used to form larger electrodes. The nickel anode sustains extended chlorine evolution without evident deterioration; the nickel cathode may slowly alloy with deposited magnesium. The magnesium product forms both as solid and liquid (Mg mp 649°C). The liquid magnesium is less dense than the electrolyte, floats upward, and eventually needs to be separated and removed to prevent an interelectrode short, or to prevent a reaction with chlorine that is evolved at the anode. In a scaled-up cell configuration (not shown in Fig. 6.13), a larger Ni cathode (200-cm^2 cylindrical nickel sheet; McMaster 9707K35) was employed, sandwiched between two coupled cylindrical Ni sheet anodes (total 200 cm^2 of area across from the cathode) in a 250-ml alumina (Adavalue) crucible, and sustains multi-amp large ampere currents. The potential at constant current is initially stable, but this cell configuration leads to electrical shorts unless liquid magnesium is removed.

One salt source for the STEP generation of magnesium and chlorine from $MgCl_2$ are via chlorides extracted from salt water, with the added advantage of the generation of less saline water as a secondary product. In the absence of effective heat exchanger, CPVs heat up to > 100°C, which decreases cell performance. Heat exchange with the (nonilluminated side of) CPVs can vaporize seawater for desalinization and simultaneously prevent overheating of the CPV. The simple concentrator STEP mode (coupling super-bandgap electronic charge with solar thermal heat) is applicable when sunlight is sufficient to both generate electronic current for electrolysis and sustain the electrolysis temperature. In cases requiring the separation of salts from aqueous solution followed by molten electrolysis of the salts, a single source of concentrated sunlight can be insufficient to both drive water desalinization and heat and drive electrolysis of the molten salts.

Figure 6.13 includes a schematic representation of a Hy-STEP process with separate (i) solar thermal and (ii) photovoltaic field to drive both desalinization and the endothermic CO_2-free electrolysis of the separated salts, or water splitting, to useful products. As illustrated, the separate thermal and electronic sources may each be driven by insolation, or alternatively, can be (i) solar thermal and (ii) (not illustrated) wind-, water-, nuclear-, or geothermal-driven electronic transfer.

6.4 STEP CONSTRAINTS

6.4.1 STEP Limiting Equations

As illustrated on the left side of Figure 6.3, the ideal STEP electrolysis potential incorporates not only the enthalpy needed to heat the reactants to T_{STEP} from $T_{ambient}$ but also the heat recovered via heat exchange of the products with the inflowing reactant. In this derivation it is convenient to describe this combined heat in units of voltage via the conversion factor nF:

$$Q_T \equiv \sum_i H_i(R_i, T_{STEP}) - \sum_i H_i(R_i, T_{ambient}) - \sum_i H_i(C_i, T_{STEP})$$

$$+ \sum_i H_i(C_i, T_{ambient});$$

$$E_Q(V) = -Q_T(J/mol)/nF \qquad (6.26)$$

The energy for the process incorporates E_T, E_Q, and the non-unit activities, via inclusion of Eq. 6.26 into Eq. 6.4, and is termed the STEP potential, E_{STEP}:

$$E_{STEP}(T, a) = [-\Delta G°(T) - Q_T - RT \cdot \ln(\Pi_{i=1\,to\,x}\,a(R_i)^{r_i}/\Pi_{i=1\,to\,y}\,a(P_i)^{p_i})]/nF;$$

$$E°_{STEP}(a = 1) = E_T° + E_Q \qquad (6.27)$$

In a pragmatic electrolysis system, product(s) can be drawn off at activities that are less than that of the reactant(s). This leads to large activity effects in Eq. 6.27 at higher temperature [3–6,8,53–55], as the RT/nF potential slope increases with T (e.g., increasing 3-fold from 0.0592V/n at 25°C to 0.183V/n at 650°C).

The STEP factor, A_{STEP}, is the extent of improvement in carrying out a solar-driven electrolysis process at T_{STEP}, rather than at $T_{ambient}$. For example, when applying the same solar energy to electronically drive the electrochemical splitting of a molecule that requires only two-thirds the electrolysis potential at a higher temperature, then $A_{STEP} = (2/3)^{-1} = 1.5$. In general, the factor is given by:

$$A_{STEP} = E_{STEP}(T_{ambient}, a)/E_{STEP}(T_{STEP}, a); e.g., T_{ambient} = 298K \qquad (6.28)$$

The STEP solar efficiency, η_{STEP}, is constrained by both photovoltaic and electrolysis conversion efficiencies, η_{PV} and $\eta_{electrolysis}$, and the STEP factor. In the operational process, passage of electrolysis current requires an additional, combined (anodic and cathodic) overpotential above the thermodynamic potential, that is, $V_{redox} = (1 + z)E_{redox}$. Mobility and kinetics improve at higher temperature and $\xi(T > T_{ambient}) < \xi(T_{ambient})$ [63,67]. Hence, a lower limit of $\eta_{STEP(V_T)}$ is given by $\eta_{STEP-ideal(E_T)}$. At $T_{ambient}$, $A_{STEP} = 1$, yielding $\eta_{STEP}(T_{ambient}) = \eta_{PV} \cdot \eta_{electrolysis}$. η_{STEP} is additionally limited by entropy and black body constraints on maximum solar energy conversion efficiency. Consideration of a black body source emitted at the sun's surface temperature and collected at ambient earth temperature limits solar conversion to 0.933 when radiative losses are considered [90], which is further limited to $\eta_{PV} < \eta_{limit} = 0.868$ when the entropy limits of perfect energy conversion are included [91]. These constraints on $\eta_{STEP-ideal}$ and the maximum value of solar conversion are imposed to yield the solar chemical conversion efficiency:

$$\eta_{STEP} : \eta_{STEP-ideal}(T, a) = \eta_{PV} \cdot \eta_{electrolysis} \cdot A_{STEP}(T, a)$$

$$\eta_{STEP}(T, a) \cong \eta_{PV} \cdot \eta_{electrolysis}(T_{ambient}, a) \cdot A_{STEP}(T, a);$$

$$(\eta_{STEP} < 0.868) \qquad (6.29)$$

As calculated from Eq. 6.3 and the thermochemical component data [59b] and as presented in Figure 6.5, the electrochemical driving force for a variety of chemicals of widespread use by society, including aluminum, iron, magnesium, and chlorine, significantly decreases with increasing temperature.

6.4.2   Predicted STEP Efficiencies for Solar Splitting of CO_2

The global community is increasingly aware of the climate consequences of elevated greenhouse gases. A solution to rising CO_2 levels is needed, yet CO_2 is a highly stable, noncombustible molecule, and its thermodynamic stability makes its activation energy demanding and challenging. The most challenging stage in converting CO_2 to useful products and fuels is the initial activation of CO_2, for which energy is required. It is obvious that using traditional fossil fuels as the energy source would completely defeat the goal of mitigating greenhouse gases. A preferred route is to recycle and reuse the CO_2 and provide a useful carbon resource. We limit the non-unit activity examples of CO_2 mitigation in Eq. 6.15 to the case in which CO and O_2 are present as electrolysis products, which yields $a_{O2} = 0.5 a_{CO}$, and upon substitution into Eq. 6.27:

$$E_{STEP}(T,a) = E_{STEP}^{\circ}(T) - (RT/2F) \cdot \ln(N); E^{\circ}(25^{\circ}C) = 1.333\,V;$$

$$N = \sqrt{2} a_{CO_2} a_{CO}^{-3/2} \tag{6.30}$$

The example of $E_{STEP}(T, a \neq 1)$ on the left side of Figure 6.14 is derived when $N = 100$ and results in a substantial drop in the energy to split CO_2 because of the discussed influence of RT/2F. Note at high temperature conditions in Figure 6.14 $E_{STEP} < 0$ occurs, denoting the state in which the reactants are spontaneously formed (without an applied potential). This could lead to the direct thermochemical generation of products but imposes substantial experimental challenges. To date, analogous direct water splitting attempts are highly inefficient because of the twin challenges of high-temperature material constraints and the difficulty in product separation to prevent back reaction upon cooling [92]. The STEP process avoids this back reaction through the separation of products, which spontaneously occurs in the electrochemical, rather than chemical, generation of products at separate anode and cathode electrodes.

The differential heat required for CO_2 splitting, E_Q, and the potential at unit activity, E_{STEP}°, are calculated and presented in the right of Figure 6.14. E_Q has also been calculated and is included. E_Q is small (comprising tens of millivolts or less) over the entire temperature range. Hence from Eq. 6.27, E_{STEP}° does not differ significantly from the values presented for E_T° for CO_2 in Figure 6.5. ECO_2split ($25^{\circ}C$) yields $A_{STEP}(T) = 1.333V/E_{STEP}^{\circ}(T)$ with unit activity, and $A_{STEP}(T) = 1.197V/E_{STEP}(T)$ for the $N = 100$ case. Large resultant STEP factors are evident in the left of Figure 6.14. This generates substantial values of solar-to-chemical energy conversion efficiency for the STEP CO_2 splitting to CO and O_2.

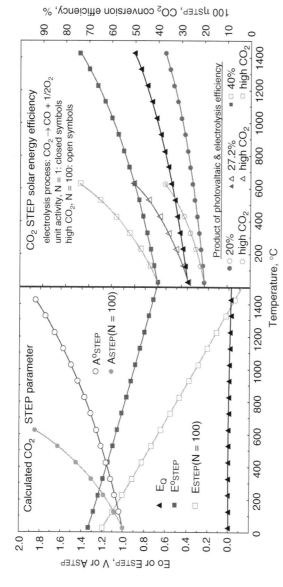

Figure 6.14 *Left:* Calculated STEP parameters for the solar conversion of CO_2. *Right:* Solar-to-chemical conversion efficiencies calculated through Eq. 6.29 for the conversion of CO_2 to CO and O_2. In the case in which the product of the photovoltaic and electrolysis efficiency is 27.2% ($\eta_{PV} \eta_{electrolysis} = 0.272$), the STEP conversion efficiency at unit activity is 35%, at the 650°C temperature consistent with molten carbonate electrolysis, rising to 40% at the temperature consistent with solid oxide electrolysis (1000°C). Non-unit activity calculations presented are for the case of $\sqrt{2} a_{CO_2} a_{CO}^{-3/2=100}$. A solar conversion efficiency of 50% is seen at 650°C when N = 100 (the case of a cell with 1 bar of CO_2 and ~ 58 mbar of CO). Modified with permission from ref. 3.

A STEP process operating in the $\eta_{PV} \cdot \eta_{electrolysis}$ range of 0.20–0.40 includes the range of contemporary 25–45% efficient CPVs [69] and electrolysis efficiency range of 80–90%. From these, the CO$_2$ solar splitting efficiencies are derived from Eqs. 6.29 and 6.30 and are summarized on the right side of Figure 6.14. The small values of E$_{STEP}$(T) at higher T generate large STEP factors and result in high solar-to-chemical energy conversion efficiencies for the splitting of CO$_2$ to CO and O$_2$. As one intermediate example from Eq. 6.30, we take the case of an electrolysis efficiency of 80% and a 34% efficient PV ($\eta_{PV} \cdot \eta_{electrolysis} = 0.272$). This will drive STEP solar CO$_2$ splitting at molten carbonate temperatures (650°C) at a solar conversion efficiency of 35% in the unit activity case, and at 50% when N = 100 (the case of a cell with 1 bar of CO$_2$ and $\sim$ 58 mbar of CO).

6.4.3   Scalability of STEP Processes

STEP can be used to remove and convert CO$_2$. As with water splitting, the electrolysis potential required for CO$_2$ splitting falls rapidly with increasing temperature (Fig. 6.4), and we have shown here (Fig. 6.5) that a PV, converting solar to electronic energy at 37% efficiency and 2.7 V, may be used to drive three CO$_2$ splitting lithium carbonate electrolysis cells, each operating at 0.9V and each generating a 2-electron CO product. The energy of the CO product is 1.3 V (Eq. 6.1), even though generated by electrolysis at only 0.9V because of the synergistic use of solar thermal energy. As seen in Figure 6.8, at lower temperature (750°C rather than 950°C), carbon, rather than CO, is the preferred product, and this 4-electron reduction approaches 100% faradaic efficiency.

The CO$_2$ STEP process consists of solar-driven and solar thermal-assisted CO$_2$ electrolysis. Industrial environments provide opportunities to further enhance efficiencies; for example, fossil-fueled burner exhaust provides a source of relatively concentrated, hot CO$_2$. The product carbon may be stored or used, and the higher-temperature product, carbon monoxide, can be used to form a myriad of industrially relevant products including conversion to hydrocarbon fuels with hydrogen (which is generated by STEP water splitting in Section 6.3.1), such as smaller alkanes, dimethyl ether, or the Fischer-Tropsch-generated middle-distillate-range fuels of C11–C18 hydrocarbons including synthetic jet, kerosene, and diesel fuels [93]. Both STEP and Hy-STEP represent new solar energy conversion processes to produce energetic molecules. Individual components used in the process are rapidly maturing technologies including wind electric [94], molten carbonate fuel cells [67], and solar thermal technologies [95–100].

It is of interest whether material resources are sufficient to expand the process to substantially impact (decrease) atmospheric levels of CO$_2$. The buildup of atmospheric CO$_2$ levels from 280 to 392 ppm occurring since the Industrial Revolution comprises an increase of 1.9×10^{16} mole (8.2×10^{11} metric tons) of CO$_2$ [101] and will take a comparable effort to remove. It would be preferable if this effort resulted in useable, rather than sequestered, resources. We calculate below a scaled-up STEP capture process that can remove and convert all excess atmospheric CO$_2$ to carbon.

In STEP, 6 kWh m^{-2} of sunlight per day, at 500 suns on 1 m^2 of 38% efficient CPV, will generate 420 kAh at 2.7 V to drive three series-connected molten carbonate electrolysis cells to CO or two series-connected series connected molten carbonate electrolysis cells to form solid carbon. This will capture 7.8×10^3 moles of CO_2 day-1 to form solid carbon (based on 420 kAh · 2 series cells/4 Faraday mol^{-1} CO_2). The CO_2 consumed per day is threefold higher to form the carbon monoxide product (based on 3 series cells and 2 F mol^{-1} CO_2) in lieu of solid carbon. The material resources to decrease atmospheric CO_2 concentrations with STEP carbon capture appear to be reasonable. From the daily conversion rate of 7.8×10^3 moles of CO_2 per m^2 of CPV, the capture process, scaled to 700 km^2 of CPV operating for 10 years, can remove and convert all the increase of 1.9×10^{16} mole of atmospheric CO_2 to solid carbon. A larger current density at the electrolysis electrodes will increase the required voltage and would increase the required area of CPVs. While the STEP product (chemicals rather than electricity) is different from contemporary concentrated solar power (CSP) systems, components including a tracker for effective solar concentration are similar (although an electrochemical reactor replaces the mechanical turbine). A variety of CSP installations, which include molten salt heat storage, are being commercialized, and costs are decreasing. STEP provides higher solar energy conversion efficiencies than CSP, and secondary losses can be lower (for example, there are no grid-related transmission losses). Contemporary concentrators, such as based on plastic Fresnel or flat mirror technologies, are relatively inexpensive but may become a growing fraction of cost as concentration increases [100]. A greater degree of solar concentration, for example, 2000 suns rather than 500 suns, will proportionally decrease the quantity of required CPV to 175 km^2, while the concentrator area will remain the same at 350,000 km^2, equivalent to 4% of the area of the Sahara desert (which averages ~6 kWh m^{-2} of sunlight per day), to remove anthropogenic carbon dioxide in 10 years.

A related resource question is whether there is sufficient lithium carbonate, as an electrolyte of choice for the STEP carbon capture process, to decrease atmospheric levels of carbon dioxide; 700 km^2 of CPV plant will generate 5×10^{13} A of electrolysis current and require ~ 2 million metric tons of lithium carbonate, as calculated from a 2 kg/l density of lithium carbonate, and assuming that improved, rather than flat, morphology electrodes will operate at 5 A/cm^2 (1000 km^2) in a cell of 1 mm thickness. Thicker, or lower-current density, cells will require proportionally more lithium carbonate. Fifty, rather than ten, years to return the atmosphere to preindustrial CO_2 levels will require proportionally less lithium carbonate. These values are viable within the current production of lithium carbonate. Lithium carbonate availability as a global resource has been under recent scrutiny to meet the growing lithium battery market. It has been estimated that the current global annual production of 0.13 million tons of (LCE) will increase to 0.24 million tons by 2015 [102]. Potassium carbonate is substantially more available but, as noted in the main portion of chapter, can require higher carbon capture electrolysis potentials than lithium carbonate.

6.5 CONCLUSIONS

To ameliorate the consequences of rising atmospheric CO_2 levels and their effect on global climate change, there is a drive to replace conventional fossil fuel-driven electrical production by renewable energy-driven electrical production. In addition to the replacement of the fossil fuel economy by a renewable electrical economy, we suggest that a renewable chemical economy is also warranted. Solar energy can be efficiently used, as demonstrated with the STEP process, to directly and efficiently form the chemicals needed by society without CO_2 emission.

The portfolio of STEP processes continues to grow and develop [103-110], iron, a basic commodity, currently accounts for the release of one-quarter of worldwide CO_2 emissions by industry, which may be eliminated by replacement with the STEP iron process. The unexpected solubility of iron oxides in lithium carbonate electrolytes, coupled with facile charge transfer and a sharp decrease in iron electrolysis potentials with increasing temperature, provides a new route for iron production. Iron is formed without an extensive release of CO_2 in a process compatible with the predominant naturally occurring iron oxide ores, hematite, Fe_2O_3, and magnetite, Fe_3O_4. STEP can also be used in direct carbon capture and the efficient solar generation of hydrogen and other fuels.

In addition to the removal of CO_2, the STEP process is shown to be consistent with the efficient solar generation of a variety of metals, as well as chlorine via endothermic electrolyses. Commodity production and fuel consumption processes are responsible for the majority of anthropogenic CO_2 release, and their replacement by STEP processes provides a path to end the root cause of anthropogenic global warming, as a transition beyond the fossil fuel, electrical, or hydrogen economy to a renewable chemical economy based on the direct formulation of the materials needed by society. An expanded understanding of electrocatalysis and materials will advance the efficient electrolysis of STEP's growing portfolio of energetic products.

ACKNOWLEDGMENTS

The author is grateful to Baouhui Wang and Hongun Wu for excellent experimental contributions to refs. 4 and 5, and to NSF grant 1230732 for partial support of this work.

REFERENCES

1. L. Vayssieres (2009). *On Solar Hydrogen & Nanotechnology*, John Wiley and Sons, US.
2. K. Rajeshwar, S. Licht, R. McConnell (2008). *The Solar Generation of Hydrogen: Towards a Renewable Energy Future*, Springer, New York, USA.
3. S. Licht (2009). *J. Phys. Chem., C*, 113: 16283.
4. S. Licht, B. Wang, S. Ghosh, H. Ayub, D. Jiang, J. Ganely (2010). *J. Phys. Chem. Lett.*, 1: 2363.

5. S. Licht, B. Wang (2010). *Chem. Comm.*, 46: 7004.

6. S. Licht, H. Wu, Z. Zhang, H. Ayub (2011). *Chem. Comm.*, 47: 3081.

7. S. Licht, O. Chityat, H. Bergmann, A. Dick, S. Ghosh, H. Ayub (2010). *Int. J. Hyd. Energy*, 35: 10867.

8. S. Licht, B. Wang, H. Wu (2011). *J. Phys. Chem. C*, 115: 11803.

9. G. Ohla, P. Surya, S. Licht, N. Jackson (2009). *Reversing Global Warming: Chemical Recycling and Utilization of* CO_2. Report of the National Science Foundation sponsored 7–2008 Workshop, available at: http://www.usc.edu/dept/chemistry/loker/ReversingGlobalWarming.pdf

10. C. Graves, S. Ebbsen, M. Mogensen, K. Lackner (2011). *Renewable Sustainable Energy Rev.*, 15: 1.

11. J. Barber (2009). *Chem. Soc. Rev.*, 38: 185.

12. A. Stamatiou, P. G. Loutzenhiser, A. Steinfeld (2010). *Energy Fuels*, 24: 2716.

13. S. Abanades, M. Chambon (2010). *Energy Fuels*, 24: 6677.

14. L. J. Venstrom, J. H. Davidson (2010). *J. Solar Energy Eng. Chem.*, 133: 011017–1.

15. W. Chueh, S. Haile (2010). *Phil. Trans. Roy. Soc. A*, 368: 3269.

16. J. Miller, M. Allendorf, R. Diver, L. Evans, N. Siegel, J. Stueker (2008). *J. Mat. Sci.*, 43: 4714.

17. S. Licht (1987). *Nature*, 330: 148.

18. S. Licht, D. Peramunage (1990). *Nature*, 345: 330.

19. B. Oregan, M. Gratzel (1991). *M. Nature*, 353: 737.

20. S. Licht (1998). *J. Phys. Chem.*, 90: 1096.

21. S. Licht (2002). *Semiconductor Electrodes and Photoelectrochemistry*, Wiley-VCH, Weinheim, Germany.

22. S. Licht, G. Hodes, R. Tenne, J. Manassen (1987). *Nature*, 326: 863.

23. S. Licht, B. Wang, T. Soga, M. Umeno (1999). *Appl. Phys. Lett.*, 74: 4055.

24. S. Yan, L. Wan, Z. Li, Z. Zou (2011). *Chem. Comm.*, 47: 5632.

25. H. Zhou, T. Fan, D. Zhang (2011). *Chem. Cat. Chem.*, 3: 513.

26. R. Huchinson, E. Holland, B. Carpenter (2011). *Nature Chem.*, 3: 301.

27. E. E. Barton, D. M. Rampulla, A. B. Bocarsly (2008). *J. Am. Chem. Soc.*, 130: 6342.

28. S. Kaneco, H. Katsumata, T. Suzuki, K. Ohta (2006). Chem. Eng. *J*, 92: 363.

29. P. Pan, Y. Chen (2007). *Catal, Comm.*, 8: 1546.

30. A. B. Murphy (2008). *Solar Energy Mat.*, 116: 227.

31. A. Currao (2007). *Chimia*, 61: 815.

32. (a) S. R. Narayanan, B. Haines, J. Soler, T. I. Valdez (2011). *J. Electrochem. Soc.*, 158: A167; (b) C. Delacourt, J. Newman (2010). *J. Electrochem. Soc.*, 157: B1911.

33. E. Dufek, T. Lister, M. McIlwain (2011). *J. Appl. Electrochem.*, 41: 623.

34. M. Gangeri, S. Perathoner, S. Caudo, G. Centi, J. Amadou, D. Begin, C. Pham-Huu, M. Ledoux, J. Tessonnier, D. Su, R. Schlögl (2009). *Catal. Today*, 143: 57.

35. B. Innocent, D. Liaigre, D. Pasquier, F. Ropital, J. Leger, K. Kokoh (2009). *J. Appl. Electrochem.*, 39: 227.

36. A. Wang, W. Liu, S. Cheng, D. Xing, J. Zhou, B. Logan (2009). *Intl. J. Hydrogen Energy*, 39: 3653.

37. N. Dong-fang, X. Cheng-tian, L. Yi-wen, Z. Li, L. Jiz-xing (2009). *Chem. Res. Chinese U.*, 34: 708.

38. D. Chu, G. Qin, X. Yuan, M. Xu, P. Zheng, J. Lu (2008). *ChemSusChem*, 1: 205.

39. J. Yano, T. Morita, K. Shimano, Y. Nanmi, S. Yamsaki (2007). *J. Sol. State Electrochem.*, 11: 554.

40. Y. Hori, H. Konishi, T. Futamura, A. Murata, O. Koga, H. Sakuri, K. Oguma (2005). Electrochim. *Acta*, 50: 5354.

41. K. Ogura, H. Yano, T. Tanaka (2004). *Catal. Today*, 98: 414.

42. (a)H. Chandler, F. Pollara (1966). *AICHE Chem. Eng. Prog. Ser.: Aerospace Life Support*, 62: 38; (b) L. Elikan, D. Archer, R. Zahradnik, *ibid*, 28.

43. (a)M. Stancati, J. Niehoff, W. Wells, R, Ash (1979). *AIAA*, 79–0906: 262; (b) R. Richter, *ibid*, 1981, 82–2275: 1.

44. (a)J. Mizusaki, H. Tagawa, Y. Miyaki, S. Yamauchi, K. Fueki, I. Koshiro (1992). *Solid State Ionics*, 126: 53; (b) G. Tao, K. Sridhar, C. Chan (2004). *ibid*, 175, 615; (c) *ibid*, 621; R. Green, C. Liu, S. Adler,; (d) *ibid* 2008, 179, 647.

45. C. Meyers, N. Sullivan, H. Zhu, R. Kee (2011). *J. Electrochem. Soc.*, 158: B117.

46. P. Kim-Lohsoontorn, N. Laosiripojana, J. Bae (2011). Current Appl. *Physics*, 11: 5223.

47. S. Ebbesen, C. Graves, A. Hausch, S. Jensen, M. Mogensen (2010). *J. Electrochem. Soc.*, 157: B1419.

48. S. Jensen, X. Sun, S. Ebbesen, R. Knibbe, M. Mogensen (2010). *Intl. J. Hydrogen Energy*, 35: 9544.

49. Q. Fu, C. Mabilat, M. Zahid, A. Brisse, L. Gautier (2010). *Energy Environ. Sci.*, 3: 1382.

50. C. M. Stoots, J. E. O'Brien, K. G. Condie, J. Hartvigsen (2010). *Intl. J. Hydrogen Energy*, 35: 4861.

51. Q. Fu, C. Mabilat, M. Zahid, A. Brisse, L. Gautier (2010). *Energy Environ. Sci.*, 3: 1382.

52. D. Lueck, W. Buttner,J. Surma (2002). *Fluid System Technologies*, at: http://rtreport .ksc.nasa.gov/techreports/2002report/600%20Fluid%20Systems/609.html.

53. S. Licht (2002). *Electrochem. Comm.*, 4: 789.

54. S. Licht (2003). *J. Phys. Chem. B*, 107: 4253.

55. (a)S. Licht, L. Halperin, M. Kalina, M. Zidman, N. Halperin (2003). *Chem. Comm.*, 3006; (b) S. Licht (2005). *Chem. Comm.*, 4623.

56. A. Fujishima, K. Honda (1972). *Nature*, 238: 37.

57. Z. Zou, Y. Ye, K. Sayama, H. Arakawa (2001). *Nature*, 414: 625.

58. (a) S. Licht, B. Wang, S. Mukerji, T. Soga, M. Umento, H. Tributsh (2000). *J. Phys. Chem. B*, 104: 8920; (b) S. Licht (2001). *J. Phys. Chem. B*, 105: 6281.

59. (a) A. J. deBethune, T. S. Licht (1959). *J. Electrochem. Soc.*, 106: 616; (b) M. W. Chase (1998). *J. Phys. Chem. Ref. Data*, 9: 1; data available at: http://webbook.nist.gov/ chemistry/form-ser.html

60. W. E. Wentworth, E. Chen (1976). *Solar Energy*, 18: 205.

61. J. O'M. Bockris (1980). *Energy Options*, Halsted Press, NY.

62. T. S. Light, S. Licht, A. C. Bevilacqua (2005). *Electrochem & Sol State Lett.*, 8: E16.

63. T. S. Light, S. Licht (1987). *Anal. Chem.*, 59: 2327.

64. S. Licht (1985). *Anal. Chem.*, 57: 514.

65. S. Licht, K. Longo, D. Peramunage, F. Forouzan (1991). *J. Electroanal. Chem.*, 318: 119.

66. C. Elschenbroich, A. Salzer (1992). *Organometallics*. 2nd Ed., Wiley-VCH, Weinheim, Germany.

67. K. Sunmacher (2007). *Molten Carbonate Fuel Cells*, Wiley-VCH, Weinheim, Germany.

68. J. L. Pellegrino (2000). *Energy & Environmental Profile of the U.S. Chemical Industry*, available at: http://www1.eere.energy.gov/industry/chemicals/tools_profile.html

69. (a) R.R. King, D. C. Law, K. M. Edmonson, C. M. Fetzer, G. S. Kinsey, H. Yoon, R. A. Sherif, N. H. Karam (2007). *Appl. Phys. Lett.*, 90: 183516; (b) M. Green, K. Emery, Y. Hishikawa, W. Warata (2011). *Prog. Photovoltaics*, 19: 84.

70. J. E. Miller, J. E.; Allendorf, M. D.; Diver, R. B.; Evans, L. R.; Siegel, N. P.; Stuecker, J. N. (2008). *J. Mat. Sci.*, 43: 4714.

71. Y. Woolerton, Y., W.; Sheard, S.; Reisner, E.; Pierce, E.; Ragsdale, S. W.; Armstrong, F. A. (2010). *J. Amer. Chem. Soc.*, 132: 2132.

72. E. Benson, C. P. Kubiak, A. J, Sathrum, J. M. N. Smieja (2009). *Chem. Soc. Rev.*, 38: 89.

73. Z. Zhang, Z. Wang (2006). *Principles and Applications of Molten Salt Electrochemistry*, Chemical Industry Press, Beijing, p. 191.

74. T. Kojima, Y. Miyazaki, K. Nomura, K. Tanimoto, K. Density (2008). *J. Electrochem. Soc.*, 155: F150.

75. V. Kaplan, E. Wachtel, K. Gartsman, Y. Feldman, I. Lubormirsky (2010). *J. Electrochem. Soc.*, 157: B552.

76. L. Andrieux, G. Weiss (1944). *Comptes Rendu*, 217: 615.

77. G. M. Haarberg, E. Kvalheim, S. Rolseth, T. Murakami, S. Pietrzyk, S. Wang (2007). *ECS Transactions*, 3: 341.

78. S. Wang, G. M. Haarberg, E. Kvalheim, E. (2008). *J. Iron and Res. Int.*, 15: 48.

79. G. M. Li, D. H. Wang, Z. Chen (2009). *J. Mat. Sci. Tech.*, 25: 767.

80. B. Y. Yuan, O. E. Kongstein, G. M. Haarberg (2009). *J. Electrochem. Soc.*, 156: D64.

81. W. Palmaer, J. A. Brinell (1913). *Chem. Metall. Eng.*, 11: 197.

82. F. A. Eustis (1922). *Chem. Metall. Eng.*, 27: 684.

83. E. Mostad, S. Rolseth, S. Thonstad (2008). *J. Hydrometallurgy*, 90: 213.

84. L. Qingeng, F. Borum, I. Petrushina, N. J. Bjerrum (1999). *J. Electrochem, Soc.*, 146: 2449.

85. R. Collongues, G. Chaudron (1950). *Compt. Rend.*, 124: 143.

86. A. Wijayasinghe, B. Bergman, C. Lagergren (2003). *J. Electrochem. Soc.*, 150: A558.

87. A. Lykasov, M. Pavlovskaya (2003). *Inorg. Mat.*, 39: 1088.

88. H. Q. Li, S. S. Xie (2005). *J. Rare Earths*, 23: 606.

89. G. Demirci, I. Karakaya (2008). *J. Alloys & Compounds*, 465: 255.

90. C. S. Solanki, G. Beaucarne (2006). *Advanced Solar Cell Concepts*, AER India-2006, 256.

91. A. Luque, A. Marti (2003). *Handbook of Photovoltaic Sci. & Eng.*, (Eds. A. Luque, S. Haegedus), Wiley-VCH, Weinheim, Germany, 113.

92. A. Kogan (1998). *Intl. J. Hydrogen Energy*, 23: 89.

93. A. Andrews, J. Logan (2008). Fischer-Tropsch Fuels from Coal, Natural Gas, and Biomass: Background and Policy. *Congressional Research Service Report for Congress*, RL34133, (March 27, 2008); available at: http://assets.opencrs.com/rpts/RL34133_20080327.pdf.

94. E. Barbier (2010) *Nature*, 464: 832.

95. Power tower solar technologies are described at: brightsourceenergy.com; ausra.com; esolar.com; bengoasolar.com/corp/web/en/our_projects/solana/.

96. Siemens to build molten salt solar thermal test facility in Portugal, siemens.com, 2011, at: http://www.siemens.com/press/pool/de/pressemitteilungen/2011/renewable_energy/ERE201102037e.pdf.

97. Solarreserve.com, 2011, at: http://www.solarreserve.com/projects.html.

98. Parabolic solar concentrator technologies are described at: stirlingenergy.com.

99. Fresnel solar concentrator technologies are described at: amonix.com, energy innovations.com/sunflower.

100. R. Pitz-Paal (2007). High Temperature Solar Concentrators, in *Solar Energy Conversion and Photoenergy Systems*, Eds. Galvez, J. B.; Rodriguez, S. M., *EOLSS* Publishers, Oxford, UK.

101. P. Tans (2009). *Oceanography*, 22: 26.

102. Tahil, W. "The Trouble with Lithium 2; Under the Microscope", 2008, 54 pages, Meridan International Research, Martainsville, France.

103. S. Licht (2011). *Advanced Materials*, 47, 5592.

104. S. Licht, H. Wu, Z. Zhang, H. Ayub (2011). *Chem. Comm.*, 47; 3081.

105. S. Licht, H. Wu (2011). *J. Phys. Chem., C*, 115: 25138.

106. S. Licht, H. Wu, C. Hettige, B. Wang, J. Lau, J. Asercion, J. Stuart (2012). *Chem. Comm.*, 48: 6019.

107. B. Wang, H. Wu, G. Zhang, S. Licht (2012). *ChemSusChem*, 5: 2000.

108. S. Licht, B. Cui, B. Wang (2013). *J. CO2 Utilization*, 2: 58.

109. S. Licht, B. Cui, J. Stuart, B. Wang, J. Lau (2013). *Energy & Environmental Sci.*, 6: 3646.

110. B. Cui, S. Licht (2013). *Green Chem.*, 15: 881.

CHAPTER 7

Electrocatalytic Reduction of CO_2 in Methanol Medium

M. MURUGANANTHAN, S. KANECO*, H. KATSUMATA, T. SUZUKI, and M. KUMARAVEL

7.1   INTRODUCTION

The atmosphere can be divided vertically into different layers such as troposphere, stratosphere, mesosphere, and thermosphere. The troposphere extends above the Earth to a distance of 10–16 km. The atmosphere protects the living things on Earth from the sun's-cancer causing UV radiation, and it also moderates the earth's climate. Otherwise, the Earth would experience extremes of hot and cold. The percentage of CO_2 in the atmosphere is extremely small, but has been found to play an important role in maintaining the Earth's heat balance. The biological processes of CO_2 production and consumption by respiration and photosynthesis is a cyclic process. The green environment has been gradually destroyed because of the progressive increase in industrial activity. There is a growing concern now that the Earth's protective ozone layer is being destroyed by certain gases and chemicals

*Indicates the corresponding author.

Green Carbon Dioxide: Advances in CO₂ Utilization, First Edition.
Edited by Gabriele Centi and Siglinda Perathoner.
© 2014 John Wiley & Sons, Inc. Published 2014 by John Wiley & Sons, Inc.

that include CO_2, which are produced by industrial activities. It is obvious that a huge quantity of coal has been burned for the fuel needs of industries. This could be a major source of CO_2 emission. Also, our continued dependence on fossil fuels for energy is resulting in the emission of CO_2 in large quantities to the atmosphere. It is noteworthy that CO_2 is the ultimate by-product of all chemical processes in the oxidation of carbon compounds.

These practices have resulted in the greenhouse effect in the Earth's atmosphere which is suspected to be causing global warming. The contribution of CO_2 is considerably higher in the greenhouse effect. CO_2 is a very stable molecule, as its standard free energy $(\Delta G°)$ of formation is -394.359 kJ mol^{-1}. It is the most oxidized form of carbon; therefore the only possible chemical transformation is to reduce it. The structure of CO_2 changes from linear to bent shape by transfer of an electron, which results in an irreversible reduction product. Furthermore, hydrogen has emerged as one of the best energy sources for future energy systems such as fuel-cell vehicles and fuel-cell power generation. The production of hydrogen from hydrocarbons that are abundant in all kinds of fossil fuels, biomass, and organic waste would be an alternate energy source. Along with the gaseous products, during the gasification of hydrocarbon with water CO and CO_2 are also expected to form and pollute the atmosphere. CO is one of the raw materials in the hydroformylation reaction, which plays an important role in the chemical industry. Formic acid has been used industrially in various processes including textile dyeing and finishing, leather tanning, and electroplating [1].

Owing to the increase of atmospheric CO_2, many environmental scientists have suggested that CO_2 itself can be used as a source of carbon for production of petroleum-like products. The CO_2 can be converted into fuel, feedstock for chemical manufacturing, and useful products that are consumed by humans and animals. Accordingly, the carbon energy cycle can be written as follows:

$$CO_2 + Energy \rightarrow Hydrocarbon/Fuel \rightarrow CO_2 + Energy \qquad (7.1)$$

Many utilization techniques are available, including biochemical, chemical, and electrochemical methods. Since 1870, there have been many attempts to reduce CO_2 to useful hydrocarbons by nonbiological methods [2]. Many techniques have been adopted for the conversion of CO_2, such as chemical, thermochemical, radiochemical, photochemical, electrochemical, and biochemical processes. Of late, many researchers have actively studied electrochemical reduction of CO_2 using a spectrum of electrode materials in nonaqueous solvents, because electrochemical conversion of CO_2 in organic solvent is advantageous in many ways as follows. CO_2 can be dissolved in organic solvent in greater volume than in water solvent [3]. The important fact is that a parasite side reaction of H_2 evolution takes place less efficiently in nonaqueous electrolyte [4]. Also, several metals have been found to work well in nonaqueous electrolyte rather than water. It has been reported that low reduced products containing formic acid, CO, and oxalic acid were produced by the electrochemical reduction of CO_2 in acetonitrile, dimethyl sulfoxide, N,N-dimethyl formamide, and propylene carbonate.

One of the protic nonaqueous solvents, methanol is a better choice for CO_2 than the above-mentioned aprotic solvents. The solubility of CO_2 in methanol is approximately five times that in water at ambient temperature and more than eight times that in water at temperatures below 273 K. This is the reason methanol has been predominantly used as a physical absorbent of CO_2 in the Rectisol method at low temperature. Currently about 70 large-scale plants practice the Rectisol process in Japan. Therefore, the direct electrochemical conversion of CO_2 in methanol is an advantageous choice, especially when the process is carried out under energetically efficient conditions.

In general, power consumption is not a consideration in electrochemical reactions and processes. However, a major constraint for the utilization of CO_2 is energy consumption, which can be determined by the number of electrons required to reduce a single molecule of CO_2 to its reduction products. The current efficiency is often called faradaic efficiency. Faradaic efficiency is the current utilized for the reduction to a particular desired product. In general, the hydrogen evolution competes with the CO_2 reduction; thus the suppression of the hydrogen evolution reaction is the major task during the electrochemical reduction process. The supplied energy is dissipated for the hydrogen evolution reaction rather than the CO_2 reduction at higher current densities.

The parameters that can control the energy consumption are high current densities, high faradaic efficiency, low specific electricity consumption, low potential difference, and stable electrode materials. The higher current densities result in lower faradaic efficiency as the parallel undesired reactions like hydrogen evolution are accelerated. The optimization of current densities is thus the most important parameter. When increasing the electrolyte pressure, the concentration of CO_2 can be increased. The mole fraction of CO_2 is 0.34 at 4.0 MPa and 0.94 at 5.8 MPa [5]. In such a highly concentrated CO_2-containing electrolyte, the transport of the CO_2 molecule to the electrode surface is easy; therefore the reduction process is achieved with higher faradaic current efficiency.

7.2 ELECTROCATALYTIC REDUCTION OF CO_2 IN METHANOL MEDIUM

Ever since the first work on electrochemical reduction of CO_2 in a methanol-based electrolyte was done in 1993 [6] many research groups have been actively working on it, varying parameters such as different groups of cathode materials ("sp" and "d" groups), electrolyte salt, temperature, and potential. The electrochemical conversion of CO_2 in methanol medium-based electrolyte salts is summarized in Table 7.1 [7–38]. Obviously, the roles of electrodes and electrolytes containing salt are the most important parameters that influence the selective formation of reduction products. The metal electrode has been classified on the basis of its electrocatalytic behavior, caused by its electronic configuration. The modification has also been done based on its alloying properties, that is, chemically modified electrodes. The modification could be due to a combination of changes in electronic structure and surface crystallographic characteristics including introduction of dislocation and vacancies. This leads to major differences in selective product formation and reaction rates. A metal-hydrogen electrode and the Cu-Re alloy

TABLE 7.1 Electroreduction of CO_2 on a Spectrum of Electrode Materials and Electrolytes Containing Salts in CH_3OH Medium

Ref.	Electrodes	Electrolyte Conditions	Products Formed with Priority	Findings
7	P-InP	Cu particle suspended CH_3OH	In presence of Cu: HCOOH and CO In absence of Cu: CH_4 and C_2H_4	Current Efficiency (CE) of hydrocarbon was better at lower temperature (273 K), and the maximum was observed at around −2.5 V. The addition of Cu particles shifts the onset potential to a more positive value than that obtained without Cu particles.
8	Zn particle mixed/pressed with CuO, Cu_2O disc plate electrode	KOH/CH_3OH	Without CuO and Cu_2O – HCOOH, CO; with CuO and Cu_2O – Hydrocarbon	Copper oxide effective for hydrocarbon. Faradaic efficiency (FE) for C_2H_4 was better than CH_4 at CuO/zinc, whereas FE for CH_4 was better than C_2H_4 at Cu_2O.
9	Cu	CsOH / cold CH_3OH	CO, HCOOH, CH_4, C_2H_4 The product distribution is due to the hydrophobic nature of the electrode surface created by Cs ions	FE for CO formation was a maximum of 84% at −3.5 V. H_2 evolution was suppressed to <2% at more negative potential than −3.0 V.
10	Cu deposited Ni	$KHCO_3/H_2O$ and CH_3OH (8:2)	C_2H_4, CH_4, CO, HCOOH	Cu-modified Ni electrode found to be effective for the formation of C_2H_4.
11	Cu	$NaOH/CH_3OH$	CH_4, C_2H_4, CO, HCOOH CE of CH_4 was 80.6% at −2.5 V to −5.0 V.	Effect of Na^+ was studied, and CH_4 formation is so effective because of the hydrophilic environment near the electrode surface.
12	Cu	Cold $CH_3OH/LiCl$, under high pressure	CO, CH_4, HCOOH, C_2H_4 The maximum CE for CH_4 was 20% at −3.0 V, and the FE increases gradually as the temperature goes down.	The selective formation of reduction products is explained in terms of the hydrophobic nature of the supporting salt used. CO formation is favored in a hydrophobic environment, and CH_4 formation is favored in a less hydrophobic environment. Li cation provides a less hydrophobic atmosphere.

13	P-GaAs, P-InP	CH_3OH medium Na_2CO_3, Na_2SO_4, Na_3PO_4, $LiClO_4$	FE for CO: 41.5% on P-InP cathode at -2.4 V HCOOH: 12–15% on both P-GaAs and p-InP cathode	Onset potential of P-GaAs -1.2 V and P-InP -1.8 V; P-InP was effective in achieving maximum CE for CO formation. The total efficiency of CO_2 reduction was always higher at all working potentials comparing H_2 evolution.
14	Pb, Zn foil	Copper particle suspended CH_3OH	With Cu particle:CH_4, C_2H_4 Without Cu particle: no hydrocarbon (HCOOH, CO)	FE of hydrocarbon increases with increase in the quantity of the Cu particles, whereas formation efficiency of HCOOH and CO decreased. FE of CH_4 was 6% at Pb and 12% at Zn cathode
15	P-InP deposition of Pb, Ag, Au, Pd, Cu, Ni	$LiOH$-CH_3OH,	CO, HCOOH, CH_4,C_2H_4 The maximum CE (80.4%) for CO formation was achieved using Ag deposited P-InP. As Pb deposition increases, FE of HCOOH increases up to 29.9%.	The selective formation of the reduction products and its efficiency are dependent on the catalytic property of the metals deposited over the P-InP electrode. CE of CO formation is considerably increased with Ag- and Au-deposited P-InP rather than intrinsic electrode.
16	Cu	KOH and $RbOH/CH_3OH$ medium	CH_4, C_2H_2, CO, HCOOH; C_2H_2 $-$ 37.5% at 4.0 V	H_2 evolution minimized <3.3%; FE ratio between C_2H_2 and CH_4 increases in $RbOH/CH_3OH$ medium.
17	Cu	CH_3COONa, NaCl, NaBr, NaI, NaSCN, $NaClO_4/CH_3OH$	CH_4, C_2H_2, CO, HCOOH CH_4 formation was predominant in all Na-containing salt electrolytes.	FE of CH_4–70.5% in $NaClO_4$ at -3.0 V; H_2 evolution suppressed to <18% in the electrolyte salts except CH_3 COONa and especially <1% in NaSCN

(continued)

TABLE 7.1 (*Continued*)

Ref.	Electrodes	Electrolyte Conditions	Products Formed with Priority	Findings
18	Cu	LiCl, LiBr, LiI, LiClO$_4$, CH$_3$COOLi/ CH$_3$OH	Ambient pressure: CH$_4$, CO, HCOOH, C$_2$H$_4$ High pressure and low temperature: CO, CH$_4$, HCOOH, and C$_2$H$_4$	FE decreases with increasing Temperature H$_2$ evolution was suppressed to <12% and especially <1% in LiI salt-containing electrolyte.
19	Cu	CH$_3$ COOCs, CsCl, CsBr, CsI, CsSCN/CH$_3$OH T-243 K	CH$_4$, C$_2$H$_4$, C$_2$H$_6$, CO, HCOOH	FE of C$_2$H$_4$ greater than CH$_4$, CO$_2$ reduction is dependent on anionic species. The order of selectivity formation of C$_2$H$_4$ over CH$_4$ is Br > I > Cl > SCN
20	Cu	CH$_3$COOK, KBr, KI, KSCN/ CH$_3$OH T-243 K	CH$_4$, C$_2$H$_4$, C$_2$H$_6$, CO and HCOOH, The maximum FE was 27% for CH$_4$ at −3.0 V in CH$_3$COOK and 19.9% for C$_2$H$_2$ at −3.0 V in KI.	H$_2$ evolution <8.1%; total CE of hydrocarbon was relatively lower (2%).
21	Cu	LiOH/CH$_3$OH	CH$_4$,C$_2$H$_4$, CO, HCOOH	CE of CH$_4$ was 63% at −4.0V and FE of CH$_4$ and C$_2$H$_4$ was 78% at −4.0 V. H$_2$ evolution depressed to 2% at relatively negative potentials.
22	Cu	CsOH/CH$_3$OH	CH$_4$, C$_2$H$_4$, C$_2$H$_6$, CO and HCOOH; the maximum FE for CH$_4$ was 8.3% at −4.0 V and for C$_2$H$_4$ was 32.3% at −3.5 V.	The onset potential of electrolyte containing CsOH is more negative than other salts such as CsCl, CsBr, and CsI.
23	Ti, hydrogen-storing Ti(Ti + H)	KOH/CH$_3$OH	HCOOH, CO	(Ti + H) electrode favored the formation of CO at relatively positive potential (−2.2 V), and the efficiencies of formic acid and CO formation on both Ti and (Ti + H) electrodes generally decreased with increasing temperature.

#	Electrode	Electrolyte	Products	Remarks
24	Cu wire	Tetraethyl ammonium perchlorate (TEAP) and benzalkonium chloride/CH_3OH T- 243 K	CH_4, C_2H_4, CO, HCOOH	The partial current density for CO_2 reduction at Cu wire is better than the Cu sheet used in earlier studies.
25	Ag	KOH/CH_3OH	CO, HCOOH; FE for CO is larger than for HCOOH.	The CE of CO increased at relatively negative potential with decreasing temperature.
26	Au	KOH/CH_3OH	Products formed are CO and HCOOH. FE for CO increases and H_2 decreases as the temperature goes down.	From Tafel plot, high mass transfer of CO_2 to the electrode has been confirmed even at lower temperature. H_2 evolution was found to be suppressed at temperatures below 273 K
27	Pb wire	KOH/CH_3OH	HCOOH, CO, CH_4	The formation of HCOOH was effective in the potential range of -1.8 V to 2.5 V. The partial current density of reduction products was 20 times higher than the H_2 evolution reaction. From Tafel plot, it is concluded that the formation of HCOOH and CO are independent from each other and reduction product formation is not limited by mass transfer.
28	Polyaniline-coated Pt	$CH_3OH/LiClO_4$	CH_3COOH and HCOOH with a maximum FE of 78% and 12%, respectively.	The electrolysis potential has been decreased to -0.4 V with the polyaniline-modified electrode.

(continued)

TABLE 7.1 (*Continued*)

Ref.	Electrodes	Electrolyte Conditions	Products Formed with Priority	Findings
29	Polyaniline-coated Pt	Electrochemical impedance spectroscopic (EIS) study in CH_3OH $LiClO_4$, a small amount of H_2O and H_2SO_4	HCHO, CH_3COOH, HCOOH. The formation of HCOOH and CH_3COOH are competitive.	Film thickness was optimized at 0.6 μm. It was observed that there is no direct electron transfer to CO_2 and only formation of H_{ad} was the initial step. Transfer coefficient and exchange current density were calculated.
30	Polypyrole (Ppy)-coated Pt	Under high pressure, CH_3OH $LiClO_4$	Products formed: HCHO, HCOOH, CH_3COOH with maximum CE of 1.8%, 40.5%, and 62.2%, respectively.	An increase of applied pressure results in the higher CE of reductant product formations with the polypyrrole-modified electrode.
31	Polyaniline (pAn)	$LiClO_4 - CH_3OH$	Products formed: HCHO, HCOOH, CH_3COOH with maximum CE of 26.5%, 13.1%, and 37%, respectively.	pAn can be used as cathode in electroreduction of CO_2 with higher CE and lower overpotential.
32	W, Ti, Pt, Sn, Pb, Ag, Zn, Pd, Cu, Ni	Tetrabutyl ammonium perchlorate (TBAP)	HCOOH production is suppressed and CO formation is enhanced due to the supporting electrolyte effect.	The metals, which failed to show better efficiency in aqueous medium, have been shown to enhance production of hydrocarbon along with H_2 evolution.
33	Cu wire	Tetrabutyl ammonium tetrafluoroborate (TBABF$_4$), TBAP-TBAClO$_4$, LiBF$_4$, LiClO$_4$, NH$_4$ClO$_4$	CO: predominant with TBA salt HCOOCH$_3$: major product with Li salt. Li ion prevents H_2 evolution.	The hydrophobic environment of 'TBA$^+$ results in the formation of CO and CO_2. Li ion suppresses the CO_2 reduction, more pronounced under higher pressures.

#	Electrode	Electrolyte	Products	Remarks
34	Cu wire	$TBABF_4$, TBAP, tetraethyl ammonium perchlorate (TEAP)	CO, CH_4, C_2H_4 formed. Formation of H_2 increases in presence of TBAP, and formation of hydrocarbon increases when TBAP is replaced by TEAP	The CO_2 can be converted into hydrocarbon instead of CO by changing the alkyl chain length of tetraalkylammonium cation.
35	Cu	$CH_3OH^-TBABF_4$	CO, CH_4, C_2H_4, $HCOOCH_3$. Pressure varied from 1 to 60 atm and classified into 3 regions: 1–20 atm, 20–40 atm, and 40–60 atm. Resistance increases as pressure goes up.	CE for CO_2 reduction is achieved up to 87% at −2.3 V under 40 atm. Mass transfer of CO_2 was sufficiently high above a pressure of 20 atm. The formation efficiency of reduction products increases from 23% at 1 atm to 92% at 20 atm.
36	Cu	$TBABF_4$, liquid CO_2 under high pressure	CO, CH_4, C_2H_4, $HCOOCH_3$	CO_2 reduction was carried out at higher current density, which is equal to the conditions used in industries such as chlor-alkali and Al refinement.
37	Au/Ppy/Re Au/Ppy/Cu-Re	$LiClO_4$ CH_3OH	CO and CH_2 formed: Au/Cu CH_2 : Au/Cu-Re	FE value absolutely depends on the nature of substrate and the operating potential.
38	Au and Re electrodes	CH_3OH	CO and CH_4 formed	FE value absolutely depends on operating potential and hydrodynamic conditions of electrolyte.

are interesting examples. It was observed that electroreduction of CO_2 is highly enhanced using alloy metal [37,38]. Furthermore, the metal electrodes in the presence of electrocatalytic material dissolved in the electrolyte are responsible for the selective product formation.

7.2.1 Effect of Electrolyte Containing Salt

Media containing cationic salts are found to play a vital role in forming reduction products. Recently, CO_2 conversion was achieved by dispersing Cu particles in CH_3OH medium on a P-InP electrode [7]. The selective nature of the reduction product changed because of the presence of Cu particles. Only HCOOH and CO were found to be formed in the absence of Cu particles on the P-InP electrode, and the hydrocarbon formation could be achieved by adding Cu particles in methanol electrolyte. The faradaic efficiency increased with increase in the quantity of Cu particles, and the maximum efficiency was achieved for CH_4 and C_2H_4 at 300 and 400 mg of Cu particles, respectively. Total hydrocarbon efficiency was found to have a maximum value at 300 mg of Cu particle suspension, and the efficiency was decreased beyond 400 mg of Cu particles. The addition of Cu particles shifts the onset potential to a more positive value than that obtained without Cu particles. It is clear that the reduction of CO_2 to hydrocarbon could be achieved by adding the electrocatalytic Cu metal particle in electrolyte.

The effect of electrolytic salts such as sodium, potassium, lithium, cesium, and rubidium in methanol medium has been studied in detail for electroreduction of CO_2. The lithium salts were found to be suitable for CH_4 formation, and an efficiency of 78% was shown with $LiClO_4$ as supporting salt [18]. Methane formation tends to increase in the order Cs^+, K^+, Li^+, the decreasing order of cation size. In Li^+/CH_3OH-based electrolyte, the faradaic efficiency of H_2 evolution on Cu at 243 K was observed to be <12%, except for the case of CH_3COOLi.

Electrochemical reduction of CO_2 with a Cu electrode in $CsOH/CH_3OH$-based electrolyte was studied [9,22]. The main products from CO_2 were methane, ethylene, ethane, CO, and HCOOH. The maximum faradaic efficiency of C_2H_4 was 32.3% at −3.5 V. The CH_4 formation efficiency was 8.3% at −4.0 V. Current efficiency ratio of ethylene and methane was in the range of 2.9–7.9. in the $CsOH/CH_3OH$ electrolyte. The efficiency of H_2 formation, being a competitive reaction against CO_2 reduction, was depressed to below 23%. The faradaic efficiency for C_2H_4 was larger than that of methane at all potential ranges. A trace quantity of ethane <1% was formed. The formation efficiency of ethylene was 23.7–32.3%. Thus in alkaline medium, CO and HCOOH are the main products at the Ag electrode and the ethylene formation is predominant at the Cu electrode. With the Cs salt as supporting electrolyte, the reduction of CO_2 was carried out under high pressure in cold methanol [9]. The main reduction products were CO, HCOOH, CH_4, and C_2H_4. The maximum current efficiency for CO was 84%, unlike the above-described case.

This shows that high pressure and low temperature decide the selectivity of the reduction products and faradaic efficiency. Furthermore, in the potential region more negative than -3.0 V, the efficiency of H$_2$ formation was suppressed to <2%. This could be achieved by changing the pressure and temperature and shifting the potential to the more negative side. The current efficiency for CO increased from 43% to 84% while increasing the potential in the negative direction, and the same phenomenon was observed for ethylene. In contrast, faradaic efficiency for CH$_4$, HCOOH, H$_2$ decreases with more negative potential. This leads to two groups of reduction products, (i) CO and C$_2$H$_4$ and (ii) CH$_4$, HCOOH, and H$_2$.

The hydrophobic electrode surface is expected to permit hydrophobic cations such as Cs$^+$ to adsorb on the electrode surface, which produces a hydrophobic atmosphere in vicinity of the electrode. In this case, Cs ion provides a hydrophobic environment and hydrogen-poor conditions near the electrode surface during a cathodic polarization. The protonation reaction of CO$_2$ reduction intermediates for the formation of CH$_4$, HCOOH, and H$_2$ formation reaction therefore seems to proceed less efficiently at relatively more negative potential, and, on the other hand, the formation of CO may be favorable. The difference in electrochemical reduction of CO$_2$ between Li and Cs salts can be explained. For example, in the electrochemical reduction of CO$_2$ in cold methanol at ambient pressure, Li salt was suitable for CH$_4$ formation (71.8% in LiClO$_4$) whereas Cs salt was favorable for ethylene formation (32.3% in CsOH salt).

Under high pressure, CO$_2$ reduction in cold methanol with LiCl salt yields CO, CH$_4$, HCOOH, and C$_2$H$_4$ as reduction products. The current efficiency (20%) of CH$_4$ was better than ethylene (2.3%). Conversely, the trend is reversed in Cs$^+$ salt under high pressure at low temperature. The large cation Cs$^+$ is adsorbed on the electrode surface, thus making the electrode vicinity more hydrophobic and creating a hydrogen ion-poor environment. Therefore, the dimerization of intermediate Cu = CH$_2$ to ethylene may be favored at lower surface hydrogen coverage in the case of Cs$^+$ salt, because its reaction does not require the presence of adsorbed proton. In a similar trend, the electrochemical reduction of CO$_2$ was carried out on Cu electrode with K$^+$ and Rb$^+$ salt cations instead of Li$^+$ and Cs$^+$. The main reduction products were CH$_4$, C$_2$H$_4$, CO, and HCOOH. The maximum current efficiency for C$_2$H$_4$ was 37.5% at -4.0 V, and it was higher in K$^+$ salt electrolyte than Rb$^+$ salt electrolyte. As the K$^+$ and Rb$^+$ are adsorbed on the surface of electrode, since these are less hydrated bulky cations, the conversion of C$_2$H$_4$ is favored.

The electrochemical reduction of high-pressure CO$_2$ with a Cu electrode in cold methanol was studied [12]. In the electrolysis of high-pressure CO$_2$ at low temperature, the reduction products formed in the order of CO, CH$_4$, HCOOH, and C$_2$H$_4$. The faradaic efficiency of CH$_4$ decreases with increase in temperature. The increase in concentration of the supporting electrolyte affects the current efficiency to a certain extent. Its curve is convex for C$_2$H$_4$ and concave for CO. However, the predominant product was CO irrespective of the concentration of the supporting electrolyte. The faradaic efficiency of the CO was drastically decreased from 71% to 12% as the potential became more negative, and for H$_2$ it increased from 7% to 98%. It has been proven that the formation efficiency of CH$_4$ in cold CH$_3$OH at high pressure is effective compared to electrochemical reduction of high-pressure CO$_2$ at ambient temperature in CH$_3$OH with other electrolytes such as TBABF$_4$,

$TBAClO_4$, $LiBF_4$, and $LiClO_4$. In the electrochemical reduction of high-pressure CO_2 in methanol at ambient temperature, the reduction of CO_2 was suppressed with Li as supporting salt because of the stability of the ion pair of single electron reduced species (CO_2^-) with the cation of the supporting salt $[Li^+ - (CO_2^-)]$. The formation of $[Li^+ - (CO_2^-)]$ is favored under high pressure at ambient temperature. Although CO formation is predominant when tetrabutylammonium (TBA) salts are used, the formation efficiency slightly favors the CH_4 formation. One of the major differences between TBA and Li ion is their hydrophobic nature. A hydrophobic surface is thought to allow hydrophobic cations such as TBA^+ to adsorb on the cathode surface, thus producing a hydrophobic atmosphere. In contrast, Li cation provides a less hydrophobic environment near the electrode surface. CO formation can be expected to be favorable in the hydrophobic environment, whereas CH_4 formation may proceed efficiently in the less hydrophobic atmosphere because its reaction does require the presence of adsorbed proton.

In 1995, Fujishima and co-workers developed an electrochemical system for reduction of CO_2 in methanol electrolyte containing tetrabutyl ammonium tetrafluoroborate ($TBABF_4$) to overcome the mass transfer problem. In other words, the concentration of the CO_2 was maintained at 17 M at 293 K, 64 atm; thus it became a liquid CO_2-containing methanol electrolyte. The aim of the work was accomplished by achieving current efficiency of 85% and the major reduction products CO, methane, ethylene, and methyl formate [36]. The reduction product formation is often based on the electrodes and metal cation of the supporting electrolyte that is presented in the CO_2/methanol medium.

Methanol is known to be a very good solvent. It is well known that metal electrodes themselves serve as catalyst, and the reduction product distribution strongly depends on the electrode material used. It was proved that when Cu is used as cathode hydrophobic cation (TBA) leads to CO as main product, while the hydrophilic cation such as Li^+ leads to formate and hydrogen as main products. The electrotransfer products, namely, CO, HCOOH, and $HCOOCH_3$, are formed at most of the metals, even at Cu in the presence of TBA or Li salts. The main products of this electroreduction in the presence of TBA—a hydrophobic environment—were CO, CH_4, C_2H_4, and $HCOOCH_3$, with faradaic efficiency of 48.1%, 40.7%, 9%, and 34.6%, respectively.

On the other hand, the Li salt-containing supporting electrolyte-hydrophilic environment enhanced the formation of methyl formate. The formation of methyl formate was explained by two consecutive processes. The two-electron reduction of CO_2 to HCOOH, according to the usual mechanism in protic media, is given here:

$$(CO_2)_{ad} + e^- \rightarrow (CO_2^{-\cdot})_{ad} \qquad (7.2a)$$

$$(CO_2^{-\cdot})_{ad} + 2H^+ + e^- \rightarrow HCOOH \qquad (7.2b)$$

Methanol serves as the protic solvent in this system and for the subsequent chemical esterification of formic acid with methanol:

$$HCOOH + CH_3OH \rightarrow HCOOCH_3 + H_2O \qquad (7.3)$$

When the high current density electroreduction on a Cu cathode was conducted at elevated pressure [35], the formation of CO was no longer limited by the mass transfer of CO$_2$. At 40 atm and −2.3 V, the total current density was 436 mA/cm^2, while the faradaic efficiency of CO$_2$ reduction reached 87%. In the gas phase CO, CH$_4$, C$_2$H$_4$, and H$_2$ were detected as products, while the main components present in the liquid phase were methyl formate and dimethoxymethane (CH$_3$O$^-$CH$_2$$^-OCH_3$) [35].

When the electrolyte salt was replaced by tetraethyl ammonium perchlorate, efficient formation of methane and ethylene was observed with a current efficiency of 89.7% [34]. It has also been proven that the total current efficiency of CO$_2$ reduction is higher in the CO$_2$/methanol system than in the CO$_2$/H$_2$O/ system. Experimental results had proven that formation of hydrocarbon is highly favored at Cu and Ni electrodes and Cu is the only metal at which CO$_2$ is reduced efficiently to hydrocarbons in an aqueous system. Among various electroactive metals, Cu is always found to be effective to form hydrocarbons as reduction products in aqueous electrolyte. It has been reported that CO$_2$ reduction in CH$_3$OH medium using TBA salt yields CO as major product while the same process with Li salt gives HCOOCH$_3$. It was suggested that the hydrophobic atmosphere of the TBAA ion and the hydrophilic atmosphere of the Li ion are advantageous to CO and HCOOCH$_3$ formation, respectively [33]. On the other hand, it is obvious that metal cathodes suffer corrosion or passivation problems that prevent their use as electrocatalyst during long-term electrolysis. To overcome this problem and look for versatile and cheaper systems, a number of metal transition complexes have been studied as electrocatalysts for the reduction of CO$_2$.

It is more interesting to note that the reduction product selectivity of CO$_2$ greatly depends on the anionic species present in the supporting salt [20]. Electrochemical conversion of CO$_2$ on Cu using CH$_3$OH containing various potassium salts (CH$_3$COOK, KBr, KI, KSCN) was investigated at 243 K. The reduction products were CH$_4$, C$_2$H$_2$, C$_2$H$_4$, CO, and HCOOH. The best current efficiency for C$_2$H$_4$ was obtained in KI- CH$_3$OH-based electrolyte at −3.0 V. The maximum efficiency of CH$_4$ formation was 27% in CH$_3$COOK-CH$_3$OH electrolyte at −3.0 V. The formation of H$_2$ evolution was suppressed to <8.1%. In contrast, higher efficiency (53.9%) for H$_2$ formation was observed with CH$_3$COOK salt. The formation efficiency of CO was very poor in CH$_3$COOK salt-containing electrolyte compared to other salts such as KBr, KI, and KCN. H$_2$ formation tends to increase in the order of I < SCN < CH$_3$COO. The faradaic efficiency of hydrocarbon is independent of the above-mentioned supporting salt.

The selective formation of reduction product is often based on the concentration of CO$_2$ at the vicinity of the electrode. Generally, the hydrocarbon is expected to form only when the applied current density is high enough to meet the total flux of CO$_2$ toward the electrode surface that is, the reduction reaction undergoes mass transfer control. On the other hand, when a lower current density is employed, the main products are always HCOOH and CO. The surface of the electrode is covered with adsorbed CO, and other parasite reaction is diminished. The situation persists until a higher concentration of CO$_2$ is maintained regardless of applied potential and current. The hydrocarbon formation may be due to hydrogen presence at the vicinity of the electrode. This is possible when the solvent is protic, and the surface of the electrode should have strong interaction with hydrogen.

7.2.2 Effect of Electrode Materials

It is very obvious that metal electrodes themselves serve as catalyst and the reduction product distribution is strongly dependent on the electrodes employed. The different electrocatalytic behavior of metals can be related to their electronic configuration, which can be grouped into "sp" and "d" metals. These configurations can further be modified on the basis of alloying properties. This modification could be due to a combination of changes in the electronic structure and surface morphological characteristics, including the introduction of dislocation and vacancies. It leads to the major changes in product distributions and reaction rates. Metal hydrogen alloys such as Pd-H are an interesting example. It was observed that CO_2 reduction is highly enhanced with alloy metal [39,40].

The electrocatalytic behavior of the metals has been classified in three categories based on response under high pressure of CO_2 in electrolytic solution. The first category of electrodes (Fe, Co, Ni, Pt, C, W, Rh, Ir) shows a high faradaic efficiency for CO and HCOOH in which hydrogen formation is also predominant. In the second category (Ag, Au, Zn, In, Pb, Sn, Bi), faradaic efficiency of CO and/or HCOOH increases under high pressure of CO_2 and the selectivity of the product remains unchanged. However, in the third category(Cu), selective formation of hydrocarbon from CO and HCOOH changes.

Cu metal is well known for its specific electrocatalytic nature of yielding hydrocarbon during the electroreduction of CO_2. Ohya et al. [8] investigated zinc particles mixed with CuO and Cu_2O powder and used a disk plate electrode for the CO_2 conversion. Hydrocarbon formation was only observed in the presence of copper oxides, and zinc material alone led to the formation of CO and HCOOH. This again proves that Cu has an electrocatalytic nature for forming hydrocarbon molecules. However, the formation efficiency of hydrocarbon was observed to be poor at CuO and Cu_2O electrodes compared to intrinsic copper electrodes [8]. Similarly, a Cu-modified Ni electrode was fabricated by an electrodeposition method for the CO_2 reduction [10]. The nickel substrate with low coverage of Cu coating was evaluated. The faradaic efficiency of hydrocarbon was very low (<0.3%) at an intrinsic Ni cathode, and the efficiency could be considerably improved with the use of the Cu-deposited Ni electrode. Also, the H_2 evolution reaction was not favored with the Cu-modified Ni electrode. However, hydrocarbon formation is highly enhanced when Ni is deposited on a P-InP electrode.

The Gibbs energy of $CO_{(g)}$ on different metal electrodes has been calculated by Frese [41]. According to this calculation, $CO_{(g)}$ is adsorbed as such onto Ag, Pd, and Au, which does not allow $CO_{(g)}$ to be dissociated. However, $CO_{(g)}$ undergoes dissociation on a Cu electrode only at high pressure and never in an ultra-high vacuum. Ni has the largest value of Gibbs energy for $CO_{(g)}$ splitting among these metals and is predicted to dissociate $CO_{(g)}$ at any reasonable pressure. Hence, the

Gibbs energy may be causing the formation of hydrocarbons only at Ni-modified p-InP electrode material [15].

With the aim of reducing overpotential and increasing the faradaic efficiency of reduction product, a spectrum of Cu base alloy materials such as Cu-Ni, Cu-Sn, Cu-Pb, Cu-Zn, Cu-Cd, and Cu-Ag have been employed [42]. The formation of CH$_3$OH was achieved by Cu-Ni and Cu-Cd alloys, whereas the intrinsic Cu does not produce CH$_3$OH. Also, the formation rate of CO was dramatically increased with Cu-Zn and Cu-Cd alloy compared to Cu-alone material. The formation of double carbon reduction products such as C$_2$H$_4$, C$_2$H$_5$OH, and CH$_3$CHO was achieved when Cu and Ag were put together as an alloy. On the other hand, the intrinsic Cu or Ag does not produce such double carbon products in electroreduction of CO$_2$.

Our research group has employed Pb electrode in a KOH-methanol-based electrolyte at ambient temperature and pressure [27]. It has been found that the solubility of the CO$_2$ is considerably increased comparing pure methanol and water. The solubility of CO$_2$ was 330 µmol/cm^3 of methanol solvent at 288 K. The increase in solubility in KOH-based methanol was explained by the fact that the formation of bicarbonate helps to accumulate CO$_2$ in larger quantity. Bubbling of CO$_2$ gas for several minutes in KOH-methanol electrolyte results in the conversion of KOH to KHCO$_3$. Among the reduction products, formic acid was predominantly formed in the potential range -1.8 to -2.5 V at the Pb cathode. The faradaic efficiency for hydrogen formation at the Pb electrode in KOH-methanol catholyte was found to be <3.5%. The formation efficiency of CH$_4$ at a Zn electrode is better than that obtained with a Pb electrode [32]. It could be suggested that the Zn electrode is favorable for production of the reaction intermediate CO for the formation of CH$_4$. It is quite possible to change and control the reduction product distributions in electrochemical reduction of CO$_2$ by adding metal particles to the electrolyte.

Earlier work [24] on the use of a benzalkonium/methanol catholyte at ambient temperature found that the faradaic efficiencies at Ag, Au, Co, Cu, Fe, Ni, Pt, Sn, Ti, and Zn for H$_2$ evolution were 41%, 16%, 84%, 54%, 86%, 95%, 94%, 10%, 114%, and 21%, respectively. In addition, the formation of formic acid and CO is proved as an independent process. Ti metal has tremendous affinity toward hydrogen at higher temperatures, and the adsorption of hydrogen on Ti metal is a reversible process. It has been proven that concurrent desorption of hydrogen on Pd metal in aqueous electrolyte caused enhancement of the electrochemical reduction of CO$_2$. Hence, Mizuno et al. [23] made a comparison study between Ti and hydrogen-storing Ti (Ti-H) metal in KOH/CH$_3$OH electrolyte, varying the potential and temperature. The total faradaic efficiency for the products by CO$_2$ electroreduction on the Ti-H electrode was larger than that on the Ti electrode at more negative potential and lower temperature. Among the main products HCOOH and CO, the formation of the latter was larger at the Ti-H electrode than on the Ti electrode. In the case of an aqueous electrolyte, oxalic acid formation was observed at the Ti electrode with 18.7% faradaic efficiency at 273 K [43].

The same authors extended the work by lowering the temperature of the supporting electrolyte to 243 K. Under optimal conditions, the faradaic efficiency for methane exceeded 42%. Furthermore, the efficiency of the competing hydrogen evolution reaction was controlled to <8% at such a low temperature [44]. Despite the fact that oxalic acid formation was not observed in KOH/CH_3OH medium, the faradaic efficiency was found to be better. Under similar experimental conditions, cathodes such as Ag and Au electrodes were employed, varying the potential and temperature for faradaic efficiency, by the same authors [25,26].

It was stated that the formation efficiency of CO was increased at negative potential and lower temperature, with a lesser formation of hydrogen. However, the selective formation of CO_2 reduction products was enhanced at lower temperature. These effects were attributed to the poor reactivity of the electrode surface with hydrogen ion in the methanol at low temperature. Also, the reduction current density was found to be lower in CO_2-saturated methanol medium than in N_2-saturated methanol medium at constant potential. The electrolyte pH of the former was found to be 8.5, whereas the latter was 13.9. This extreme pH may affect the onset potential and the reduction process.

Saeki et al. [32] investigated the electrochemical conversion of CO_2 in CH_3OH, using various electrodes. The results were compared with the electrodes' efficiency in aqueous system. The W, Ti, and Pt electrodes showed poor activity in the reduction of CO_2, and H_2 evolution was predominant at these electrodes at room temperature. Nevertheless, the CO_2 reduction was enhanced. In aqueous solution, Pb and Sn selectively produced HCOOH. It has also been proven that CO is the predominant product when TBA salt is used as supporting electrolyte. When Ag, Zn, Pd, and Cu electrodes are used, CO is the principal product. In aqueous system CO_2 is selectively reduced to CO at an Ag electrode. At a Zn electrode both CO and formate were produced in aqueous system with comparable current efficiency. It has also been proven that current efficiency of CO_2 reduction is higher in the CO_2/CH_3OH system than in a CO_2 aqueous system. The experimental results show that formation of hydrocarbon is highly favored at Cu and Ni electrodes and Cu is the only metal at which CO_2 can be reduced to hydrocarbon effectively. Ni has the largest value of Gibbs energy for $CO_{(g)}$ splitting among these metals and is predicted to dissociate $CO_{(g)}$ at any reasonable pressure. Hence, the Gibbs energy may be causing the formation of hydrocarbons only at Ni modified electrode materials.

Ortiz et al. [46] investigated the electrochemical reduction of CO_2 on Sn, Cu, Au, In, Ni, and Pt electrodes in CH_3OH containing 0.1 M $NaClO_4$ by cyclic voltammetry and in situ FTIR spectroscopy. It was observed from the FTIR spectra of electrolysis products that there was no reduction of CO_2 on any of the electrodes employed and CO_3^{2-} was formed because of the reaction of CO_2 with hydroxide ions produced in the electroreduction of residual water molecule. To identify the reaction products, the FTIR spectra of sodium oxalate and sodium carbonate in methanol were obtained. These solutions were prepared by the electroreduction of oxalic acid and alkaliniation of CO_2 saturated methanol, respectively. It has been proven that electroreduction of carboxylic acids to carboxylate anions in organic solvents does not require either H-chemisorbing metal electrode or the presence of water in the solvent [45].

F. Koleli and co-workers studied the electrochemical conversion of CO_2 at a conducting polymer-coated Pt cathode under ambient conditions and high pressure in $CH_3OH/LiClO_4$ electrolyte under high pressure. The conducting polymers were polyaniline [28,31], polypyrole [30], and polythiophene. To achieve the process at lower potential, the mediated electrode system was prepared by forming a double layer, that is, the inorganic conductor was the inner layer over which the conducting polymer was coated, because the bare metal electrodes are reported to be effective in reducing CO_2 only at relatively lower negative potential less than -1.7 V in both protic and aprotic solvents. The reduction process was carried out in divided, undivided, and zero gap membrane cells. The advantages of membrane cells have been described as less potential drop in electrolyte and transfer of the cation from the anodic compartment to the cathodic compartment.

The pressure applied plays a major role in that the solubility of the CO_2 increases and it affects the faradaic value of the reduction products. The reduction products were formic acid, formaldehyde, and acetic acids. Earlier the Prussian blue-laminated polyaniline electrode was used for the reduction of CO_2 under high pressure, and it was found that CO_2 can be effectively reduced at around -0.8 V [45]. Formic acid was the main product when polypyrole was used as cathode, whereas acetic acid formation was predominant when a polyaniline cathode was used. However, the maximum faradaic efficiency of 62.2% was achieved only for CH_3COOH in both cases under similar experimental conditions. Furthermore, the thickness of the film was observed to play an important role, and adsorption of the substrate molecule and thereby transfer of H_{ad} atom to the CO_2 molecule is easier.

The thickness of the film was optimized to be between 0.3 and 0.6 μm. No reaction was observed at -0.4 V when the film was thin, and the reaction reflects the characteristic behavior of Pt cathode. The current density is decreased when an excessively thick polymeric film on Pt is used, because of the long diffusion path of H_{ad}, generated on the metal electrode Pt, to reach the CO_2 that is adsorbed on the film. The formation of H_{ad} was further enhanced by changing to metal electrodes such as Au or indium tin oxide coated with polymeric film. The current density was decreased to an extent of 100 μA/cm^2 and 10 μA/cm^2 for polyaniline-coated Au and polypyrole-coated Au cathodes, respectively. In continuation of this work, the authors carried out an electrochemical impedance spectroscopic (EIS) study on CO_2 reduction on polyaniline-coated Pt cathode at different applied DC potentials and characterized the Pt/polyaniline depletion layer, the charge transfer across the polymeric film, and the electrochemical reduction of CO_2 by using different equivalent electrical circuits corresponding to the obtained EIS results [29]. The above-described mechanism for the electrochemical reduction of CO_2 has been proposed when introducing a small quantity of H_2SO_4. According to this mechanism, a proton initially accepts an electron and gets adsorbed on the electrode surface as H atom. This H_{ad} combines with CO_2 in solution to form $HCOO_{ad}$ and with OH_{ad} to form H_2O. The $HCOO_{ad}$ formed further reacts with H_{ad} and CH_3OH to form $HCOOH$ and CH_3COOH along with OH_{ad}, respectively. It has been shown that the formation of $HCOOH$ and CH_3COOH are competitive reactions for CO_2

reduction. According to the mechanism, there is no direct transfer of electron to CO_2 and only the formation of H_{ad} atom was the initial step. From the EIS data, they were able to calculate the transfer coefficient and exchange current density. These results show that in the presence of protic or partially protic electrolyte the formation of H_{ad} is the very first step.

The electrocatalytic materials Re and Au were compared in the electroreduction of CO_2, and is was found that reduction of CO_2 begins at -0.5 V and -0.8 V on Re and Au, respectively [37]. Although CO was a major reduction product in both cases, formation of hydrocarbon (CH_4) was quite possible while extending the reduction process only in the case of the Re electrocatalyst system. Faradaic efficiency was found to be dependent on operating potential and hydrodynamic condition of the electrolyte.

The same research group studied the electroreduction of CO_2 using a polypyrole film electrode modified by Re and Re-Cu alloy electrocatalytic materials [38]. The results were compared with Au modified by the same electrocatalytic materials of Re and Re-Cu. The three electrolytic systems (Au/Cu, Au/Cu-Re, Au/Ppy-Cu-Re) reduced the CO_2 at more positive potential than other electrodes. Furthermore, among the three, the best energy efficiency for CO_2 reduction was achieved only with the Cu-Re electrode system. Also, the initial potential at which reduction process starts is shifted to more positive potential with the Au/Cu-Re system. The presence of Re in the Cu-Re alloy causes a synergic effect in CO formation but not in CH_4 formation. However, the intrinsic Cu shows no synergic effect on either of the products. In the Au/Cu electrode, the formation of CH_4 was predominant rather than CO irrespective of the operating potential. This indicates that CO molecule remains strongly adsorbed, which leads to the CH_4 formation. On the other hand, in Au/Cu-Re the formation of CO is always higher than that of CH_4. This shows that the presence of Re in the system makes the CO the principal product, possibly because of weak metal-CO interaction. Conclusively, Re displaces the adsorption equilibrium of CO toward the solution. Mass change measurements reveal that CH_2 is formed at the Cu-Re alloy, whereas a predominant formation of CO and COOH is achieved at the Re system [38].

7.2.3   Effect of Potential

The current potential curves with metal electrodes in CO_2-saturated methanol and nitrogen-purged methanol were studied. CO_2 reduction was evident on voltammogram recorded in CO_2-saturated methanol. When the electrolysis was carried out under N_2 atmosphere, H_2 evolution was observed. Hence, the cathodic current under N_2 atmosphere is attributed to reduction of H_2O molecule. The aim of the current-potential curve is to find out the onset (starting) potential for the particular electrochemical system; thus the electrochemical reduction of CO_2 can be carried out by cathodic polarization at more than the onset potential value.

In Tables 7.2 and 7.3, the onset potential for the cathodic current of various metal electrodes, that is, the potential value at which the minimum current density (0.1 mA cm^{-2}) was varying with respect to the electrolysis conditions and nature

TABLE 7.2   Influence of Anion on Onset Potential at Cu Electrode in CH_3OH Medium

Cation of Supporting Salts	Anion of Supporting Salts						
	OH^-	Cl^-	Br^-	I^-	SCN	ClO_4	CH_3COO^-
Li	−0.5	−1.3	−1.5	−1.5	—	−1.4	−1.0
Na	−1.5	−1.5	−1.8	−1.8	−1.5	−1.8	−0.8
K	−1.4	—	−1.4	−1.4	−1.5	—	−1.3
Rb	−1.3	—	—	—	—	—	—
Cs	−1.5	−1.3	−1.3	−1.3	−1.1	—	−1.3

TABLE 7.3   Onset Potential Value on Various Electrodes and CH_3OH-Based Electrolyte Salts

Electrode	Salt Presented in CH_3OH Medium	Onset Potential in CO_2 Saturated
Pb	KOH	−1.6
Ti	—	−1.8
Ti-H	—	−1.7
P type InP	—	−0.8
P type GaAs	KOH	−1.2
Zn	Cu_2O particle	−1.2
Ni-Cu deposited	H_2O	−1.1
Pb	Cu particle	−1.9
Zn	Cu particle	−0.9
Ni	—	−1.4
Ag	—	−1.7
In	—	−1.7
Pb	—	−1.7
Ti	—	−1.8
Zn	—	−1.5
Au	—	−1
Pb-deposited P-InP	—	−0.4
Ag-deposited P-InP	—	−0.75
Au-deposited P-InP	—	−0.75
Pd-deposited P-InP	—	−0.8
Cu-deposited P-InP	—	−0.5 (−0.1)
Ni-deposited P-InP	—	−1.25

of the metal electrodes is shown. Generally, the cathodic current in CO_2-saturated methanol is higher than that in N_2-purged methanol. Ohya et al. [8] reported that the current efficiency of non-hydrocarbon product formation was almost independent of the operating potential and, on the other hand, the maximum formation efficiency of CH_4 and C_2H_4 was observed at around −3.0 V. The best faradaic efficiency of methane and the total current efficiency of hydrocarbon formation were also observed at the same potential (−3.0 V). The onset potential of the cathodic current was -0.5 and -1.2 V for CuO and Cu_2O mixed zinc electrodes, respectively. The supporting salts containing Li, Cs, Na, K, Rb, etc., were examined for electrochemical reduction of CO_2 on the Cu electrode. The onset potential of cathodic current for CO_2-saturated KOH/CH_3OH and $RbOH/CH_3OH$ was -1.4 V and -1.3 V, respectively. In the Tafel plot for LiOH and CsOH supporting electrolytes under similar conditions, the onset potentials were around -0.5 V and -1.5 V, respectively.

With the use of P-InP, the current density increased almost double the time as the potential became more negative. The maximum current efficiency for methane and ethylene was obtained at -2.5 and -2.6 V, respectively. In the photo electrochemical reduction of CO_2 at the P-InP electrode in the methanol-based electrolyte formic acid was predominantly formed [13]. However, in Cu particle suspended methanol, CO was predominant irrespective of operating potentials. The hydrocarbon formation and suppression of H_2 evolution were accomplished with the use of a P-InP electrode at -2.5 V in Cu particle suspended methanol. The Cu-deposited Ni electrode has shown better faradaic efficiency for CH_4, while increasing the potential in the more negative direction and the potential was optimized at -1.9 V [10], because the current efficiency for H_2 evolution increases as the negative potential is increased.

The faradaic efficiency for CH_4 increases as temperature decreases down to 268 K. Beyond this range, the formation of HCOOH increases and H_2 and CH_4 formation is decreased. The onset potential of Cu electrode with various electrolyte salt dissolved in CH_3OH medium is shown in Table 7.2. The onset potential observed on Cu electrode with $LiOH/CH_3OH$-based electrolyte at 243 K was -0.5 V. Ortiz et al. observed onset potential at around -0.6 V with a Cu electrode and $LiClO_4$-CH_3OH electrolyte [46]. Generally, the onset potential for CO_2 reduction in CH_3OH tends to increase with increasing size of the cation. The onset potential is affected by various parameters such as the pH of the catholyte and the nature of the cation and anion and their concentrations.

It is interesting to note that the methane and ethylene were formed in the electrochemical reduction of CO_2 in methanol medium at a relatively more negative potential region (-4.5 to -5.0 V) [22]. This may be attributed to the fact that the diffusion of CO_2 to the cathode in methanol-based electrolyte is more effective than in aqueous solution since the solubility of CO_2 in methanol at 243 K is almost 19 times that in water at ambient temperature. The onset potential of electrolyte containing CsOH is more negative than other salts such as CsCl, CsBr, and CsI.

7.3 MECHANISMS OF CO_2 REDUCTION IN NONAQUEOUS PROTIC (CH_3OH) MEDIUM

Generally, the electroreduction products (CH_3COOH, HCOOH, CH_4, C_2H_4, and CO) decide the pathway mechanism of electroreduction of CO_2. The product formation depends on the nature of the electrode material, the operating potential, the electrolyte salt, and the temperature. Also, the CO_2 concentration at the vicinity of the electrode and the nature of the electrolyte (aqueous or nonaqueous) determine the product formation. In nonprotic solvents, the strength of CO_2 binding is weakly correlated with the dielectric constant of the solvent. This is attributed to the difference in the solubility of CO_2 and ion pairing in the most nonpolar solvents. It has also been found that CO, $(COOH)_2$, and HCOOH were mainly produced from CO_2 in organic solvents such as propylene carbonate and dimethyl sulfamide. In aqueous solutions, the usual product is formic acid, which can be further reduced to HCHO followed by methane and methanol. However, protic solvents like CH_3OH allow the formation of CO. It has been proven that the products of the electrochemical reduction of CO_2 in deuterated methanol electrolyte did not contain any deuterated compounds. This clearly confirms that the no products are formed from methanol electrolyte medium. The conversion of CO_2 into hydrocarbons is a complex multistep reaction with adsorbed intermediates, particularly CO_{ads}. In methanol medium, the usual pathway involving single-electron reduction is assumed. Most studies have indicated that first steps in the reduction of CO_2 at high hydrogen over potential cathodes are as follows:

$$CO_{2(ads)} + e^- \rightarrow \cdot CO_2^-{}_{(ads)} \qquad (7.4a)$$

$$CO_2^-{}_{(ads)} + H^+ + e^- \rightarrow HCOO^-{}_{(ads)} \qquad (7.4b)$$

The presence of adsorbed anion radicals has been confirmed by electrochemical and spectroscopic techniques [47]. These steps are followed by second electron transfer or protonation to yield HCOOH and then by parallel disprotonation of the $\cdot CO_2^-{}_{(ads)}$ radical anions to neutral CO molecules. It is believed that the $\cdot CO_2^-{}_{(ads)}$ radicals produced are adsorbed and remain on the electrode surface. As the $\cdot CO_2^-{}_{(ads)}$ adsorbed on the electrode surface tends to be attacked by CO_2 molecules, the formation of CO is greater than the formation of HCOOH. Hydrocarbons are produced by a series of simultaneous or consecutive electronation/protonation steps. The adsorbed $\cdot CO_2^-{}_{(ads)}$ radical anion formed in the first electronation steps undergoes a second electronation/protonation to yield adsorbed CO as the key intermediate. By four successive electronation/protonation steps an adsorbed reactive methylene group forms, and this may stabilize as a methane molecule by a subsequent double electronation or protonation sequence to dimerize to form ethylene and ethane. The most common pathway in methanol medium can be proposed as shown in Figure 7.1.

7.4 CONCLUSIONS

The electroreduction of CO_2 on metal and polymeric film electrode materials in methanol-based electrolyte salts has been summarized. This chapter has mainly focused on the role of operating parameters that affect the nature of reduction product formation and the faradaic efficiency value. It is quite possible to enhance the faradaic efficiency of reduction products by changing the electrode materials and

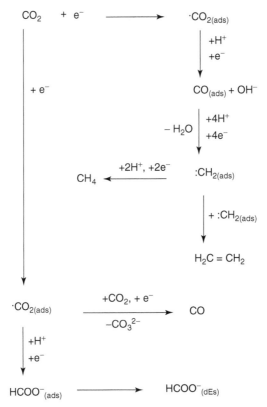

Figure 7.1   Reaction pathway for electroreduction of CO_2 on a metal cathode using CH_3OH as the electrolyte medium.

employing suitable electrolyte salts. Obviously, methanol medium is a better solvent for electroreduction of CO_2 under high pressure and low temperature.

The hydrocarbon formation depends on the size and nature of the cation salts presented in the methanol medium. Alloy materials such as Re and Cu-Re influence the faradaic efficiency of hydrocarbon formation. The hydrophobic nature of TBA^+ and Li^+ plays a significant role in the selective formation of reduction products. Data in the literature reveal that formation of hydrocarbon is highly favored at Cu and Ni electrodes and Cu is the only metal at which CO_2 can be reduced to hydrocarbon effectively. The efficiency of H_2 formation, being a competitive reaction against CO_2 reduction, was depressed <1% at lower temperature with LiI salt. The higher current densities result in lower faradaic efficiency as the H_2 evolution is accelerated in parallel. The onset potential in methanol medium tends to increase with increasing size of the cations employed as electrolyte salt. As shown in the literature, the reduction reaction involves simultaneous electronation/protonation reaction steps to yield hydrocarbons. The main outcome of future work will be in the direction of saving energy, which is the most important factor affecting the economic viability of scaling up the process.

REFERENCES

1. S. Kaneco, H. Katsumata, T. Suuki, K. Ohta (2006). *Chem. Eng. J.*, 116: 227.
2. M. Ulman, B. Aurian-Blajeni, H. Halman (1984). *Chemtech.*, 14: 235.
3. A.Gennaro, A.Isse, E.Vianello (1990). *J. Electroanal. Chem.*, 289: 203.
4. S. Ikeda, Y. Saito, M. Yoshida, H. Noda, M. Maeda, K. Ito (1989), *J. Elecctroanal. Chem.*, 260: 335.
5. E. Brunner, W. Hultenschmidt, G. Schichtharle (1987). *J. Chem. Thermodynam.*, 19: 273.
6. A. Naitoh, K. Ohta, T. Mizuno, H. Yoshida, M. Sakai, H. Noda (1993). *Electrochim. Acta*, 38: 2177.
7. S. Kaneco, Y. Ueno, H. Katsumata, T. Suzuki, K. Ohta (2009). *Chem. Eng. J.*, 148: 57.
8. S. Ohya, S. Kaneco, H. Katsumata, T. Suzuki, K. Ohta (2009). *Catal. Today*, 148: 329.
9. S. Kaneco, K. Iiba, H. Katsumata, T. Suzuki, K. Ohta (2007). *Chem. Eng. J.*, 128: 47.
10. S. Kaneco, Y. Sakaguchi, H. Katsumata, T. Suzuki, K. Ohta (2007). *Bull. Catal. Soc. India*, 6: 74.
11. S. Kaneco, K. Iiba, H. Katsumata, T. Suzuki, K. Ohta (2007). *J. Solid State Electrochem.* 11: 490.
12. S. Kaneco, K. Iiba, H. Katsumata, T. Suzuki, K. Ohta (2006). *Electrochim. Acta*, 51: 4880.
13. S. Kaneco, H. Katsumata, T. Suzuki, K. Ohta (2006). *Chem. Eng. J.*, 116: 227.
14. S. Kaneco, Y. Ueno, H. Katsumata, T. Suzuki, K. Ohto (2006). *Chem. Eng. J.*, 119: 107.
15. S. Kaneco, H. Katsumata, T. Suzuki, K. Ohta (2006). *Appl. Cataly. B: Environ.* 64: 139.
16. S. Kaneco, H. Katsumata, T. Suzuki, K. Ohto (2006). *Electrochim. Acta*, 51: 3316.
17. S. Kaneco, H. Katsumata, T. Suzuki, K. Ohta (2006). *Energy & Fuels*, 20: 409.
18. S. Kaneco, K. Iiba, M. Yabuuchi, N. Nishio, H. Ohnishi, H. Katsumata, T. Suzuki, K. Ohto (2002). *Ind. Eng. Chem. Res.*, 41: 5165.
19. S. Kaneco, K. Iiba, K. Ohto, T. Mizuno (2000). *Energy Sources*, 22: 127.
20. S. Kaneco, K. Iiba, K. Ohto, T. Mizuno (1999). *J. Solid State Electrochem.* 3: 424.
21. S. Kaneco, K. Iiba, H. Katsumata, S. Suzuki, K. Ohta, T. Mizuno (1999). *J. Phys. Chem. B*, 103: 7456.
22. S. Kaneco, K. Iiba, N. Kimura, K. Ohto, T. Mizuno, T. Suzuki (1999). *Electrochim. Acta*, 44: 4701.
23. T. Mizuno, M. Kawamoto, S. Kaneco, K. Ohta (1998). Electrochim. *Acta*, 43: 899.
24. S. Kaneco, K. Iiba, K. Ohto, T. Mizuno (1999). *Energy Sources*, 21: 643.
25. S. Kaneco, K. Iiba, K. Ohto, T. Mizuno, A. Saji (1998). *Electrochim. Acta*, 44: 573.
26. S. Kaneco, K. Iiba, K. Ohta, T. Mizuno, A. Saji (1998). *J. Electroanal. Chem.*, 441: 215.
27. S. Kaneco, R. Iwao, K. Iiba, K. Ohto, T. Mizuno (1998). *Energy*, 23: 1107.
28. F. Koleli, T. Ropke, C.H. Hamann (2004). *Synthetic Metals*, 140: 65.
29. F. Koleli, T. Ropke, C.H. Hamann (2003). *Electrochim. Acta*, 48: 1595.
30. R. Aydin, F. Koleli (2004). *Synthetic Metals*, 144: 75.
31. R. Aydin, F. Koleli (2002). *J. Electroanal. Chem.* 535: 107.
32. T. Saeki, K. Hashimoto, N. Kimura, K. Omata, A. Fujishima (1996). *J. Electroanal. Chem.*, 404: 299.

33. T. Saeki, K. Hashimoto, N. Kimura, K. Omata, A. Fujishima (1995). *J. Electroanal. Chem.* 390: 77.

34. T. Saeki, K. Hashimoto, N. Kimura, K. Omata, A. Fujishima (1995). *Chem. Lett.* 361.

35. T. Saeki, K. Hashimoto, A. Fujishima, N. Kimura, K. Omata (1995). *J. Phys. Chem.*, 99: 8440.

36. T. Saeki, K. Hashimoto, Y. Noguchi, K. Omata, A. Fujishima (1994). *J. Electrochem. Soc.*, 141: L130.

37. R. Schrebler, P. Cury, F. Herrera, H. Gome, R. Cordova (2001). *J.Electroanal. Chem.*, 516: 23.

38. R. Schrebler, P. Cury, C. Suare, E. Munoz, H. Gomz, R. Cordova (2002). *J. Electroanal. Chem.*, 533: 167.

39. K. Ohkawa, K. Hashimoto, A. Fujishima, Y. Noguchi, S. Nakayama (1993). *J. Electroanal. Chem.*, 345: 445.

40. K. Ohkawa, Y. Noguchi, S. Nakayama, K. Hashimoto, A. Fujishima (1994). *J. Electroanal. Chem.*, 369: 247.

41. K.W. Frese Jr. (1993). in: *Electrochemical and Electrocatalytic Reduction of Carbon Dioxide*, B.P. Sullivan (Ed.), Elsevier, Amsterdam, 145–216.

42. M. Watanabe, M. Shibata, A. Kato, M. Auma, T. Sakata (1991). *J. Electroanal. Chem.*, 305: 319.

43. M. Azuma, K. Hashimoto, M. Hiramoto, M. Watanabe, T. Sakata (1990). *J. Electrochem. Soc.*, 137: 1772.

44. T. Mizuno, A. Naitoh, A. K. Ohta (1995). *J. Electroanal. Chem.*, 391: 199.

45. K. Ogura, M. Nakayama, C. Kusumoto (1996). *J. Electrochem. Soc.*, 143–11: 3606.

46. R. Ortiz, O.P. Marque, J. Marque, C. Gutierre (1995). *J. Electroanal. Chem.*, 390: 99.

47. A. Aylmer-Kelly, A. Bewick, P. Cantrill, A. Tuxford (1974). *Faraday Discuss Chem. Soc.*, 56: 96.

CHAPTER 8

Synthetic Fuel Production from the Catalytic Thermochemical Conversion of Carbon Dioxide

NAVADOL LAOSIRIPOJANA*, KAJORNSAK FAUNGNAWAKIJ, and SUTTICHAI ASSABUMRUNGRAT

8.1   INTRODUCTION

Carbon dioxide (CO_2) is generally known as an important poisonous greenhouse gas, which makes a significant negative impact on current global warming. This gaseous compound usually is emitted by the combustion of hydrocarbon fuels (i.e., coal, oil, and natural gas) and requires costly treatment, disposal, capture, and/or sequestration to prevent several global impacts from its release. Nevertheless,

*Indicates the corresponding author.

Green Carbon Dioxide: Advances in CO₂ Utilization, First Edition.
Edited by Gabriele Centi and Siglinda Perathoner.
© 2014 John Wiley & Sons, Inc. Published 2014 by John Wiley & Sons, Inc.

because of its cheapness and abundant carbon resources, the image of this gaseous compound has recently changed, and CO_2 is now viewed as a business opportunity rather than a useless compound or harmful waste, which requires high cost for treatment, disposal, or storage. Currently, there are several attempts to utilize carbon dioxide for several beneficial purposes; one attractive route is the conversion of CO_2 to valuable chemicals and synthetic fuels. Important examples of chemicals that can be produced from the conversion of CO_2 to chemicals are carbonates, carbamates, urethanes, lactones, pyrones, and formic acid, while examples of synthetic fuels from CO_2 conversion include synthesis gas ("syngas," which is a mixture of gaseous carbon monoxide and hydrogen at various ratios depending on its production route), hydrogen, methanol, dimethyl ether (DME), and alkane fuel.

Syngas and hydrogen are considered promising fuels to replace conventional oil in the near future because of their environmental friendliness and economic advantages. Hydrogen is commonly used as the primary fuel for low-temperature fuel cells (i.e., proton exchange membrane fuel cell, PEMFC), while syngas can be widely used in several applications [e.g., as the fuel for combustion in gas turbines, as the primary fuel for high-temperature fuel cells (i.e., solid oxide fuel cell, SOFC), and as the reactant for methanol, DME, and alkane syntheses]. Hydrogen is also widely used as feedstock in several refinery and petrochemical industries. Another synthetic fuel from CO_2 conversion, methanol is the smallest type of alcohol and can efficiently replace conventional gasoline fuel in transportation applications; it can be efficiently produced from the reaction of carbon monoxide and hydrogen in syngas, which is known as the methanol synthesis reaction ($CO + 2H_2 \rightarrow CH_3OH$). DME is the simplest ether compound that can be synthesized from the dehydration of methanol; this compound can replace diesel fuel in the transportation sector and liquefied petroleum gas (LPG) in the household sector.

Generally, syngas and hydrogen can be readily produced from the reaction of hydrocarbon compounds (i.e., methane, natural gas, biogas, LPG, methanol, ethanol, gasoline, and other oil derivatives) with an oxygen-containing co-reactant (i.e., CO_2 or steam). The reaction between hydrocarbon and CO_2 is called CO_2 reforming, while the reaction between hydrocarbon and steam is called steam reforming. The steam reforming process is the most widely employed route for hydrogen or syngas industrial production, and is well-established technology. The production from steam reforming could be a wide range of gas products depending on operating conditions, which are the operating temperature, pressure, steam to carbon ratio, and the carbon to hydrogen ratio of the hydrocarbon feed. The variation of these parameters is the primary means applied to enable the steam reforming to produce the different gas streams. Typically, the steam reforming reaction (8.1) and the associated water-gas shift reaction (8.2) are always carried out over a catalyst at elevated temperatures.

$$CH_4 + H_2O \rightleftharpoons CO + 3H_2 \qquad \Delta H = 206.2 \text{ kJ/mol} \qquad (8.1)$$

$$CO + H_2O \rightleftharpoons CO_2 + H_2 \qquad \Delta H = -41.1 \text{ kJ/mol} \qquad (8.2)$$

Both reactions are reversible, and the reaction rates are very fast at conditions close to their equilibrium. The steam reforming reaction is strongly endothermic, while the water-gas shift reaction is mildly exothermic. Therefore, the steam reforming reaction is favored by high operating temperature with low pressure. In contrast, the water-gas shift reaction is favored by low temperature, but this reaction is largely unaffected by pressure. To maximize the overall production of hydrogen, carbon monoxide, and CO_2, the steam reforming reaction needs to be operated at high operating temperature. As seen from reaction 8.1, the stoichiometric requirement for the steam to carbon ratio (methane) is 1.0. However, it has been demonstrated that carbon deposition reactions easily occur under this condition. In order to prevent carbon deposition, steam is normally added in excess of the stoichiometric requirement. The minimum steam to carbon ratio is in the region of 1.5–2.0; nevertheless, it is well established that the steam reforming reaction is promoted by the excess of steam; hence, the steam to carbon ratio is commonly used between 3.0 and 5.0. If steam is completely replaced by CO_2, the reaction is called CO_2 reforming. In fact, the steam and CO_2 reforming reactions have similar thermodynamic characteristics except that in case of CO_2 reforming there is a greater potential for carbon deposition on the catalyst surface due to the lower H/C ratio of this reaction. Nickel and cobalt are usually applied as the catalyst for this reaction. One advantage of the CO_2 reforming reaction over the steam reforming reaction is the lower hydrogen to carbon monoxide ratio production, which is preferred in the synthesis of oxo-alcohols and oxygenated compounds. The excess of hydrogen gas, which is formed by conventional steam reforming reaction, can suppress chain growth and decrease the selectivity to higher hydrocarbons. Therefore, CO_2 reforming provides a more suitable hydrogen to carbon monoxide ratio for these purposes compared with steam reforming. It is noted that the combination of steam and CO_2 reforming reactions introduces the possibility of getting a range of hydrogen to carbon monoxide ratios, depending on the final use. One of the most attractive features of the CO_2 reforming reaction is the utilization of CO_2, which is the greenhouse effect gas. More details of the CO_2 reforming reaction are given in Section 8.2.

This chapter mainly focuses on the conversion of CO_2 to synthetic fuels via a thermochemical process, which is the most common process for commercial applications. Importantly, the main challenge for CO_2 conversion is that it is a thermodynamically stable compound that requires high energy for processing. Technically, the processes for converting CO_2 to synthetic fuel can be classified into two pathways, including (i) hydrogenation of CO_2 to form oxygenates (e.g., alcohol and DME) and/or hydrocarbons and (ii) reforming of CO_2 with hydrocarbons to form syngas. The hydrogenation of CO_2 has been intensively studied recently; for example, the process for methanol synthesis from the reaction between CO_2 and hydrogen is currently being developed at the pilot scale, while the further process for converting methanol to DME for replacing conventional diesel fuel has also been widely investigated. In addition, the production of ethanol via the homologation of methanol and the conversion of CO_2 to formic acid are also known as promising technologies for the near future. Nevertheless, the major barrier to CO_2 hydrogenation is the availability of hydrogen. It has been reported that this reaction route is feasible when hydrogen is produced from renewable resources or hydrogen is excess from other processes (i.e., petrochemical dehydrogenation process) and the production plant is located near the source of hydrogen [1]. For the second reaction route, CO_2 reforming, this process is applied industrially to convert the CO_2-containing feedstock (e.g., biogas and natural gas) to syngas for further processing. Importantly, alcohol and DME can also be produced from this reaction route by further converting of syngas (generated from CO_2 reforming) via the methanol synthesis and methanol dehydration reactions. In addition, alkane fuel can be efficiently produced from this syngas via a Fischer–Tropsch process. Also, this syngas can be converted to pure hydrogen by water-gas shift/CO oxidative processes and applied efficiently in fuel cell application.

Considering the technological feasibility point of view, this review focuses mainly on the second route of CO_2 conversion, from which details of CO_2 reforming, catalyst selection, possible operation, and potential application are provided. In addition, research approaches to the conversion of syngas to methanol, DME, and alkane fuel (which is commonly known as gas-to-liquid or GTL) are also given.

8.2   GENERAL ASPECTS OF CO_2 REFORMING

CO_2 reforming (or dry reforming) is the reaction between hydrocarbons (C_nH_m) and CO_2 to form synthesis gas or syngas (the mixture of carbon monoxide and hydrogen), Eq. 8.3.

$$C_nH_m + nCO_2 \, P \, 2nCO + 1/2mH_2 \tag{8.3}$$

Theoretically, the hydrocarbon reactant can be any compound including light and heavy hydrocarbons; nevertheless, the use of heavy hydrocarbon compounds can easily result in rapid catalyst deactivation from the carbon deposition. Among all types of CO_2 reforming, CO_2 reforming with methane is attractive because of several advantages from an environmental viewpoint because this reaction utilizes greenhouse gases, that is, methane and view point, which make significant contributions to the global warming problem. Furthermore, this reaction causes relatively lower carbon deposition compared to the view point reforming of other hydrocarbons. In general, the CO_2 reforming of methane reaction (Eq. 8.4) is typically accompanied by the simultaneous occurrence of the reverse water-gas shift (RWGS) reaction (Eq. 8.5).

$$CH_4 + CO_2 \rightleftharpoons 2CO + 2H_2 \tag{8.4}$$

$$CO_2 + H_2 \rightleftharpoons CO + H_2O \tag{8.5}$$

From the reaction, the hydrogen to carbon monoxide production ratio (H_2/CO ratio) is always <1.0. Bradford and Vannice [2] presented the apparent activation energies for the consumption of methane and CO_2, as well as the production of carbon monoxide, hydrogen, and water, in order to investigate the influence of the RWGS reaction. The apparent activation energy for hydrogen formation is always greater than that for the formation of carbon monoxide, which supports the strong influence of the RWGS reaction on the reaction mechanism. The steam and CO_2 reforming reactions have similar thermodynamic characteristics except that the carbon deposition in CO_2 reforming is more severe than in steam reforming because of the lower H/C ratio of this reaction. The carbon deposition is a critical problem in the reforming process. The negative effect of carbon deposition over the catalyst surface on reaction activity generally includes the poisoning of the metal surface, the blockage of catalyst pores, and the physical disintegration of the catalyst support; in consequence, deactivation and loss of catalyst activity can occur rapidly. Therefore, the understanding and controlling of this effect are of great importance.

According to Rostrup-Nielsen [3], three different carbon species are formed during the reforming process, including

1. Whiskerlike carbon, which forms at a temperature greater than 450°C
2. Encapsulating film, which forms by the polymerization at a temperature less than 500°C;
3. Pyrolytic carbon, which forms by the cracking of hydrocarbon at a temperature above 600°C

The form of carbon deposited depends on the reaction conditions. For instance, the formation of amorphous carbon always occurs at moderate temperatures and high pressures, whereas the formation of graphitic carbon is more likely to occur at low pressures and very high temperatures. Regarding the possible carbon formation during the CO_2 reforming process, the following reactions are theoretically the most probable reactions that could lead to carbon formation:

$$2CO \rightleftharpoons CO_2 + C \tag{8.6}$$

$$CH_4 \rightleftharpoons 2H_2 + C \tag{8.7}$$

$$CO + H_2 \rightleftharpoons H_2O + C \tag{8.8}$$

$$CO_2 + 2H_2 \rightleftharpoons 2H_2O + C \tag{8.9}$$

At low operating temperature, reactions 8.8 and 8.9 are favorable, while reaction 8.6 is thermodynamically unfavored. The Boudouard reaction (Eq. 8.6) and the decomposition of methane (Eq. 8.7) are the major pathways for carbon formation at high operating temperature as they show the largest change in Gibbs energy. Apart from the concern over carbon deposition, water formation is another factor that must be considered since this formation consequently results in the decrease of total hydrogen production yield. According to thermodynamic analysis of the CO_2 reforming of methane at the temperature range 400–1000°C using a CO_2/CH_4 ratio of 1.0, water generally forms at low reaction temperatures (400–800°C) and mostly disappears at temperatures above 900°C, where the conversions of methane and carbon dioxide are close to 100% [4]. It is notable that CO_2/CH_4 ratio also plays an important role in reaction performance; the use of a low CO_2/CH_4 ratio leads to the low water formation, but it promotes the degree of carbon deposition. In contrast, the use of a high CO_2/CH_4 ratio efficiently minimizes the amount of carbon deposition, but higher water yield would be found because of the RWGS reaction.

Overall, the main concern for the CO_2 reforming reaction is to overcome the carbon deposition problem, which results in catalyst deactivation over long-term operation. In order to minimize this problem, selection of suitable operating conditions (i.e., temperature and CO_2/CH_4 ratio) is one possibility, while the other approaches are (i) the use of a catalyst with high resistance to carbon deposition and (ii) the use of specific operation modes. Details of catalyst selection for the CO_2 reforming reaction is presented in detail in Section 8.3. Regarding the details of the specific operation for CO_2 reforming, one alternative mode of operation is periodic operation by feeding methane and carbon dioxide alternately. According to this process, the first step is the catalytic cracking of methane, in which methane is converted to hydrogen and carbon species deposited on the surface of the catalyst (Eq. 8.7).

Carbonaceous deposition typically causes plugging and catalytic deactivation problems, and thus the second step is then applied by feeding CO_2 in order to regenerate the catalyst via the reverse Boudouard reaction. Under periodic operation, pure hydrogen can be produced separately from further gaseous product, for example, carbon monoxide. In addition, this operation also offers another potential benefit for heat integration within the reactor, as the exothermic heat from the catalyst regeneration can supply heat for endothermic methane cracking during the hydrogen generation cycle. Another alternative mode is the tri-reforming operation of feeding steam along with methane and CO_2, which results in of steam reforming occurring simultaneously with the CO_2 reforming. The presence of steam (as well as hydrogen) along with CO_2 could substantially reduce the formation of carbon from the CO_2 reforming, since the adsorbed steam (and hydrogen) react with adsorbed atomic carbon (from dissociation of CO or decomposition of CH_4) and thus remove these carbon precursors.

For instance, the simultaneous steam and CO_2 reforming of methane over Ni-CaO catalyst was studied at 700–850°C, and it was found that the carbon deposition on the catalyst surface was significantly reduced [5]. Furthermore, the combination of CO_2 reforming with steam reforming can produce syngas with a controllable H_2/CO ratio for several utilizations, for example, later conversion to methanol/DME, alkane (via Fischer–Tropsch process). More details of the periodic and tri-reforming operations are provided below in this chapter.

In addition to these operations, the influence of oxygen added to the CO_2 reforming on the amount of carbon formation has also been investigated. It was found that addition of oxygen could significantly reduce the carbon formation rate; furthermore, it also showed several other advantages including a lower energy requirement, higher energy efficiency, and lower operating temperature. Nevertheless, a major drawback of adding oxygen along with hydrocarbon and CO_2 is the large investment required for an oxygen production unit. Although air can be used directly instead of oxygen, the high content of inert nitrogen in air causes a large gas volume, and consequently the system requires larger associated components (such as feed/effluent heat exchangers and compressors). In addition, the use of air lowers the hydrogen concentration in the gas product because of the partial combustion of hydrogen produced from the reforming with addition of oxygen.

8.3 CATALYST SELECTION FOR CO$_2$ REFORMING REACTION

Importantly, the CO$_2$ reforming reaction requires a suitable catalyst to catalyze the reaction. It is noted that the role of the catalyst is not only to activate the overall reaction but also to minimize the catalyst deactivation. Generally, catalyst deactivation mainly results from two causes, carbon deposition and catalyst poisoning. Carbon deposition and sulfur poisoning are the major concerns when heavy hydrocarbons and/or sulfur-containing feeds, for example, natural gas, biogas, and LPG, are used. Typically, pre-reforming and/or desulfurization units are required to reform these feedstocks; however, both installations reduce flexibility and hence the potential for practical and economic application of hydrogen/fuel cell technologies. Research is therefore continuing to develop catalysts with high resistance to carbon formation and sulfur interaction. Theoretically, high carbon deposition is usually observed in CO$_2$ reforming compared to steam reforming, particularly when heavy hydrocarbons, for example, propane, butane, and gasoline, are applied even under oxidant-rich conditions. It has been widely established that the carbon deposition can be minimized by proper selection of catalyst type, catalyst support, and catalyst modifier. In addition, it has been reported that the carbon-free operation during CO$_2$ reforming of methane can be enhanced if the reaction is kinetically controlled by using suitable catalysts. For sulfur poisoning, it is known that most of the metal catalysts are very susceptible to sulfur compounds. Nickel is one of the metals most susceptible to sulfur poisoning. At low sulfur concentrations (<100 ppm) in the feed, the catalyst lifetime significantly decreases even with the operating temperature as high as 900°C. Doping of some metals such as rhenium or lanthanum can improve sulfur tolerance. For instance, 5 wt% Re−Sr/Ni/ZrO$_2$ was reported to provide high stability for the autothermal reforming of sulfur-containing fuels [6]. Therefore, it is clear that the selection of catalyst is an important criterion for enhancing good reaction activity. Recently, a number of publications have reported on the development of catalysts for the CO$_2$ reforming reaction. The development of catalysts for this reaction might be categorized into two main groups: development of the active component and investigations of the catalyst support and modifier. Details of these developments are presented below.

8.3.1 Active Components

Metallic catalysts are known to be active for the CO$_2$ reforming reaction. Numerous supported transition metal catalysts (Ni, Ru, Rh, Pd, Pt, and Ir) have been tested for catalyzing the carbon dioxide reforming of methane. Rostrup-Nielsen and Bak-Hansen [7] investigated the activity of this reaction over several metals; their order of activity for this reaction is

$$Ru > Rh > Ni \sim Ir > Pt > Pd$$

which is similar to their proposed order for steam reforming. They also observed that the replacement of steam with CO_2 gave similar activation energies, which indicated a similar rate-determining step in these two reactions. An order of catalyst activity similar to that of Rostrup-Nielsen and Bak-Hansen [7] was reported by Takayasu et al. [8], who presented the order of catalyst activity for the CO_2 reforming of methane as

$$Ru > Rh > Pt > Pd$$

However, several researchers have reported a different order of catalyst activity for the CO_2 reforming of methane. For instance, Solymosi et al. [9] reported the order of catalyst activity (supported by alumina) for the CO_2 reforming of methane as

$$Rh > Pt \sim Pd > Ru > Ir$$

whereas Ashcroft et al. [10] reported high activity for Ni, Ir, and Rh and low activity for Pd and Ru. Nevertheless, in light of these previous claims, it is generally agreed that Rh is the most highly active catalyst for this reaction.

Apart from the use of precious metals (Ru, Rh, Pd, Pt, and Ir), the use of Ni as catalyst for the CO_2 reforming reaction has also been widely investigated since it is much cheaper than all precious metal catalysts. For example, Takayasu et al. [8] indicated that the use of 5 mol% Ni showed slightly higher activity than the use of 1 mol% Ru (91% methane conversion compared to 90% conversion). Nevertheless, it is well agreed that the major drawback of Ni as catalyst is its high sensitivity to carbon deposition. Therefore, numerous researchers have widely investigated approaches to minimize the degree of carbon deposition from the CO_2 reforming over Ni-based catalyst. For instance, Wang et al. [12] indicated that the loading amount of Ni as well as its preparation method have great influence on the catalytic activity. In addition, the size and morphology of the Ni particle also affect its catalytic performance, in that the use of smaller size of Ni particle results in less carbon deposition. Importantly, the interaction between Ni and its support also plays an important role on the degree of carbon deposition. Yuliang et al. [11] reported that Ni supported by Al_2O_3 and SiO_2 (Ni/Al_2O_3 and Ni/SiO_2) treated by glow discharge plasma exhibits high CO_2 reforming activity with low carbon deposition, which is mainly due to the improvement of Ni dispersion and the formation of $Ni-Al_2O_3$ and $Ni-SiO_2$ interaction by plasma treatment.

In addition to studies on the metallic-based catalysts, recently several investigations have also focused on the development of other alternative reforming catalysts, for example, ceria-based catalysts and perovskite-based catalysts. Cerium oxide (or ceria) is an important catalyst for a variety of reactions involving oxidation of hydrocarbons. It is also used as catalyst promoter and catalyst support in several industrial processes and as a key component in the formulation of catalysts for the control of noxious emissions in the transportation sector. In detail, this material contains a high concentration of highly mobile oxygen vacancies, which act as local sources or sinks for oxygen involved in reactions taking place on its surface. This high oxygen mobility, high oxygen storage capacity, and ability to be modified render the ceria-based material very interesting for a wide range of catalytic applications. Recently, one of the great potential applications of ceria-based material is in the solid oxide fuel cell (SOFC) as an in-stack reforming catalyst (IIR-SOFC). Ceria is a candidate catalyst for the CO_2 reforming reaction since it is much less active and more resistant to carbon deposition compared to Ni. Recently, successful tests of ceria for methane cracking, CO_2 reforming of methane, and steam reforming of methane have been reported, from which it has been proposed that the reforming reactivity of ceria-based materials involves the gas-solid reaction between hydrocarbons and the lattice oxygen ($O_O{}^x$) on the ceria surface. Some previous researchers have proposed a redox mechanism to explain the CO_2 reforming of methane over ceria-based catalysts, suggesting that the CH_4 reaction pathway for ceria-based materials involves the reaction between absorbed CH_4 (forming intermediate surface hydrocarbon species) with the lattice oxygen ($O_O{}^x$) at the CeO_2 surface, as illustrated schematically below.

CH_4 *adsorption*:

$$CH_4 + 2^* \rightarrow CH_3{}^* + H^* \tag{8.10}$$

$$CH_3{}^* + {}^* \rightarrow CH_2{}^* + H^* \tag{8.11}$$

$$CH_2{}^* + {}^* \rightarrow CH^* + H^* \tag{8.12}$$

$$CH^* + {}^* \rightarrow C^* + H^* \tag{8.13}$$

Co-reactant adsorption:

$$CO_2 + 2^* \rightleftharpoons CO^* + O^* \tag{8.14}$$

Redox reactions of lattice oxygen (O_{O^x}) with C^* and O^*:

$$C^* + O_O{}^x \rightarrow CO^* + V_{O\bullet\bullet} + 2e' \tag{8.15}$$

$$V_{O\bullet\bullet} + 2e' + O^* \rightleftharpoons O_O{}^x + {}^* \tag{8.16}$$

where * is the surface active site of ceria-based materials. $V_{O\bullet\bullet}$ denotes an oxygen vacancy with an effective charge 2^+, and e' is an electron that can be either more or less localized on a cerium ion or delocalized in a conduction band. During the reactions, CH_4 adsorbed on *, forming intermediate surface hydrocarbon species (CH_x^*) (Eqs. 8.10–8.13) and later reacted with the lattice oxygen ($O_O{}^x$) (Eq. 8.15). The steady-state reforming rate is due to the continuous supply of the oxygen source by CO_2 (Eq. 8.14) that reacted with the reduced-state catalyst to recover lattice oxygen ($O_O{}^x$) (Eq. 8.16). An identical reaction rate for steam and CO_2 reforming at similar methane partial pressures, as well as a stronger linear dependence of the reforming rate on methane partial pressure with the positive fraction value of reaction order in this component and the independent effects of steam and CO_2 on the reaction rate have been reported, which provides evidence that the sole kinetically relevant elementary step is the reaction of intermediate surface hydrocarbon species with the lattice oxygen (O_O^x), and that oxygen is replenished by a significantly rapid surface reaction of the reduced state with the oxygen source from either steam or CO_2; this fast step maintains the lattice oxygen (O_O^x) essentially unreduced by adsorbed intermediate surface hydrocarbon. In kinetics studies, negative effects of CO and H_2 were found, which could be due to the reactions between these adsorbed components (CO^* and H^*) with the lattice oxygen (O_O^x) (Eqs. 8.17–8.20), and consequently results in the inhibition of methane conversion.

$$H_2 + 2^* \leftrightharpoons H^* + H^* \tag{8.17}$$

$$H^* + O_O^x \leftrightharpoons OH^* + V_{O\bullet\bullet} + 2e' \tag{8.18}$$

$$CO + {}^* \leftrightharpoons CO^* \tag{8.19}$$

$$CO^* + O_O^x \leftrightharpoons CO_2{}^* + V_{O\bullet\bullet} + 2e' \tag{8.20}$$

In addition to the study on the CO_2 reforming of hydrocarbons (i.e., methane) over ceria-based catalyst, the CO_2 reforming of oxyhydrocarbons (i.e., methanol) over ceria-based catalysts was also reported [13]. Good reforming activity with a low amount of carbon deposition was indicated by the study; furthermore, the mechanism of CO_2 reforming of CH_3OH over ceria-based materials was reported as shown in Eqs. 8.21–8.25, in which CH_3OH first adsorbed on *, forming intermediate surface hydrocarbon species (CH_x^*) and OH* (Eq. 8.21). Similar to the CO_2 reforming of methane, the intermediate surface hydrocarbons then adsorbed on the active surface site and reacted with the lattice oxygen (O_O^x). The steady-state reforming rate is due to the continuous supply of the oxygen-containing compounds present in the system (i.e., CO_2) to regenerate the lattice oxygen (O_O^x).

$$CH_3OH + 2^* \rightarrow CH_3{}^* + OH^* \tag{8.21}$$

$$CH_3{}^* + 3^* \rightarrow C^* + 3H^* \tag{8.22}$$

$$C^* + O_O^x \rightarrow CO^* + V_{O\bullet\bullet} + 2e' \tag{8.23}$$

$$OH^* + {}^* \leftrightharpoons H^* + O^* \tag{8.24}$$

$$V_{O\bullet\bullet} + 2\,e' + O^* \leftrightharpoons O_O^x + {}^* \tag{8.25}$$

The main drawback of ceria is its low specific surface area and high deactivation due to thermal sintering, particularly when used at a high temperature. Research has been conducted on overcoming these constraints by developing a synthesis method that can enhance the surface area of ceria, for example, the use of homogeneous precipitation techniques with different precipitating agents and additives, hydrothermal synthesis, spray pyrolysis methods, inert gas condensation of cerium followed by oxidation, thermal decomposition of carbonates, microemulsion, and electrochemical methods. In addition to the investigation of preparation methods, the addition of doping elements (i.e., Gd, Zr, Y, Nb, La, and Sm) has also been reported to improve the specific surface area, oxygen storage capacity (OSC), redox property, thermal stability, and catalytic activity for steam and CO$_2$ reforming using ceria. Apart from the use of ceria as a catalyst, several researchers have also investigated the use of this material as a catalyst support for several reforming reactions. These investigations a presented in Section 8.3.2.

In addition to the ceria-based catalyst, there has also been an interest in the use of perovskite-based material with the general formula of ABO$_3$ as an alternative reforming catalyst for hydrogen production and fuel cell technologies. Perovskite materials with La at the A-site and a first-row transition metal at the B-site, namely, Cr, Ti, Fe, or Co, can provide a good reforming reaction in terms of high resistance to carbon formation. Among these, LaCrO$_3$-based perovskite material has been widely investigated for SOFC applications, that is, as an anode component and as an internal reforming catalyst (IR-SOFC). Nevertheless, it is well known that pure lanthanum chromite shows a decrease in mechanical strength under reducing conditions as well as phase segregation in the microstructure due to the evaporation of gaseous CrO$_3$ from LaCrO$_3$ particles at high temperature; the partial substitution of Cr on the B-site by Ni (Cr/Ni of 0.9/0.1) has been reported to improve the structural stability without a significant decrease in its catalytic reactivity. Furthermore, the partial substitution of the A-site cation with alkaline earths (i.e., Sr and Ca) has been found to increase the catalytic reactivity of LaCrO$_3$-based perovskite material because of the stabilizing of the B-site cation as well as the introduction of structural defects, for example, oxygen vacancies, while the addition of Ca and Sr elements has been reported to increase the expansion of perovskite materials [14,15]. It is well established that perovskite materials expand when exposed to reducing conditions. The expansion correlates with change in the defect structure of the chromites and results in the reduction of Cr^{4+} to Cr^{3+} as well as the formation of oxygen vacancies as presented below:

$$2Cr^{\bullet}{}_{Cr} + O^x{}_o \leftrightharpoons 2Cr^x{}_{Cr} + V_o{}^{\bullet\bullet} + 1/2\,(O_2) \qquad (8.26)$$

The expansion of several perovskite materials at high temperatures and reducing conditions was observed to increase linearly with increasing temperature, acceptor dopant concentration, and oxygen nonstoichiometry. In addition, the catalyst expansion decreases with increasing oxygen partial pressure. Under the same conditions, Ca doping resulted in higher expansion than Sr doping.

The reaction with dry reforming of methane over La$_{0.8}$Ca$_{0.2}$CrO$_3$ was studied, and it was reported that the reaction of methane on this material had two pathways depending on the operating temperature: (i) At intermediate temperature, the complete oxidation reaction occurs as the lattice oxygen is unlikely to be mobile over this temperature range. Methane reacts with surface oxygen and produces CO$_2$, and H$_2$O as the main products with only a small amount of H$_2$, and CO. (ii) At high temperature, methane dissociation occurs, as the lattice oxygen is likely to be mobile over this temperature range. Methane adsorbs on the oxide surface and forms unsaturated carbon and monatomic hydrogen. The carbon atom oxidizes to produce CO, while monatomic hydrogen combines rapidly to form H_2. The main products in this temperature range are CO and H_2. It was also reported that carbon can be deposited on the perovskite surface because of methane dissociation but with a much lower rate than that of metallic-based catalyst in the same operating condition.

Recently, $La_{0.6}Sr_{0.4}Co_{0.2}Fe_{0.8}O_3$ was investigated as a SOFC anode operating directly with methane. It was observed that CO_2 and H_2O were produced at temperatures above 200°C and did not show any carbon formation. Methane steam reforming activities for $La_{0.8}Sr_{0.2}Cr_{0.97}V_{0.03}O_3$, $La_{0.8}Sr_{0.2}Cr_{0.8}Mn_{0.2}O_3$, and $La_{0.8}Sr_{0.2}Cr_{0.8}Fe_{0.2}O_3$ were also investigated for SOFC application [16]. It was reported that $La_{0.8}Sr_{0.2}Cr_{0.8}Fe_{0.2}O_{3-\delta}$ cannot be applied as the anode material in SOFC because of its high methane cracking activity, whereas $La_{0.8}Sr_{0.2}Cr_{0.97}V_{0.03}O_{3-\delta}$ presented a good electrochemical oxidation of hydrogen and $La_{0.8}Sr_{0.2}Cr_{0.8}Mn_{0.2}O_{3-\delta}$ exhibited electrochemical behavior similar to that of nickel without any carbon formation at 800°C. Lanthanum calcium chromite was also investigated as the anode in SOFC, and it was found that that this material presents lower reforming activity than Ni-YSZ cermet anode but higher resistance to carbon deposition than Ni-YSZ cermet. Recently, $La_{0.8}Sr_{0.2}Cr_{0.9}Ni_{0.1}O_3$ was reported to have high CO_2 reforming of methane activity with high resistance to carbon formation. At low inlet CO_2/CH_4 ratios, its reforming reactivity was as high as that of a precious metal catalyst (i.e., Rh/Al_2O_3).

8.3.2 Support and Promoter

Until now, several contradictory theories have been reported on the role of the catalyst support in the reforming reaction. Some studies have concluded that catalyst supports have no effect on catalyst reactivity, whereas some investigations have reported a strong impact of the supports on the reforming performance in terms of reactivity and kinetics. Recently, some studies have indicated that the type of catalyst support has no effect on the activity and the turnover frequency (TOF) of catalysts in the CO_2 reforming of methane and the key parameter that affects the activity is the dispersion of the metal on the supports. However, there are also several reports showing the strong dependency of the catalyst support on activity and an inverse relationship as well as noncorrelation of the metal dispersion with catalyst activity. They reported that the activity dependency of the supports is related to the electronic interactions between the metal and its support. Some reports compared the activities for the partial oxidation and CO_2 reforming of methane over rhodium catalysts supported on a reducible oxide, for example, CeO_2, niobium pentoxide (Nb_2O_5), tantalum pentoxide (Ta_2O_5), titanium dioxide (TiO_2), and zirconium dioxide (ZrO_2) with a nonreducible oxide, for example, Al_2O_3, lanthanum (III) oxide (La_2O_3), MgO, SiO_2, and yttrium oxide (Y_2O_3). It was found that the catalysts supported on reducible oxides, except for Ta_2O_5, exhibited significantly lower methane conversion than those supported on nonreducible oxides. The difference in catalytic reactivity between catalysts supported by reducible and nonreducible oxides is explained by the partial coverage of the metal by reducible oxides and also by the ability of the reducible oxides to partic- ipate in oxidation reactions. Furthermore, some reports also claim that the acidity of the support strongly affects the reforming activity by promoting the dissocia- tion of hydrocarbon during the reforming reactions. Alternatively, in some cases, the supports are also active for the reforming reactions. For instance, as described above, it was reported that ceria-based supports, for example, CeO_2 and $Ce-ZrO_2$, have good reactivity for CO_2 reforming of methane, methanol, LPG, and ethanol. Furthermore, the improvement in reforming reactivity as well as resistance to car- bon deposition of catalysts supported over ceria-based materials compared to cat- alysts supported over conventional aluminum oxide support has also been widely investigated.

Although there are contradictory theories about the role of the catalyst support in the reforming reaction, it is well agreed that the important criterion for enhancing high reaction rate of the catalyst is that the metal be well dispersed over the catalyst support. Therefore, the important properties of the catalyst support to obtain a high reaction rate are the high specific surface area and good resistance to thermal sin- tering, particularly for the high-temperature reaction. Generally, α-Al_2O_3, γ-Al_2O_3, MgO, and CaO are the common catalyst supports for the CO_2 reforming reaction because of their high specific surface area and high thermal stability. In addition, the mixed-oxide supports including MgO- Al_2O_3, CaO- Al_2O_3, and SiO_2-Al_2O_3 have also been reported to improve some properties of the single-oxide supports, that is, the resistance to carbon deposition. For instance, the problem of deactivation of Ni/Al_2O_3 and Ni/SiO_2 due to carbon deposition can be minimized by substituting Al_2O_3 and SiO_2 supports with MgO (as MgO- Al_2O_3 and MgO- SiO_2). Importantly, the use of ceria-based materials as the support and promoter for the catalytic CO_2 reforming reaction has recently been investigated by several researchers.

Ceria-based materials have been reported to be promising supports among α-Al_2O_3, γ-Al_2O_3, and γ-Al_2O_3 with alkali metal oxide and rare earth metal oxide and $CaAl_2O_4$. In addition, the addition of zirconium oxide (ZrO_2) to pure ceria as ceria-zirconia ($CeO_2 - ZrO_2$ or $Ce - ZrO_2$) has also been reported to provide a good catalyst support for the reforming reaction since it can improve oxygen storage capacity, redox property, thermal stability, and catalytic activity. This high oxygen storage capacity was associated with enhanced reducibility of cerium (IV) in $Ce - ZrO_2$, which is a consequence of the high O^{2-} mobility inside the fluorite lattice. The possible reason for the increasing mobility might be related to the lattice strain, which is generated by the introduction of a smaller isovalent Zr cation into the CeO_2 lattice (Zr^{4+} has a crystal ionic radius of 0.84 Å, which is smaller than 0.97 Å for Ce^{4+} in the same coordination environment) [17]. Because of the high thermal stability of this material, $Ce - ZrO_2$ would be a good candidate to be used as the catalyst support for a high-temperature reforming reaction.

Ceria-based materials have also been reported to be a good promoter for dry methane reforming at intermediate temperature. Wang and Lu [18a] prepared CeO_2-doped Ni/Al_2O_3 by adding CeO_2 on γ-Al_2O_3 powder before impregnated Ni on CeO_2-Al_2O_3 support and tested the dry methane reforming reactivity at

500–800°C. They found that the doping of CeO_2 significantly improved the resistance of catalyst to the carbon deposition. According to the range of temperature for the CO_2 reforming reaction, carbon would be formed via the decomposition of hydrocarbon and Boudouard reactions. By applying ceria as catalyst support and promoter, both reactions could be inhibited by the redox reaction between the surface carbon (C) forming via the adsorptions of hydrocarbon and CO (produced during the reforming process) with the lattice oxygen ($O_O{}^x$) at ceria surface (Eq. 8.27).

$$C + O_O{}^x \rightarrow CO + V_{O\bullet\bullet} + 2e' \qquad (8.27)$$

In addition to the use of ceria-based materials as catalyst promoter, several alkaline earth oxides, such as calcium oxide and magnesium oxide, and rare earth oxides are also known as good modifier or promoter materials for Ni-based catalysts (e.g., Ni/Al_2O_3 and Ni/ZrO_2) to reduce the amount of carbon formation by suppressing the decomposition of hydrocarbon feeds to coke and preventing the formation of nickel aluminate ($NiAl_2O_4$) (in the case of Ni/Al_2O_3), which has been known as a cause of catalyst deactivations. Furthermore, in the case of Ni/ZrO_2, the incorporation of calcium oxide with zirconium dioxide as $Ni/CaO–ZrO_2$ was reported to show a high reforming activity by preventing the formation of nickel oxide (NiO) and reducing the degree of carbon formation.

In addition to the use of metal oxide-based support, carbon-based support is another alternative route for the CO_2 reforming reaction that has also been receiving much attention recently. Among several carbon materials, activated carbon and carbon black have been mostly applied as the catalyst support for several reactions including the CO_2 reforming reaction [18b]. Its great benefit is the relatively high surface area and porosity compared to the metal oxide-based support, which allows the high dispersion of metallic material over its surface and results in high catalyst reaction activity. Another advantage of the carbon-based support is the potential adjustment of its pore size distribution; hence, the catalyst pore size can be well-designed to match with the selected catalytic reactions. Nevertheless, the main barrier to the carbon-based support is the fact that it is easily gasified under reducing and oxidizing conditions, which makes it difficult to apply in hydrogenation and oxidation reactions.

8.4 REACTOR TECHNOLOGY FOR DRY REFORMING

The development of reactors is another important issue for CO_2 reforming technology. CO_2 reforming is similar in nature to a well-established steam reforming technology, and therefore conventional reactors for steam reforming can be applied to CO_2 reforming. However, design of the reactors must take into account the stronger endothermicity and higher potential for carbon formation of CO_2 reforming. Typically, water could be included with CO_2 to operate the reactor under combined steam and CO_2 reforming for suppression of carbon formation and adjustment of a H_2/CO ratio. Autothermal reforming achieved by combined feeding of CO_2 and oxygen (or air) is also an attractive operation, as the exothermic partial oxidation directly generates heat required for the endothermic dry reforming, leading to several benefits such as a simplified system, better temperature control, lower operating temperature, easier start-up, and less coking [19]. The same concept extends to tri-reforming (CO_2, steam, and partial oxidation reforming) with the use of CO_2, steam, and oxygen (or air). At present, there are many well-known commercial processes for large-scale hydrogen or syngas production from steam reforming and autothermal reforming of hydrocarbons, in particular methane, a main component in natural gas. The processes are based on the applications of conventional fixed-bed or fluidized bed reactors. For the CO_2 reforming of methane, the same reactors/processes could be employed. As the details of the commercial reactors/processes are available in the literature they are not discussed here, but this chapter focuses on the development of other innovative reactors such as the membrane reactor, plasma-assisted reactor, microreactor, and periodic operation reactor. Although most of these novel reactors are at the stage of research and development in laboratories except for some technologies that are being tested at pilot plant scale, they show some promising performances compared with conventional reactors and further efforts are required to bring the technologies to commercial application.

A number of efforts have been devoted to developing membrane reactor technology for hydrogen/syngas production. Thermodynamic analysis of dry reforming indicates that the reaction is thermodynamically favorable at high operating temperatures. The use of hydrogen-selective membranes such as palladium and its alloys has been demonstrated to shift the forward equilibrium-limited reaction according to Le Chatelier's principle, thus allowing the reactor to operate at a lower operating temperature with less of a catalyst sintering problem as well as offering a pure hydrogen product. The major challenges remain the long-term stability and high cost of the membrane as well as catalyst deactivation. A number of efforts have been devoted to developing high-performance palladium-based membrane and module design for membrane reactor. Recently, a novel class of membrane reactor for the dry reforming of methane was proposed [20]. The membrane reactor employs a CO_2-selective membrane consisting of mixed-conducting oxide and molten carbonate phases to extract CO_2 from hot flue gas to react with methane fed in the other chamber packed with a reforming catalyst. Heat available in the flue gas can be utilized for the endothermic reactor. The operation is also attractive in terms of utilization of CO_2 available in waste flue gas from combustion. Plasma-assisted reforming technology is an interesting subject nowadays. In a plasma reactor, the energy and free radicals used for reactions are provided by plasma. This offers several benefits over a conventional reactor in terms of cost, catalyst deactivation, and weight requirement. A number of studies have employed the plasma reactor for hydrogen/syngas production from various hydrocarbons [21–24]. The operation can take place with or without the presence of catalysts. Some studies have experimentally demonstrated a synergistic effect when combining plasma with thermal catalysis for hydrogen production. Plasma can be broadly categorized into thermal (high temperature) plasma and nonthermal (cold) plasma. Most reforming studies are based on a nonthermal plasma because of its higher efficiency. Among various types of plasma technologies such as plasmatron, gliding arc, dielectric barrier discharge, corona, microwave, and pulsed discharge, it was evaluated that arc discharge-based reactors are among the most attractive technologies as they offer high energetic densities, relative simplicity of setup, and the ability to create a large reactive volume. However, a number of studies on the dry reforming of methane still focus on thermal plasma technology because the challenge is to improve energy conversion efficiency. Generally, the main disadvantages of a plasma reactor arise from the requirements of electrical power and high electrode erosion, particularly at elevated pressures.

The microreactor is another potential reactor for syngas production. One side of the microreactor is coated with a reforming catalyst and the other with an oxidation catalyst where fuel combustion takes place to provide enough heat for the endothermic reaction. Excellent utilization of catalyst and a high heat/mass transfer characteristic allow the system to be very compact and highly efficient. Selection of a catalyst with high coking resistance is important for the microreactor technology to avoid microchannel plugging. The technology is also considered to be rather expensive for practical operation compared to the conventional reformers. The periodic operation reactor is based on transient operation. Two reaction steps are involved. In the first step, methane is fed to the reactor where endothermic methane decomposition takes place, yielding pure hydrogen and solid carbon on catalyst surface. Then CO_2 is fed to the catalyst bed to convert carbon to carbon monoxide and heat. Oxygen can be added along with CO_2 to enhance the catalyst regeneration and provide sufficient energy for the first step. The catalyst bed serves as a regenerative heat exchanger with excellent internal heat transfer between gas and packing. Although the concept of periodic operation is rather simple, successful operation depends significantly on system design and control.

8.5 CONVERSION OF SYNTHESIS GAS TO SYNTHETIC FUELS

As discussed above, global warming has caused serious problems worldwide because of its effects on the environment and human life. CO_2 is the principal greenhouse gas, and its emissions are largely derived from the burning of fossil fuels for transportation (vehicles) and the generation of electricity (power plants). Conversion of CO_2 to other value-added products (synthetic fuels in particular) would help suppress the global warming effect. A sustainable solution to the energy crisis must include the generation of energy from renewable resources and a reduction in both pollution emissions and gross consumption of raw materials. A number of synthetic fuels such as biomass-to-liquid (BTL), gas-to-liquid (GTL), methanol, and DME have attracted considerable attention as promising alternative fuels. The production of these fuels can be obtained via catalytic reaction of syngas. The catalysts play crucial roles in the production processes, where the optimization of reaction condition would provide optimized performance with minimal effect on the environment. In this section, the focus is paid to three major renewable fuels: GTL, methanol, and DME. The production process and catalyst used are presented along with a variety of applications of these synthetic chemicals.

8.5.1 Gas-to-Liquid

8.5.1.1 Introduction to GTL Technology and Its Applications.

A new technology that is being developed and applied to synthesize liquid hydrocarbon fuels ranging from gasoline to middle distillates has been receiving increasing attention because of a variety of environmental issues, societal benefits and economics. Gas-to-liquid (GTL) technology is one alternative technology to make clean, versatile liquid fuel and chemicals. GTL is an integrated refinery process to convert natural gas or other gaseous hydrocarbons into longer-chain hydrocarbons. Therefore, the process involves the manufacture of synthesis gas (syngas) by feeding natural gas (methane) into a reformer or generator. The two major components of syngas are hydrogen and carbon monoxide, along with a trace amount of CO_2. Conversion of the syngas into hydrocarbon (synthetic crude or syncrude) via the Fischer–Tropsch (FT) reaction by a liquid processing unit follows [25]. Typical output products for a GTL process consist of about 70% ultra-clean diesel fuel, 25% naphtha, and a few percent LPGs, lubes, and waxes. In addition, synthetic crude can also be produced by the same process from coal, biomass, or any carbon-containing material [26]. Three main steps typically include

1. Parallel syngas generation using natural gas, coal, and biomass gasifiers
2. Hydrocarbon production via Fischer–Tropsch
3. Product upgrading units

There are two major technologies for GTL to produce syncrudes: a direct conversion from natural gas (or CH_4-based feedstock), and an indirect conversion via syngas (Fig. 8.1). The direct conversion refers to processes in which feedstock is converted directly into intermediate or final products, without any conversion to syngas. This process can eliminate the cost of production of syngas but requires high activation energy and is difficult to control [27]. Direct conversion is not yet possible in commercial quantities, being economically unattractive. The indirect conversion process refers to a process in which feedstocks are first converted to syngas either through gasification or steam/dry reforming of methane. That syngas is processed into a liquid transportation fuel using Fischer–Tropsch synthesis or methanol (Mobil process) depending on the desired end product [28]. Indirect conversion has the most deployment worldwide, with global production estimated at around 980,000 barrels per day by the year 2010, and many additional projects under active development [29,30].

The synthetic fuels from GTL have a practical low aromatic content and are sulfur free, reflecting a worldwide trend toward the reduction of sulfur and aromatics in fuel. These technologies have also empowered the production of ultra-clean

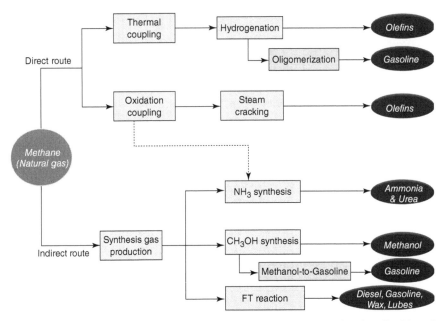

Figure 8.1 Conceptual routes for the conversion of methane (natural gas) to liquid products.

diesel with a higher cetane number of 75–80 compared with the typical conventional diesel range of 45–50, permitting a superior performance engine design [30]. Moreover, GTL not only adds value but is also capable of producing products that could be blended into refinery stock as superior products with fewer pollutants. With the expected rise in demand for diesel, GTL technology provides the option to make a fuel with qualities that can enable significant reductions in emissions. The GTL process provides various end products, including GTL-based LPG, naphtha, jet fuels, diesel constitutes, lubes, and wax. In addition, GTL fuels can used as a blending component for upgrading other fuels to reach the specifications of the oil standard [25]. The GTL technology has a tendency to produce more of the lighter petroleum fractions (kerosene and diesel) compared to the refined oil barrel distillate fraction.

8.5.1.2 *GTL Production Processes and Catalysts.* GTL process can be divided into three basic process steps, syngas generation, FT synthesis, and refining of the synthetic crude. For syngas production, natural gas is converted to syngas by various processes. The most widely practiced commercial processes for production of syngas are steam reforming, CO_2 reforming, autothermal reforming, and partial oxidation [27]. The next process of producing synthesis fuels through indirect conversion is often referred to FT synthesis. FT synthesis is the catalytic hydrogenation of CO to give a range of products, which can be used for the production of high-quality diesel fuel, gasoline, linear chemicals, and oxygenated hydrocarbons [25]. The FT chemistry can be described by the following set of reactions:

Methanation reactions: $CO + 3H_2 \rightarrow CH_4 + H_2O \, \Delta H_{298}° = -247 \text{ kJ/mol}$

$$(8.28)$$

Hydrocarbon synthesis: $CO + 2H_2 \rightarrow (1/n)(C_n H_{2n}) + H_2O$ (8.29)

Water-gas-shift: $CO + H_2O \rightarrow CO_2 + H_2 \, \Delta H_{298}° = -41 \text{ kJ/mol}$

$$(8.30)$$

Boudouard reaction: $2CO \rightarrow C + CO_2 \, \Delta H_{298}° = -172 \text{ kJ/mol}$ (8.31)

This process is technically classified into two categories, the high-temperature Fischer–Tropsch (HTFT) and the low-temperature Fischer–Tropsch (LTFT) processes [25,27,28,31]. The principal criterion for this classification is the operation temperature (Sasol plant at Secunda, South Africa) [25].

- HTFT is operated at temperatures of 300–350°C and uses an iron-based catalyst. This process produces predominantly gasoline and light olefins. Alcohols and ketones (co-products) are either recovered or processed to become fuel components. This process consumes a lower H_2/CO ratio than LTFT, resulting in less CO_2 production.
- LTFT is operated at lower temperatures of 200–260°C and uses a cobalt-based catalyst. This process produces more paraffin and waxy products than the HTFT. The products can readily be worked up to predominantly high-quality diesel such as base oils for high-quality lubricants, automotive gas oil, and food-grade waxes.

A variety of catalysts can be used for the Fischer–Tropsch process, but the most common ones are in a transition metal class. Co, Fe, Ni, and Re have sufficiently high activities for the hydrogenation of carbon monoxide [32,33]. Among them, Ru is the most active, but its high cost and low availability rule it out for large-scale application. Ni is also very active but tends to favor methane formation; moreover, it can form volatile carbonyl nickel, resulting in continuous loss of the metal under operational conditions. Co-based catalysts are highly active, although Fe may be more suitable for syngas with low hydrogen content such as those derived from coal because of its promotion of the water-gas shift reaction. Currently, it is clear that only Co- and Fe-based catalysts are practical as FT catalysts.

Co catalysts are more active than Fe for FT synthesis when the feedstock is natural gas. Natural gas has a high hydrogen-to-carbon ratio, so the water-gas shift is not needed for Co catalysts. Co catalysts are suited for the production of high-value diesel fuel. On the other hand, Fe catalysts are preferred for lower-quality feedstocks such as coal and biomass. It rather produces the high-quality linear alkenes at high temperatures in fluidized bed reactors [33]. During the reaction, Co-, Ni-, and Ru-based catalysts remain in the metallic state, but in the case of Fe catalysts, it tends to form a number of phases including various oxides and carbides during synthesis [34]. Thus control of these phase transformations can maintain the catalytic activity and prevent the breakdown of the catalyst particles. FT catalysts are sensitive to poisoning by sulfur-containing compounds. Catalyst sensitivity to sulfur is greater for Co-based catalysts than for their Fe counterparts. Promoters also have an important influence on catalytic performance. Important promoters for FT catalysts include oxides of alkali metal (K, Ca), transition metal (Mn, V, Cr, Ti, and Cu), lanthanides (Ce and La), and actinides (Th). Noble metals are common promoters, but the formulation depends on the primary metal, Fe and Co [35–37]. It should be noted that K is a poison for Co but is the most important promoter for Fe. Bartholomew and Farrauto have summarized the functions of the promoters and catalysts in GTL synthesis [28].

Nevertheless, the literature provides considerable evidence that support, metal loading, and dispersion are influences on the activity and selectivity of catalysts, especially for CO hydrogenation [38,39]. Thus catalysts are usually supported on high-surface-area supports such as silica, alumina, or zeolites. Finally, the products of FT synthetic operations, whether they are produced from HTFT or LTFT, need upgrading operations to make them suitable for use as fuels like gasoline, kerosene, and diesel. The upgrading operations are similar to those carried on in petrorefineries, like hydrocracking, reforming, hydrogenation, isomerization, polymerization, and alkylation. For hydrocarbons of less than C_4 atoms, isomerization is needed to produce gasoline, while for heavy hydrocarbons of more than C_{20} hydrocracking is necessary to produce kerosene and diesel. For middle products, reforming is only needed to produce high-grade fuels.

8.5.2 Methanol and DME

8.5.2.1 Introduction to Methanol and DME Technologies and Their Applications.
Methanol (CH_3OH), as the simplest aliphatic alcohol, is one of the most important industrial petrochemical products. It is a colorless liquid, completely miscible with water, that is very hydroscopic. In recent years, since methanol has also attracted growing attention as a fuel for fuel cells, a co-reactant for biodiesel production, and an intermediate raw material of hydrogen and DME, which are clean energy sources, it is expected that the demand for methanol will continue increasing in the future. DME, also called methoxymethane, has the chemical formula CH_3OCH_3 and is the smallest ether. DME has long been used as an aerosol propellant in various applications because it is friendly to the environment. Nowadays, DME has been considered a promising synthetic fuel that can be employed as a LPG and diesel substitute for cooking/heating and transportation. DME can also be used as a fuel for fuel cells. DME is an oxygenated hydrocarbon that provides a high hydrogen-to-carbon ratio and is harmless. Therefore it is preferable to methanol when use is highly restricted by human and environment safety. Both methanol and DME can be produced from various sources such as biomass, natural gas, and coal. Presently, DME can be produced via two catalytic routes (Fig. 8.2):

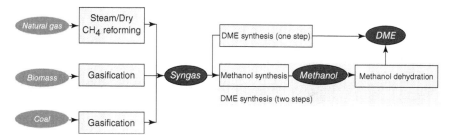

Figure 8.2 Feedstocks and typical synthesis processes of methanol and dimethyl ether (DME).

1. Methanol synthesis followed by methanol dehydration to DME (two steps)
2. Direct synthesis of DME from syngas (one step)

Because of the similar physical properties of DME and LPG, the well-developed infrastructure of LPG can be also adapted for DME, and this makes DME outstanding for practical uses as well [40–44]. Like LPG, DME is a gas at normal temperature and pressure and can readily be transformed into a liquid when it is subjected to modest pressure or cooling. As a result, DME can be easily stored and transported. Finally, thanks to its high oxygen content, ultra-clean combustion, and lack of noxious compounds such as sulfur, DME is a promising, versatile alternative to other clean, renewable, low-carbon fuels. Both DME and methanol are alternate fuels for internal combustion and other engines, either in combination with gasoline or directly. Their advantages over the fossil fuels are nonemission of particulate matter and being sulfur free upon burning. The cetane number of DME is in the range of 55–60, while that of MeOH is 5.

8.5.2.2 MeOH and DME Production Processes and Catalysts.

The reaction pathways for methanol synthesis involve natural gas or methane as a feedstock that is typically reformed to synthesis gas as the first step [45]. The commercial method of methanol production is based on synthesis gas (H_2/CO/ CO_2) in a tubular packed bed reactor at high pressure and low temperature. The conversion of synthesis gas to methanol is strongly influenced by thermodynamic factors. In the commercial process, the water-gas shift (WGS) and reverse WGS reaction can occur simultaneously at the catalyst during the synthesis.

Methanol synthesis from CO:

$$CO + 2H_2 \rightarrow CH_3OH \qquad \Delta H_{298°} = -91.0 \text{ kJ/mol} \qquad (8.32)$$

Methanol synthesis from CO_2:

$$CO_2 + 3H_2 \rightarrow CH_3OH + H_2O \qquad \Delta H_{298°} = -49.7 \text{ kJ/mol} \qquad (8.33)$$

Pure CO and H_2 are widely used for methanol production, especially in the BNL (Brookhaven National Laboratory) method, via the formation of methyl formate as two main intermediate steps [46–48]. However, this process must be free of CO_2 and H_2O in the gas feed-stream, leading to a high cost of operation. Tsubaki et al. [49] proposed a reaction route that is composed of several steps as shown below via containing of CO_2 and H_2O in synthesis gas over Cu-based catalysts.

$$CO + H_2O \leftrightharpoons CO_2 + H_2 \tag{8.34}$$

$$CO_2 + 0.5\,H_2 + Cu \leftrightharpoons HCOOCu \tag{8.35}$$

$$HCOOCu + ROH \leftrightharpoons HCOOR + CuOH \tag{8.36}$$

$$HCOOR + 2H_2 \leftrightharpoons ROH + CH_3OH \tag{8.37}$$

$$CuOH + 0.5H_2 \leftrightharpoons H_2O + Cu \tag{8.38}$$

Overall reaction:

$$CO + 2H_2 \leftrightharpoons CH_3OH \tag{8.39}$$

For methanol synthesis from CO_2 and H_2, Fan et al. [50] investigated a reaction through formic ester by using Cu-based catalysts that consisted of three steps as shown below:

$$CO_2 + H_2 \leftrightharpoons HCOOH \tag{8.40}$$

$$HCOOH + C_2H_5OH \leftrightharpoons HCOOC_2H_5 + H_2O \tag{8.41}$$

$$HCOOC_2H_5 + 2H_2 \leftrightharpoons CH_3OH + C_2H_5OH \tag{8.42}$$

Overall reaction:

$$CO_2 + 3H_2 \leftrightharpoons CH_3OH + H_2O \tag{8.43}$$

Moreover, utilizing a direct method such as using CH_4 along with CO_2 and/or O_2 as reactants in order to produce methanol can mitigate global warming issues and diminish both capital and operating cost. However, the drawback of methane partial oxidation in a thermal reactor is that the conditions that will induce methane oxidation will certainly induce methanol oxidation, leading to carbon oxide production and lowering methanol selectivity. Therefore, conventional thermal catalytic reactors for methanol production have been proposed to be replaced by several new technologies consisting of catalyst and plasma reactors as a promising approach, because of the strength of the electrical energy that can break C–H bonds of methane. Among other plasma techniques, the nonthermal dielectric barrier discharge (DBD) plasma chemical process is one of the most promising future technologies in synthesizing methanol [51–53].

Basically, a conventional methanol reactor for syngas conversion is a vertical shell/tube type in which tubes are packed with catalysts and shells are filled by the boiling water. The heat of reactions is transferred to the boiling water, and steam is produced [54]. Because of the thermodynamic equilibrium limitations of the methanol synthesis reactions, methanol conversion in the conventional methanol reactors is low. Thus most of the synthesis gas should be circulated in the process. Recently, a dual-type reactor system instead of a conventional methanol reactor was proposed [55]. The dual-type methanol reactor is an advanced technology for converting natural gas to methanol. As a solution to overcoming the thermodynamic limitations, the methanol reactor can be devised with a permselective membrane layer for H_2O removal that shifts the equilibrium in the favorable direction.

In the conventional methanol production process, which employs the gas-phase method from the standpoint of catalyst activity and plant cost, the prevalent reaction conditions are 5–10 MPaG in terms of pressure and 200–300°C for temperature. The problems in need of improvement are as follows:

1. Getting a uniform temperature distribution
2. Stabilizing the heat extraction
3. Improving the catalytic activity (lowering the reaction temperature)

In particular, it is necessary to efficiently remove the reaction heat. Therefore, a heat exchanger type or gas-quenched type reactor is used. However, the low conversion of reactants per pass through the catalytic layer makes it difficult to use a large-scale reactor. Furthermore, the development of the liquid-phase method, which is advantageous in terms of reaction heat removal, has been reported. In the conventional gas-phase reactor, unreacted gas is fed in between catalyst beds. Otherwise, the catalyst is filled into tubes, and a cooling medium is circulated outside the tubes to control temperature. This leads to a limited conversion per pass because of the limited heat removal capabilities. In a slurry reactor, gaseous reactants dissolve in the liquid and transfer to the catalyst surface and the resultant products then become gas again. The slurry medium, such as an inert mineral oil, could help enhance energy efficiency with proper reactor designs, providing high syngas conversion to methanol. Air Products and Chemicals, Inc.'s Liquid Phase Methanol (LPMEOH™) Process [56] is designed to convert synthesis gas derived from the gasification of coal into methanol for use as a chemical intermediate or as a low-sulfur dioxide- and low-nitrogen oxide-emitting alternative fuel. The liquid process is superior to the conventional gas-phase process in terms of process efficiency and flexibility of feed composition.

The first catalyst used for methanol production from syngas was $ZnO-Cr_2O_3$. It was commercialized by BASF and used in high-pressure and high-temperature synthesis. The improvements in the purification of the methanol synthesis feed (especially removal of sulfur) led to a major improvement in catalysts, as the use of a Cu/ZnO catalyst was enabled. The Cu/ZnO catalyst has a high activity, and the synthesis can be operated at the lower pressure and temperature. Industrially, methanol is produced catalytically on copper-containing catalysts such as $Cu/ZnO/Al_2O_3$ and $Cu/ZnO/Cr_2O_3$. Tsubaki et al. [49] developed methanol synthesis from carbon monoxide and hydrogen containing CO_2 and water by copper-based catalysts (Cu/ZnO and Cu/Al_2O_3) at 170°C and 3 MPa, compared to the present industrial process. CO_2 and water, which usually act as poisons to an alkali metal methoxide catalyst in the low-temperature methanol synthesis reaction, play a promotional role in this new route.

Accompanying alcohol, as a "catalytic liquid medium," with the aid of copper-based solid catalyst, altered the reaction path to a low-temperature direction. The Cu–Cr–O catalyst and the Pd–Cu–Cr–O (Pd: 5 wt%) catalyst provided a good activity for methanol synthesis from CO_2 and H_2 [50]. Farsi and Jahanmiri [54] focused on modeling and optimization of methanol production in a dual-membrane reactor. They demonstrated that coupling reaction and separation in a membrane reactor improves the reactor efficiency and reduces purification cost in the following stages. Palladium-based membranes have been widely used in hydrogen extraction because of their high permeability, selectivity, and good surface properties. The main advantages of the optimized dual-membrane reactor are higher CO_2 conversion, the possibility of overcoming the limitation imposed by thermodynamic equilibrium, improvement of the methanol production rate, and its purity.

In DME synthesis, two major processes are presently employed in industrial plants as shown in Figure 8.2. The first process is a methanol dehydration (the so-called two-step process), and the second process is the direct DME synthesis from syngas in a single reactor (the so-called single-step process).

MeOH dehydration:

$$2CH_3OH \rightarrow CH_3OCH_3 + H_2O \, \Delta H_{298°} = -23.5 \text{ kJ/mol} \qquad (8.44)$$

The two-step process is the most commonly used process at present because of the maturity of the available technology for obtaining methanol. The catalysts used for this process contain Cu/ZnO as a major component, while solid acid catalysts are the promoters. In recent years, a single component of acid catalysts such as alumina and zeolite (e.g., ZSM-5 and mordenite) has been solely used for methanol dehydration. In terms of stability, alumina outperformed the others. Among zeolites, ZSM-5 provided highest activity and stability. However, when zeolites are used, by-product formation is s serious concern. The two-step process has been developed by some companies, including Mitsubishi Gas Chemical, Toyo Engineering, Lurgi GmbH, and Haldor Topsoe.

The single-step synthesis has been developed with the aim of enhancing energy efficiency and reducing the capital cost. In this direct synthesis, a bifunctional catalyst containing metallic (such as Cu and Pd) and acidic (such as alumina and zeolite) species is generally required [57,58]. The reactions involve the water shift reaction and hydrogenation, and the overall reaction is expressed as the following equation.

Direct DME synthesis:

$$3CO + 3H_2 \rightarrow CH_3OCH_3 + CO_2 \qquad (8.45)$$

Although the reaction mechanisms for methanol formation have been clearly defined, the reaction steps in single-step DME synthesis from syngas are still unclear. The catalyst used for this process can be varied from a physical mixture of Cu-based species with solid acid to chemically prepared composites or mixed oxides. For metallic sites, oxidation state, dispersibility, particle size, and interaction with oxide composites play crucial roles in the reactions. For acid sites, acid strength, acid amount, and acid nature are three major factors bringing about the high performance of the catalysts [59–61]. The development of the bifunctional composite catalyst or hybrid materials has remained a challenging task for the new route. The direct synthesis is more thermodynamically favorable than the two-step process, resulting in enhancement of the production yield. A recycle stream of the resultant and unreacted DME and CO_2 would further allow adjustment of synthesis gas composition and hence DME yield. Like the methanol synthesis process, a slurry-type reactor can be applied to DME synthesis [62]. The slurry-phase process allows higher conversion per pass and high flexibility in syngas composition. A major problem of this process is the poor stability of the catalyst used. Therefore, a major effort must be directed to developing a catalyst with good activity and lifetime.

8.6 CONCLUSIONS

CO_2 is currently considered a poisonous greenhouse gas, which is usually emitted from the combustion of hydrocarbon fuels. Recently, the image of this gaseous compound is changing to that of to a business opportunity, with several attempts to utilize CO_2 for various beneficial purposes. One interesting route is the conversion of CO_2 via a thermochemical process to synthetic fuels including syngas, hydrogen, methanol, DME, and alkane fuel. Generally, syngas and hydrogen can be readily produced from the reaction between hydrocarbon and CO_2, known as CO_2 reforming.

This reaction requires a suitable catalyst to catalyze the reaction, in which the role of the catalyst is not only to activate the overall reaction but also to minimize catalyst deactivation mainly from carbon deposition. Numerous supported transition metal catalysts (Ni, Ru, Rh, Pd, Pt, and Ir) are commonly known to be active for the CO_2 reforming reaction. In addition, several recent investigations have also focused on the development of other alternative reforming catalysts, for example, ceria-based catalysts and perovskite-based catalysts.

Ceria is an important catalyst for a variety of reactions involving oxidation of hydrocarbons. It is also being used as promoter or support in several industrial processes and as a key component in the formulation of catalysts for the control of noxious emissions in the transportation sector. Ceria is a candidate catalyst for the CO_2 reforming reaction since it is much less active and more resistant to carbon deposition compared to Ni. Perovskite-based material with the general formula of ABO_3 is also known as an alternative reforming catalyst for hydrogen production and fuel cell technologies. Perovskite materials with La at the A-site and a first-row transition metal at the B-site, namely, Cr, Ti, Fe, or Co, are known to enhance reforming reactions in terms of high resistance to carbon formation.

At present, there are many well-known commercial processes for large-scale hydrogen or syngas production from the reforming of hydrocarbons, particularly methane. The processes are based on the applications of conventional fixed-bed or fluidized bed reactors. In addition, further developments of the CO_2 reforming reaction in various innovative reactors such as the membrane reactor, the plasma-assisted reactor, the microreactor, and the periodic operation reactor are also of interest nowadays. Although most of those novel reactors are at the stage of research and development in laboratories, except for some technologies that are being tested at pilot-plant scale, they show promising performance over the conventional technologies, and further efforts are required to bring the technologies to commercial application.

Importantly, syngas produced from CO_2 reforming can be efficiently converted to several liquid synthetic fuels: alcohols, DME, and alkane fuels. Alkane fuel, known as GTL, can be efficiently produced from the conversion of syngas via the Fischer–Tropsch process. Typical output products for a GTL process consist of about 70% ultra-clean diesel fuel, 25% naphtha, and a few percent of LPGs, lubes, and waxes, which is applicable as transportation fuel. A variety of catalysts can be used for the Fischer–Tropsch process, but the most common are in a transition metal class. Co, Fe, Ni, and Re have sufficiently high activities for the hydrogenation of carbon monoxide to warrant possible application in FT synthesis.

Currently, it is clear that only Co- and Fe-based catalysts are practical as FT catalysts. Methanol can be produced from the methanol synthesis reaction, while DME can be further synthesized by the dehydration of methanol. Industrially, methanol is produced catalytically on copper-containing catalysts such as $Cu/ZnO/Al_2O_3$ and $Cu/ZnO/Cr_2O_3$, whereas a single component of acid catalysts such as alumina and zeolite (e.g., ZSM-5 and mordenite) has been solely used for methanol dehydration.

ACKNOWLEDGMENTS

The financial support by the Thailand Research Fund (RTA5480003) is greatly appreciated.

REFERENCES

1. Centi G, Perathoner S (2009). Opportunities and prospects in the chemical recycling of carbon dioxide to fuels, *Catal. Today* 148: 191–205.
2. Bradford MCJ, Vannice MA (1999). *Catal. Rev.-Sci. Eng.*, 41: 1–42.
3. Rostrup-Nielsen JR (1979). Symposium on the science of catalysis and its application in industry, Sindri, India, 22–24.
4. Edwards JH, Maitra AM (1995). The chemistry of methane reforming with carbon dioxide and its current and potential applications, *Fuel Proc. Techn.* 42: 269–289.
5. Lemonidou AA, Vasalos IA (2002). Carbon dioxide reforming of methane over 5 wt.% $Ni/CaO-Al_2O_3$ catalyst, *Appl. Catal. A: General*, 228: 227–235.

6. Murata K, Saito M, Inaba M, Takahara I (2007). Hydrogen production by autothermal reforming of sulfur-containing hydrocarbons over re-modified Ni/Sr/ZrO$_2$ catalysts, *Appl. Catal. B*, 70: 509–514.

7. Rostrup-Nielsen JR, Bak-Hansen JH (1993). *J. Catal.*, 144: 38–49.

8. Takayasu O, Hirose E, Matsuda N, Matsuura I (1991). Partial oxidation of methane by metal catalysts supported on ultrafine single-crystal magnesium oxide, *Chem Express*, 6: 447–450.

9. Solymosi F, Kutsan G, Erdohelyi A (1991). Catalytic reaction of methane with carbon dioxide over alumina supported platinum metals, *Catal. Letters* 11: 149–156.

10. Ashcroft AT, Cheetham AK, Green MLH, Vernon PDF (1991). Partial oxidation of methane to synthesis gas using carbon dioxide, *Nature* 352: 225–226.

11. Wang YH, Liu HM, Xu BQ (2009). Durable Ni/MgO catalysts for CO$_2$ reforming of methane: Activity and metal–support interaction, *J. Mol. Catal. A: Chem.*, 299: 44–52.

12. Yuliang L, Gaihuan L, Lei S, Wei C, Xiaoyan D, Yongxiang Y (2008). Modification of Ni/SiO$_2$ catalysts by means of a novel plasma technology, *Plasma Sci. Technol.*, 10: 551–555.

13. Laosiripojana N, Assabumrungrat S (2008). Kinetic dependencies and reaction pathways in hydrocarbon and oxyhydrocarbon conversions catalyzed by ceria-based materials, *Appl. Catal. B*, 82: 103–113.

14. Pudmich G, Boukamp BA, Gonzalez-Cuenca M, Jungen W, Zipprich W, Tietz F (2000). Chromite/titanate based perovskites for application as anodes in solid oxide fuel cells, *Solid State Ionics*, 135: 433–438.

15. Mawdsley JR, Krause TR (2008). Rare earth-first-row transition metal perovskites as catalysts for the autothermal reforming of hydrocarbon fuels to generate hydrogen, *Appl. Catal. A*, 334: 311–320.

16. Vernoux P, Guindet J, Gehain E (1998). Electrochemical and catalytic properties of doped lanthanum chromite under anodic atmosphere, in Proceedings of the Third European Solid Oxide Fuel Cell Forum, Edited by P. Stevens, Nantes, France 237–247.

17. Kim D (1989). Lattice parameters, ionic conductivities, and solubility limits in fluorite-structure MO$_2$ oxide solid solutions, *J. Amer. Ceram. Soc.*, 72: 1415–1421.

18. (a) Wang S, Lu GQ (1998). Role of CeO$_2$ in Ni/CeO$_2$-Al$_2$O$_3$ catalysts for reforming of methane with carbon dioxide, *Appl. Catal. B*, 19: 267. (b) Fidalgo B, Menendez JA (2011). Carbon materials as catalysts for decomposition and CO$_2$ reforming of methane: A review, *Chin. J. Catal.*, 32: 207–216.

19. Assabumrungrat S, Laosiripojana N (2009). FUELS—HYDROGEN PRODUCTION—Autothermal reforming, in *Encyclopedia of Electrochemical Power Sources*, Vol. 3. Amsterdam: Elsevier; 238–248.

20. Tosti S (2010). Overview of Pd-based membranes for producing pure hydrogen and state of art at ENEA laboratories, *Int. J. Hydrogen Energy*, 35: 12650–12659

21. Petitpas G, Rollier JD, Darmon A, Gonzalez-Aguilar J, Metkemeijer R, Fulcheri L (2007). A comparative study of non-thermal plasma assisted reforming technologies: Review, *Int. J. Hydrogen Energy*, 32: 2848–2867

22. Chen HL, Lee HM, Chen SH, Chao Y, Chang MB (2008). Review of plasma catalysis on hydrocarbon reforming for hydrogen production–Interaction, integration, and prospects, *Appl. Catal. B: Env.*, 85: 1–9

23. Holladay JD, Hu J, King DL, Wang Y (2009). An overview of hydrogen production technologies, *Catal. Today*, 139: 244–260

24. Tao X, Bai M, Li X, Long H, Shang S, Yin Y, Dai X (2011). CH_4-CO_2 reforming by plasma–challenges and opportunities, *Progr. Energy and Combustion Science*, 37: 113–124

25. Remans TJ, Jenzer G, Hoek A (2008). in *Handbook of Heterogeneous Catalysis* (G. Ertl, H. Knözinger, F. Schüth, J. Weitkamp Eds.), 2nd ed., Vol. 6, Chap. 13, Wiley-VCH, Weinheim, 2008.

26. Samuel P (2003). *Catal. Soc. India*, 2: 82–99.

27. Al-Shalchi W (2011). Available online at http://www.scribd.com/doc/3825160/Gas-to-Liquids-GTL-Technology Date (23/11/2011).

28. Bartholomew CH, Farrauto RJ (2006). *Fundamentals of Industrial Catalytic Processes*, 2nd ed., Chap. 6, John Wiley & Sons, New Jersey 2006.

29. Velasco JA, Lopez L, Velásquez M, Boutonnet M, Cabrera S, Järås S (2010). *J. Nat. Gas Sci. Eng.* 2: 222–228.

30. Stanley IO (2009). *J. Nat. Gas Sci. Eng.* 1: 190–194.

31. Dry ME (2001). *J. Chem. Technol. Biotechnol.*, 77: 43–50.

32. Davis BH (2011). Available online at http://www.accessscience.com Date (23/11/2011).

33. Steynberg A, Dry M (2004). in *Fischer-Tropsch Technology* (M. E. Dry Ed.) Chap. 7, Elsevier, Amsterdam, 2004.

34. Rae VUS, Stiegel G, Cinquegrane GJ, Srivastava RD (1992). *Fuel Process. Technol.* 30: 83–107.

35. Lu Y, Lee T (2007). *J. Nat. Gas Chem.*, 16: 329–341.

36. Ding M, Yang Y, Wu B, Wang T, Xiang H, Li Y (2011). *Fuel Process. Technol.* 92: 2353–2359.

37. Osa AR, Lucas A, Romero A, Valverde JL, Sánchez P (2011). *Fuel*, 90: 1935–1945.

38. Dalai AK, Davis BH (2008). *Appl. Catal. A-Gen.* 346: 1–15.

39. Sakoumisa TNE, Rønninga M, Borgb Ø, Ryttera E, Holmena A (2010). *Catal. Today* 154: 162–182.

40. Faungnawakij K, Eguchi K (2011). *Catal Surv Asia* 15: 12–24.

41. Vertes A, Qureshi N, Blaschek HP, Yukawa H (Eds) (2010). *Biomass to Biofuels: Strategies for Global Industries*, John Wiley and Sons, Ltd, Publication 2010.

42. Faungnawakij K, Fukunaga T, Kikuchi R, Eguchi K (2008). *J. Catal.*, 256: 37–44.

43. Li Y, Wang T, Yin X, Wu C, Ma L, Li H, Lv Y, Sun L (2010). *Renewable Energy* 35: 583–587.

44. Pontzen F, Liebner W, Gronemann V, Rothaemel M, Ahlers B (2011). *Catal. Today* 171: 242–250.

45. Lange JP (2001). *Catal. Today*, 64: 3–8.

46. Othmer K (1964). *Encyclopedia of Chemical Technology*, 2nd Ed., Wiley, New York, 13: 390.

47. Gormley RJ, Rao VHS, Soong Y, Micheli E (1992). *Appl. Catal.*, 87: 81–101.

48. Palekar VM, Jung H, Tierney JW, Wender I (1993). *Appl. Catal.*, 102: 13–34.

49. Tsubaki N, Ito M, Fujimoto K (2001). *J. Catal.*, 197: 224–227.

50. Fan L, Sakaiya Y, Fujimoto K (1999). *Appl. Cat. A: Gen.*, 180: L11–L13.

51. Zhou LM, Xue B, Kogelschatz U, Elliason B (1998). *Plasma Chemistry and Plasma Processing*, 18: 375–393.

52. Chen L, Zhang XW, Huang L, Lei LC (2009). *Chem. Eng. Proc.*, 48: 1333–1340.

53. Larkin D, Zhou L, Lobban L, Mallinson RG (2001). *Ind. Eng. Chem. Res.*, 40: 5496–5506.

54. Farsi M, Jahanmiri A (2011). *Chem. Eng. Proc.*, 50: 1177–1185.

55. Rahimpour MR, Lotfinejad M (2008). *Chem. Eng. Proc.*, 47: 1819–1830.

56. Commercial-Scale Demonstration of the Liquid Phase Methanol (LPMEOH™) Process, A report on a project conducted jointly under a cooperative agreement between: The U.S. Department of Energy and Air Products Liquid Phase Conversion Company, L.P., 1999.

57. Maoa D, Xia J, Zhang B, Lu G (2010). *Energy Conversion and Management*, 51: 1134–1139.

58. Sierra I, Eren J, Aguayo AT, Arandes JM, Bilbao J (2010). *Appl. Catal. B: Environ.* 94: 108–116.

59. Carr RT, Neurock M, Iglesia E (2011). *J. Catal.* 278: 78–93.

60. Naik SP, Bui V, Ryu T, Miller JD, Zmierczak W (2010). *Appl. Catal. A: Gen* 381: 183–190.

61. Moradi GR, Yaripour F, Vale-Sheyda P (2010). *Fuel Proc. Technol.* 91: 461–468.

62. Jinchuan F, Chaoqiu C, Jie Z, Wei H, Kechang X (2010). *Fuel Proc. Technol.* 91: 414–418.

CHAPTER 9

Fuel Production from Photocatalytic Reduction of CO_2 with Water Using TiO_2-Based Nanocomposites

YING LI

9.1 INTRODUCTION

The global climate change associated with increased greenhouse gas concentrations in the atmosphere has revealed the urgent need for advanced technologies to control

Green Carbon Dioxide: Advances in CO_2 Utilization, First Edition.
Edited by Gabriele Centi and Siglinda Perathoner.
© 2014 John Wiley & Sons, Inc. Published 2014 by John Wiley & Sons, Inc.

CO_2 emissions from fossil fuel consumption and to develop alternative fuels that are carbon-neutral or even carbon-negative. Recently, many efforts have been made to reduce CO_2 emissions through pre- or postcombustion CO_2 capture followed by compression and geological sequestration; however, these technologies are costly and have many uncertainties in the long run [1]. In contrast, with recent innovations in photocatalysis, recycling CO_2 to fuels with sunlight as the sole energy input offers a brand new opportunity for a sustainable energy future. The technology of photocatalytic reduction of CO_2 with water not only mitigates CO_2 emissions but also produces energy-bearing compounds such as CO, methane, and methanol [2–4] that can be subsequently converted to liquid transportation fuels.

9.2 CO_2 PHOTOREDUCTION: PRINCIPLES AND CHALLENGES

Photoreduction of CO_2 has been studied since 1979, when Honda and co-workers first demonstrated conversion of CO_2 to organic compounds in an aqueous suspension of oxide and nonoxide semiconductor particles under UV irradiation [5]. To effectively achieve CO_2 photoreduction by water, the appropriate photocatalysts should meet several criteria including band gap, band edge positions, resistance to photocorrosion, and cost. With regard to band edge positions, the semiconductor conduction band (CB) edge should be located at a potential that is more negative than the reduction potentials of CO_2, while the valence band (VB) edge should be located at a potential that is more positive than the water oxidation potential [6].

A variety of semiconductor photocatalysts such as TiO_2 [2–4,7–10], ZrO_2 [11,12], CdS [13], and In TaO_4 [14] have been studied, and among them wide band-gap TiO_2 catalysts (~3.2 eV for anatase) are considered the most convenient candidates in terms of stability, cost, and nontoxicity. Possible reactions involved in photocatalytic reduction of CO_2 with water on a TiO_2 semiconductor and their reduction potentials [6,15] are illustrated in Figure 9.1. The initial step of CO_2 photoredution on TiO_2 is the generation of electron–hole pairs upon the adsorption of photons of energy greater than or equal to the band gap of the photocatalyst. The electrons, after migrating to the surface, react with CO_2 and protons to produce C1 fuels; while the holes oxidize water molecules (electron donor) to produce O_2 and protons.

There are a few challenges in CO_2 photoreduction that have been well reported in the literature. First, the timescale of electron–hole recombination is two or three orders of magnitude faster than other electron transfer processes [6]. As a result of the fast recombination, the yields of CO_2 reduction products are typically very low [16]. The second challenge is due to the wide band gap of TiO_2 that allows absorption of only ultraviolet light ($\lambda = 387$ nm, corresponding to a 3.2 eV band gap), which accounts for < 5 % of the solar energy. This directly leads to a low energy efficiency if sunlight is used as the energy source. Finally, the CO_2 photoreduction mechanism is poorly understood, and the product selectivity is typically not controlled. It is believed that CO_2 photoreduction involves multi-electron transfer steps, but the reaction pathways and reaction intermediates are difficult to identify.

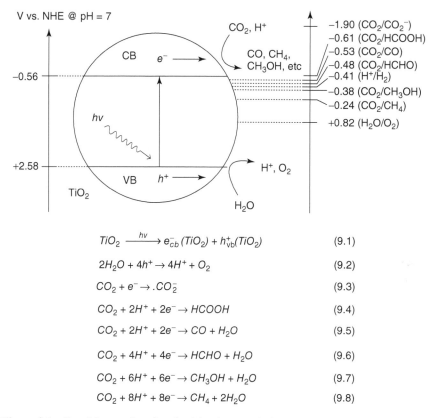

$$TiO_2 \xrightarrow{hv} e^-_{cb}(TiO_2) + h^+_{vb}(TiO_2) \tag{9.1}$$

$$2H_2O + 4h^+ \rightarrow 4H^+ + O_2 \tag{9.2}$$

$$CO_2 + e^- \rightarrow .CO_2^- \tag{9.3}$$

$$CO_2 + 2H^+ + 2e^- \rightarrow HCOOH \tag{9.4}$$

$$CO_2 + 2H^+ + 2e^- \rightarrow CO + H_2O \tag{9.5}$$

$$CO_2 + 4H^+ + 4e^- \rightarrow HCHO + H_2O \tag{9.6}$$

$$CO_2 + 6H^+ + 6e^- \rightarrow CH_3OH + H_2O \tag{9.7}$$

$$CO_2 + 8H^+ + 8e^- \rightarrow CH_4 + 2H_2O \tag{9.8}$$

Figure 9.1 Possible reactions involved in photocatalytic reduction of CO$_2$ with water on TiO$_2$ photocatalyst and their reduction potentials. Adapted from ref. 6,15.

The correlations between the catalyst properties and product selectivity are highly uncertain.

9.3 TiO$_2$-BASED PHOTOCATALYSTS FOR CO$_2$ PHOTOREDUCTION: MATERIAL INNOVATIONS

To address the above-mentioned challenges, material innovations have been developed by using nanosized photocatalyts or high-surface-area support to improve the CO$_2$ conversion efficiency and by modifying the catalyst composition or nanostructures to promote charge transfer and to enhance visible light utilization.

9.3.1 TiO$_2$ Nanoparticles and High-Surface-Area Support

Particle size is an important factor for photocatalysts that determines the surface area and photon conversion efficiency [17]. TiO$_2$ particles in the nanometer size range usually demonstrate higher photocatalytic activity than in their bulk phase.

On the other hand, studies show that there is an optimum size for TiO_2 nanoparticles in the applications of either photocatalytic oxidation of organic compounds [18] or photocatalytic reduction of CO_2 [17]. Koci et al. [17] investigated the effect of TiO_2 particle size on CO_2 photoreduction and reported that 14 nm is the optimum value for TiO_2 anatase; a particle size smaller than 14 nm leads to a higher recombination of the electron–hole pairs, while larger particles have a smaller surface area.

Other studies investigated incorporation of Ti species in silica-based micro/mesoporous materials to enhance photocatalytic reduction of CO_2. Anpo et al. [19] and Yamashita et al. [10] reported that highly dispersed tetrahedrally coordinated TiO_2 species incorporated in mesoporous silica matrix of MCM-41 and MCM-48 showed higher reactivity and selectivity for the formation of CH_3OH compared to small TiO_2 particles. Ikeue et al. [8] and Shioya et al. [20] synthesized Ti-containing porous silica thin film for CO_2 photoreduction with a quantum yield of 0.28% at 323 K for CH_3OH production, which is remarkably higher than powdered Ti-MCM-41 catalyst (quantum yield was 0.02%). The high yield and selectivity can be attributed to their high transparency and the large amount of surface OH species on the catalysts.

9.3.2 Metal-Modified TiO_2 Photocatalysts

The contact of a semiconductor and a metal generally leads to a redistribution of charges and formation of a Schottky barrier, that is, electrons migrate from the semiconductor to the metal until the two Fermi levels are aligned [21]. Hence, metal particles/clusters deposited on TiO_2 can serve as electron trappers and prohibit electron–hole recombination, thus enhancing the rate of CO_2 photoreduction [3,22–25]. Metal species can be loaded on the TiO_2 surface via various methods including incipient wetness impregnation [26], sol–gel [3], photoreduction [27], and sputter coating [23]. While various metals (e.g., Pd, Rh, Pt, Ru, Au, Ag, and Cu [28–30]) have been added to TiO_2 to enhance CO_2 photoreduction, Cu appears to be the most cost-effective choice. Tseng et al. [3] used sol–gel-derived Cu/TiO_2 catalysts for CO_2 photoreduction and found that the yield of methanol is much higher than those without Cu loading. Another study by Ishitani et al. [29] reported that CO_2 photoreduction using Pd, Rh, Pt, Au, Cu, and Ru deposited on a TiO_2 photocatalyst produces CH_4 and acetic acid, with Pd/TiO_2 exhibiting high selectivity for CH_4 production. Li et al. [30] reported increased CO_2 photoconversion efficiency by a Cu/TiO_2 catalyst compared to bare TiO_2 and selective CH_4 production due to Cu loading; they also reported Ag/TiO_2 catalysts prepared by an ultrasonic spray pyrolysis method that have the potential to simultaneously produce H_2 from water and reduce CO_2 to CO [31]. However, too high a metal loading level may result in recombination centers and lead to a reduced photocatalytic efficiency. Optimal metal concentrations have been reported for modified TiO_2 (e.g., with Ag or Cu) for CO_2 photoreduction [28,30–32]. While metal modifications on TiO_2 have apparent enhancement in charge separation, they have limited contribution to extending the photo-response to visible light region. Sasirekha et al. [33] observed that Ru-doped TiO_2 has almost the same absorption spectra as undoped TiO_2. Dholam and Patel [34] reported that Cr- and Fe-doped TiO_2 prepared by a sol–gel method had a very limited effect on inducing a red shift in TiO_2 absorption spectra compared to those prepared by a magnetron sputtering method.

9.3.3   Metal-Modified TiO₂ Supported on Mesoporous SiO₂

Since both surface modification with metals and high-surface-area support enhance the rate of CO_2 photoreduction, it is interesting to investigate the synergistic effect of combining the two strategies. Li et al. [30] used a one-pot synthesis method to prepare Cu-loaded TiO_2 (Degussa P25) nanoparticles dispersed on mesoporous silica matrix forming a Cu/TiO_2–SiO_2 catalyst; the preparation process is illustrated in Figure 9.2a. The TiO_2 mass loading was 12%, and the Cu mass loading ranged from 0.5% to 3%. The synthesized Cu/TiO_2–SiO_2 composites had a high specific surface area that is larger than 350 m²/g. X-ray diffraction (XRD) peaks for copper species were not detected, indicating that Cu is highly dispersed [4,35] or the low Cu concentration results in the formation of extremely small Cu clusters [3]. In addition, UV–vis diffuse reflectance analysis indicates that the incorporation of Cu did not change the TiO_2 absorption spectrum or narrow the band gap, implying that Cu species are loaded on the surface of TiO_2. Transmission electron microscopic (TEM) and high-resolution TEM (HR-TEM) images of the 0.5% Cu/TiO_2–SiO_2 nanocomposite are shown in Figure 9.3a and b. A homogeneous mesoporous structure of SiO_2 is clearly observed. The irregular-shaped dark particles embedded in the SiO_2 framework are P25 TiO_2 nanoparticles, which have an average size around 20 nm. However, the dispersion of TiO_2 nanoparticles is not uniform. The HR-TEM image (Fig. 9.3b) shows clear lattice fringes of TiO_2 nanoparticles and confirms the crystallinity of the P25 TiO_2. It also verifies the amorphous structure of the SiO_2 support. X-ray photoelectron spectroscopy (XPS) analysis of the samples indicated that the major Cu species on the Cu/TiO_2–SiO_2 catalysts was Cu_2O.

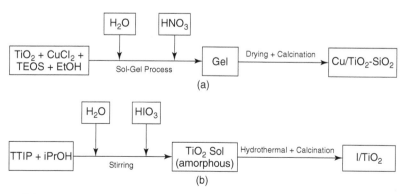

Figure 9.2   Material preparation process for Cu/TiO_2–SiO_2 catalysts (a) and I/TiO_2 catalysts (b). Adapted from refs. 30,39.

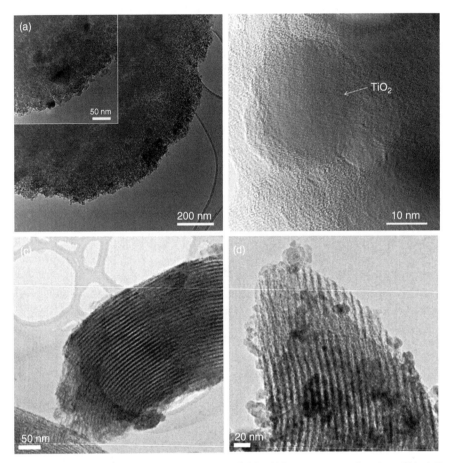

Figure 9.3 TEM and HR-TEM images of Cu/TiO_2-SiO_2 nanocomposites (a and b), 1-D mesoporous SBA-15 (c), and Ce-TiO_2/SBA-15 nanocomposites (d). Reproduced with permission from ref. 30,36.

Li and co-workers [36] also prepared Ce-doped TiO_2 dispersed on ordered mesoporous silica (SBA-15) for CO_2 photoreduction. Ce doping decreased the crystal size of TiO_2, increased the catalyst surface area, and inhibited the growth of rutile TiO_2 crystals. XPS analysis indicated that Ce species existed as a mixture of Ce^{3+}/Ce^{4+}. The specific surface areas of bare TiO_2, Ce-TiO_2, and Ce-TiO_2/SBA-15 were around 25, 130, and 400 m^2/g, respectively. Figure 9.3c demonstrates the morphology of bare SBA-15 samples, where the one-dimensional channels of ordered pores (5–7 nm) are clearly identified. Figure 9.3d demonstrates that in the Ce-TiO_2/SBA-15 sample the TiO_2 nanoparticles are dispersed on the surface, at the edge, and possibly in the pores of the SBA-15 support. Good dispersion of the TiO_2 nanoparticles on the support is believed to be important in catalytic performance in CO_2 photoreduction.

9.3.4 Nonmetal-Doped TiO$_2$ Photocatalysts

To enhance the photoactivity of TiO$_2$ under visible light, many studies have reported doping or co-doping TiO$_2$ with nonmetals (e.g., C, N, S, F) that has resulted in more significant band gap narrowing compared to metal doping. Wu et al. [37] reported that the band gaps of N-doped and N-B-co-doped TiO$_2$ were 2.16 eV and 2.13 eV, respectively, much smaller than that of pure TiO$_2$ (3.18 eV for anatase). Pelaez et al. [38] synthesized N-F-co-doped TiO$_2$ that exhibited high surface area, a low degree of agglomeration, and high activity in degradation of microcystin under visible light. In comparison to other nonmetal dopants (N, C, B, S), iodine doping is much less studied. For the first time in the literature, Li and co-workers [39] prepared I-doped TiO$_2$ through a hydrothermal and postcalcination method and tested its photocatalytic activity for CO$_2$ reduction. The material preparation process is illustrated in Figure 9.2b.

The XRD patterns of the I-doped TiO$_2$ samples indicate two major phase components: anatase and brookite TiO$_2$. As the calcination temperature increased from 375 to 450°C, the anatase phase content increased and the brookite phase decreased; when the calcination temperature increased to 550°C, the brookite content further decreased with the appearance of a small percentage of rutile. The catalyst properties including phase composition and crystal size, band gap, and BET specific surface areas are summarized in Table 9.1. Compared to undoped TiO$_2$, I-TiO$_2$ samples have smaller crystal size for anatase phase, which is in agreement with the literature that dopant favors the formation of smaller particles. The band gap of undoped I-TiO$_2$ is around 3.13 eV, and that of I-TiO$_2$ slightly decreases with iodine doping and levels off at around 3.00 eV.

Figure 9.4 shows TEM and HR-TEM images of the I-TiO$_2$ sample. Both types of images show agglomerates of TiO$_2$ nanocrystals in the size range of 6–9 nm, which is in good agreement with the average size calculated from the Scherrer equation. Similarly, selected area electron diffraction experiments (SAED inset in Fig. 9.4a) recorded from agglomerates within a selecting aperture of 450 nm confirm the phase determination of XRD and demonstrate that the anatase and brookite phases

TABLE 9.1 Phase Content and Average Crystal Size of I-Doped TiO$_2$ Samples Obtained from X-Ray Diffraction, Band Gap from Optical Spectroscopy, and Specific Surface Area from BET Analysis

Sample	Phase Content, %			Crystal Size, nm		Band Gap, eV	BET Specific Surface Area, m^2/g
	A	B	R	A	B		
TiO$_2$-375°C	66	34	0	8.8	4.7	3.13	122.9
5% I-TiO$_2$-375°C	66	34	0	7.5	5.4	3.05	137.4
5% I-TiO$_2$-450°C	71	29	0	8.4	10.1	—	99.4
5% I-TiO$_2$-550°C	76	19	5	19.1	12.7	—	43.1
10% I-TiO$_2$-375°C	70	30	0	5.5	6.3	3.00	137.6
15% I-TiO$_2$-375°C	72	28	0	5.8	5.5	3.02	137.6

A:anatase, B:brookite, R: rutile. Reproduced with permission from ref. 39.

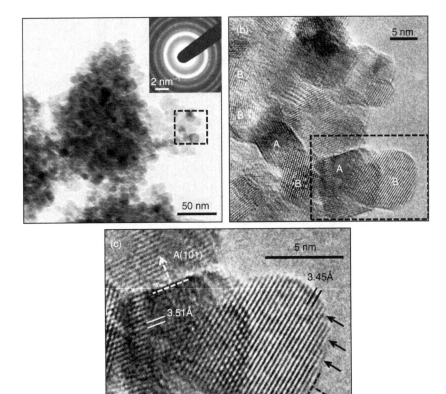

Figure 9.4   Electron microscopy of 5% I-TiO$_2$ sample: (a) TEM image and SAED (inset), (b) HRTEM image with labeled examples of anatase (A) and brookite (B) nanocrystals, and (c) HRTEM lattice spacings and dominant surface facets for A (101) and B (111) nanocrystals, with arrows pointing at steps on B (111) surface. Reproduced with permission from ref. 39.

of I-TiO$_2$ occur in close proximity. Analysis of the lattice spacing and interplanar angles from the HR-TEM image in Figure 9.4c shows clear one-dimensional lattice fringes of TiO$_2$ that correspond to brookite TiO$_2$ (111) plane (lattice spacing = 0.345 nm) and anatase (101) plane (lattice spacing = 0.351 nm).

Figure 9.5a shows the XPS survey spectrum of 10% I-TiO$_2$ calcined at 375°C, which indicates the existence of Ti, O, and I elements. Figure 9.5b–d shows the high-resolution XPS spectra scanning over the following three binding energy areas:

1. Ti 2p region (450 to 470 eV);
2. I 3d region (610 to 640 eV);
3. O 1s region (520 to 540 eV).

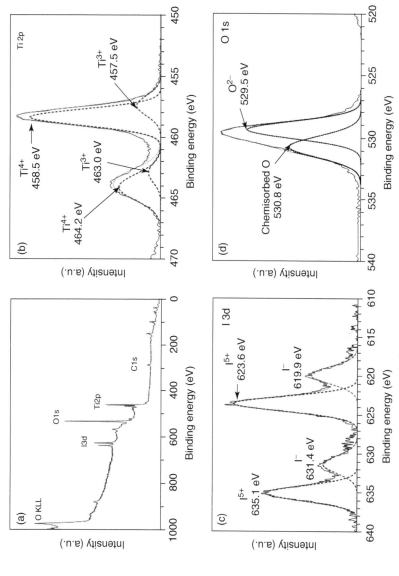

Figure 9.5 XPS spectra of 10% I-TiO$_2$ calcined at 375 °C: survey spectrum (a) and high-resolution spectra for Ti 2p (b), I 3d (c), and O 1s (d). Reproduced with permission from ref. 39.

253

As shown in Figure 9.5b, the XPS peaks of the 10% I-TiO_2-375°C sample in the Ti 2p region appear at 458.5 eV(Ti $2p_{3/2}$) and 464.2 eV(Ti $2p_{1/2}$), both of which correspond to Ti^{4+}. There are two small peaks in the lower side of Ti $2p_{3/2}$ (457.5 eV) and Ti $2p_{1/2}$ (463.0 eV), which are ascribed to Ti^{3+} that is generated to maintain electroneutrality by I^{5+} substituting Ti^{4+}, as the ionic radii of I^{5+} and Ti^{4+} are very close. The XPS spectra of the I 3d region (Fig. 9.5c) for 10% I-TiO_2-375°C show double peaks around 623.6 eV (I $3d_{5/2}$) and 635.1 eV (I $3d_{3/2}$), which infers that the oxidation state of doped iodine is I^{5+} [40,41]. Also, two weaker satellite peaks around 619.9 eV (I $3d_{5/2}$) and 631.4 eV (I $3d_{3/2}$) indicate the existence of I^- [40,42]. The XPS spectra of the O 1s region (Fig. 9.5d) show a major peak at around 529.5 eV that that corresponds to lattice oxygen O^{2-}. The other peak around 530.8 eV can be attributed to chemisorbed oxygen on the surface.

9.4 PHOTOCATALYSIS EXPERIMENTS

Figure 9.6 shows the experimental system that was used for measuring the activity of CO_2 reduction by the prepared photocatalysts. High-purity CO_2 from a gas cylinder first passed through a water bubbler to introduce a mixture of CO_2 and water vapor to a photoreactor that is made of stainless steel wall and a quartz window. The catalyst powders were dispersed on a glass fiber filter and placed inside the reactor. The light source was a 450 W Xe lamp housed in an arc source system (Oriel). It is equipped with a liquid cooler to absorb the heat and a long-pass filter that can cut off wavelengths less than 400 nm if only visible light is needed. All the experiments were conducted under visible light and/or UV + visible irradiation.

The reactor can be operated in either continuous-flow mode (when the 2-way valves are open, as in the study of Cu/TiO_2-SiO_2 catalyst [30]) or batch mode

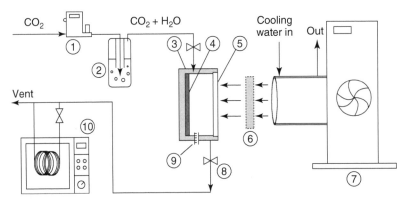

Figure 9.6 Experimental setup for CO_2 photoreduction with water vapor: 1) mass flow controller; 2) water bubbler; 3) photoreactor with a stainless steel wall; 4) catalyst powders dispersed on a glass-fiber filter; 5) quartz window; 6) long-pass UV filter; 7) Xe arc lamp with a cooling system; 8) two-way valves; 9) gas sampling port; 10) gas chromatograph (GC/TCD-FID). Adapted from ref. 30,39.

(when the 2-way valves are closed, as in the study of I/TiO$_2$ catalyst [39]). In the continuous-flow mode, the effluent gas sample from the photoreactor was taken by an automated gas valve and measured by an on-line gas chromatograph (GC) equipped with a thermal conductivity detector (TCD) and a flame ionization detector (FID). In the batch mode, the gas sample was taken by a gas-tight syringe and manually injected in the GC.

It should be noted that water vapor instead of liquid water was used as the reductant for CO$_2$. Because the solubility of CO$_2$ in water is very low, caustic NaOH is usually used to increase CO$_2$ absorption and to serve as hole scavenger in an aqueous reactor [3]. Continuously replenishing NaOH to maintain the reaction rate is not practical from a long-term perspective. Therefore, a self-sustained catalytic reaction of CO$_2$ with water vapor on TiO$_2$ is of more interest.

9.5   CO$_2$ PHOTOREDUCTION ACTIVITY

Prior to CO$_2$ photoreduction experiments, all catalyst powders were pretreated under UV irradiation in an air environment for a prolonged time to eliminate organic residues, if any, on the catalyst. GC measurements were also performed with a mixture of ultra-high-purity helium (instead of CO$_2$) and water vapor as the purging and reaction gases for the catalyst-loaded reactor; no carbon-containing compounds were produced by the catalyst under UV or visible irradiation. This verifies that the catalyst was clean (i.e., no interference from organic residues).

A series of other background tests were also conducted with a mixture of CO$_2$ and H$_2$O vapor as the purging and reaction gases for both cases of (i) empty reactor and (ii) blank glass-fiber filter in the reactor. Again, in either case no carbon-containing compounds were produced under either UV or visible irradiation. This demonstrates that the reactor and the glass-fiber filter were clean and that the CO$_2$ conversion cannot proceed without the photocatalyst. All these background tests have proven that any carbon-containing compounds produced must be originated from CO$_2$ through photocatalytic reactions.

9.5.1   Cu/TiO$_2$–SiO$_2$ and Ce-TiO$_2$/SBA-15 Catalysts

CO and CH$_4$ were identified to be the major products of CO$_2$ photoreduction with water vapor using the Cu/TiO$_2$–SiO$_2$ catalysts when the photoreactor was operated in a continuous-flow mode [30], as shown in Figure 9.7A. Catalysts without Cu or SiO$_2$ were also tested as a comparison while the mass of TiO$_2$ was kept the same for all samples in Figure 9.7Aa. For each type of catalyst, experiments were carried out three times with fresh catalysts, and the peak production rates were averaged and standard deviations reported. Bare TiO$_2$ showed the lowest production rate among the four catalysts, with 8.1 μmol g-TiO$_2{}^{-1}$ h^{-1} for CO and zero for CH$_4$. With the mesoporous SiO$_2$ support, CO production was enhanced to 22.7 μmol g– TiO$_2{}^{-1}$ h^{-1} for TiO$_2$–SiO$_2$, but CH$_4$ production was still zero. The improvement may be due to the enhanced dispersion of TiO$_2$ and improved adsorption of CO$_2$ and H$_2$O on the high-surface-area SiO$_2$ substrate. With the addition

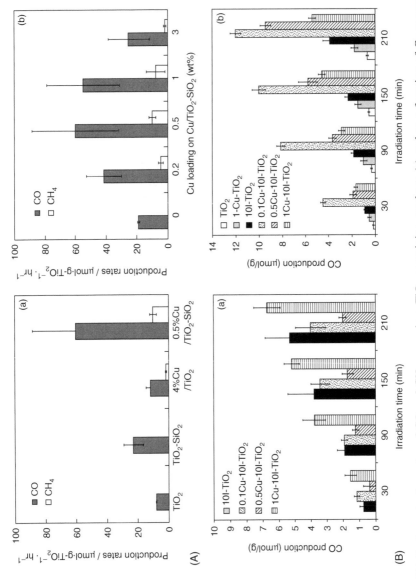

Figure 9.7 (**A**) Peak production rates of CO and CH$_4$ on various TiO$_2$-containing catalysts (a) and as a function of Cu concentration on the Cu/TiO$_2$–SiO$_2$ catalysts (b). (**B**) CO production under visible light irradiation (a) and UV/vis irradiation (b) using copper- and/or iodine-modified TiO$_2$ catalysts (xCu-yI-TiO$_2$, where x and y are weight percentages of Cu and I, respectively). Reproduced with permission from ref. 30,43.

of Cu to TiO_2, the 4% Cu/TiO_2 catalyst demonstrated a slightly higher CO production rate (11.8 $\mu mol\, g\text{-}TiO_2^{-1}h^{-1}$) than that of TiO_2, and CH_4 production occurred at a rate of 1.8 $\mu mol\, g\text{-}TiO_2^{-1}h^{-1}$. With the combination of porous SiO_2 substrate and Cu deposition, the CO_2 photoreduction rate was significantly enhanced. For the 0.5%Cu/TiO_2–SiO_2 catalyst, the average peak production rates of CO and CH_4 reached 60 and 10 $\mu mol\, g\text{-}TiO_2^{-1}h^{-1}$, respectively, indicating a truly synergistic effect (activity increased by 10 times compared to bare TiO_2 and much larger than that of the sum of Cu/TiO_2 and TiO_2/SiO_2) due to the combination of SiO_2 substrate and Cu deposition. This synergistic combination is an important finding that suggests a direction of future studies on manufacturing multicomponent nanostructured catalysts for efficient CO_2 reduction.

The effect of Cu concentration on the catalytic activity of Cu/TiO_2–SiO_2 was also investigated by varying the Cu loading from 0.2% to 3% [30], as shown in Figure 9.7Ab. It is clearly seen that 0.5%Cu had the highest production rates for both CO and CH_4. The production rates for 1%Cu were slightly lower than those of 0.5%Cu. Decreasing the Cu loading to 0.2% or increasing it to 3% reduced the production rates. At lower Cu loadings below the optimum value, Cu species can inhibit recombination of photoinduced electrons and holes by capturing the electrons, and the catalytic activity increases with Cu concentration [3,42]. However, an excess Cu loading greater than the optimum value could reduce the catalytic activity because a high concentration of Cu could mask the illuminated TiO_2 surface [3] and that excess Cu species could become recombination centers for photoinduced electrons and holes [35,42].

Quantum yield of the CO_2 photoconversion can be calculated by the following equations when CO and CH_4 are the conversion products. Two and eight electrons are required to convert CO_2 to CO and CH_4, respectively. The quantum yield based on the average peak production rates was calculated to be 0.85% and 0.56% for CO and CH_4, respectively, and the total quantum yield was 1.41% for CO_2 photoreduction using Cu/TiO_2–SiO_2.

$$\Phi_{CO}(\%) = \frac{2 \text{ mol of CO yield}}{\text{moles of photon absorbed by catalyst}} \times 100\% \qquad (9.9)$$

$$\Phi_{CH4}(\%) = \frac{8 \text{ mol of } CH_4 \text{ yield}}{\text{moles of photon absorbed by catalyst}} \times 100\% \qquad (9.10)$$

For the $Ce\text{-}TiO_2/SBA\text{-}15$ catalysts, similar synergies between the metal (e.g., Ce) and the high-surface-area support (e.g., SBA-15) were observed in CO_2 photoreduction with water [36]. The $Ce\text{-}TiO_2/SBA\text{-}15$ sample with 3% Ce loading demonstrated 8-fold enhancement in CO production and 115-fold enhancement in CH_4 production compared to the $Ce\text{-}TiO_2$ sample without SBA-15 support. This superior activity of SBA-15-supported $Ce\text{-}TiO_2$ catalysts is believed to be mainly attributable to the higher surface area and better dispersion of the $Ce\text{-}TiO_2$ nanoparticles, as verified by BET and TEM analyses [36]. Moreover, SBA-15 as the support may enhance the stability of the TiO_2 anatase phase and prevent the grain growth of TiO_2 nanocrystals. Another important finding in this study was that Ce-TiO_2 dispersed on SBA-15 was much more active (10 times higher) than that dispersed on amorphous mesoporous silica. The reason is mainly attributed to the ordered pores in SBA-15 that are more accessible to TiO_2 precursors during the material preparation (thus better TiO_2 dispersion) and could enhance CO_2 diffusion and adsorption on the catalyst [36].

9.5.2 Copper- and/or Iodine-Modified TiO_2 Catalysts

The activity of I-doped TiO_2 catalysts and Cu-I comodified TiO_2 catalysts for CO_2 photoreduction with water was measured in a batch-type photoreactor [39,43]. Experiments were conducted under visible ($\lambda > 400$ nm) and UV–vis irradiation ($\lambda > 250$ nm), respectively, and the results are illustrated in Figure 9.7B. Under visible light (Fig. 9.7Ba), no activity of CO_2 reduction was observed for either pure TiO_2 or Cu–TiO_2, which is consistent with the inability of TiO_2 or Cu–TiO_2 to absorb visible light. All I-doped TiO_2 showed visible light activity for CO_2 photoreduction to CO, and the concentration of CO increased almost linearly with illumination time. Among the three Cu–10I–TiO_2 samples with different Cu concentrations (0.1%, 0.5%, and 1%), 1%Cu appeared to be the optimum material, having a CO production of 6.7 μmol g^{-1} at 210 min. However, only the 1Cu–10I–TiO_2 sample demonstrated a higher activity than the 10I–TiO_2 sample (5.3 μmol g^{-1} at 210 min), while the other two comodified samples (0.1Cu–10I–TiO_2 and 0.5Cu–10I–TiO_2) were inferior to 10I–TiO_2. It should be noted that very few studies have reported CO_2 photoreduction with water under visible light. Varghese et al. [23] used an N-doped TiO_2 nanotube array sputtered with Cu for CO_2 photoreduction with water vapor under sunlight and reported a production rate of 0.3 μmol g^{-1} h^{-1} ascribed to the visible light portion (or 3% of the total photocatalytic activity under sunlight). The production rate of 1Cu-10I–TiO_2 catalyst under visible light (e.g., 1.9 μmol g^{-1} h^{-1}) was much higher than that reported by Varghese et al. [23], while the raw materials used were cheaper and the synthesis method was simpler.

Under UV–vis irradiation (Fig. 9.7Bb), activity results very different from those under visible light were observed. Pure TiO_2 had the lowest activity, while single ion-modified TiO_2 samples (1Cu–TiO_2 and 10I–TiO_2) exhibited enhanced, yet small activities. Cu and I comodified TiO_2 samples had remarkably higher activities, while the 0.1%Cu sample appeared to be the best among the three Cu-10I–TiO_2 samples. These results indicate that materials designed for high activity under visible light may not be necessarily optimized for UV applications. Similar findings have been reported that N- or S-doped TiO_2 has superior photo-oxidation activity to that of undoped TiO_2 under visible light irradiation but has similar or even lower photocatalytic activity in the UV region [44,45].

9.5.3 TiO_2 Polymorphs Engineered with Surface Defects

Pure phase TiO_2 polymorphs (anatase, rutile, and brookite) were prepared and engineered with surface defects such as oxygen vacancies and Ti^{3+} species by thermal

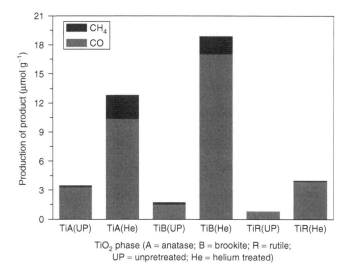

Figure 9.8 The production of CO and CH_4 on the three unpretreated and He-pretreated TiO_2 polymorphs for a 6-h period of photo-illumination. Reproduced with permission from ref. 46.

treatment in a helium environment [46]. The perfect (unpretreated) and defective (He-treated) TiO_2 polymorphs were tested for their activity of CO_2 photoreduction, as shown in Figure 9.8.

The activity of unpretreated TiO_2 follows the order of anatase > brookite > rutile. For He-treated samples, brookite had the highest CO production among all the samples, while anatase had the highest CH_4 production. Compared to the unpretreated samples, the activity of the He-pretreated samples was remarkably enhanced, and the factor of enhancement due to sample pretreatment was most significant on brookite (8–10 times), followed by anatase and rutile.

This is an important experimental finding that for the first time in the literature reveals oxygen-deficient brookite TiO_2 as a more outstanding form than anatase and rutile for CO_2 photoreduction to CO and CH_4.

9.6 REACTION MECHANISM AND FACTORS INFLUENCING CATALYTIC ACTIVITY

9.6.1 Effects of Cu and Iodine Modification on TiO_2

It is believed that Cu species on the surface of TiO_2 can trap electrons from the TiO_2 conduction band, and the trapped electrons are subsequently transferred to the surrounding adsorbed species (e.g., adsorbed CO_2), thereby avoiding electron-hole recombination and enhancing the photocatalytic reduction of CO_2. CO is the main reaction product observed when there is no Cu species loaded on the TiO_2 catalyst (i.e., bare TiO_2, TiO_2-SiO_2, and I/TiO_2). The increased selectivity of CH_4 production in the presence of Cu (i.e., Cu/TiO_2 and Cu/TiO_2-SiO_2) can be explained as

Cu species acting as electron traps and resulting in an increased probability of multielectron reactions (e.g., 8 electrons required for CH$_4$ production compared with 2 electrons required for 1 CO production).

An intriguing phenomenon observed in the study of Cu/TiO$_2$–SiO$_2$ catalysts [30] was the color change of the Cu-containing catalysts before and after the photocatalytic reaction, which has scarcely been discussed in the literature. The fresh Cu/TiO$_2$–SiO$_2$ catalyst was almost white, with very light greenish color due to the small concentration of Cu species. The color of the catalyst turned to dark gray after the CO$_2$ photoreduction experiment and changed back to white after the used catalyst was taken out of the reactor and exposed to room environment overnight.

Results of this work also indicated that the catalytic activity decreased in accordance with the color change to dark, and the activity was partially regenerated when the color returned back to white [30]. Because the XPS analysis indicated that Cu$_2$O is the Cu species in fresh Cu/TiO$_2$–SiO$_2$ catalysts, which is in line with other reports in the literature that Cu$_2$O is the active species [4,26,47], it is hypothesized that the Cu species may have undergone a redox cycle such that Cu(I) species was reduced to Cu(0) that was partially reoxidized to Cu(I) when exposed to air [30]. This Cu(I)/Cu(0) redox cycle may correlate with the observed catalyst color change. However, Irie et al. [48] point out that since XPS measurement is performed under high-vacuum conditions (10^{-10}–10^{-9} Torr), amorphous Cu(II) species could possibly be reduced to Cu(I) when Cu(II)/TiO$_2$ is introduced into the XPS chamber. Meanwhile, their X-ray absorption near edge structure (XANES) spectra indicate that Cu(II) is the major species on TiO$_2$ that facilitates electron transfer and that Cu(II) is reduced to Cu(I) in the absence of O$_2$ and reoxidized to Cu(II) when exposed to O$_2$. Given the discrepancy in the literature on the exact active Cu species, a systematic study is necessary to investigate the different Cu species [Cu(II), Cu(I), and Cu(0)] deposited on TiO$_2$ in terms of their phases (amorphous or crystalline), size (if nanoparticles formed), dispersion on TiO$_2$, activity toward CO$_2$ photoreduction, and charge transfer mechanism. It is also important to develop a strategy to stabilize the active Cu species at the catalyst surface to maintain a long-term performance of the Cu-loaded TiO$_2$ catalysts.

The effects of iodine doping are fourfold. The first and most important effect is the enhanced activity under visible light. The iodine dopant creates intra-band-gap states close to the valence band edge that induce visible light absorption at the sub-band-gap energies, which causes the calculated band gap narrowing from the analysis of UV-vis diffuse reflectance spectroscopy. The conduction band edge is not changed, so that it maintains a potential negative enough for CO$_2$ reduction as shown in Figure 9.1. Second, the substitution of Ti^{4+} with I^{5+} causes charge imbalance and results in the generation of Ti^{3+} surface states that may help trap the photoinduced electrons and inhibit charge recombination [41]. Third, the iodine atoms prefer to be doped near the TiO$_2$ surface because of the strong I–O repulsion [49], and thus the surface-doped I^{5+} will not only trap electrons but also facilitate electron transfer to the surface adsorbed species [40,49]. Finally, the continuous states caused by iodine dopant may enhance the trapping of photoinduced holes inside the TiO$_2$ particle and thus promote the separation of electron–hole pairs

[40]. On the other hand, the reduction ability of the catalyst is enhanced because of the impaired power of oxidation by holes.

9.6.2   Effect of O_2 on CO_2 Photoreduction

When O_2 is present as the surface-adsorbed species on TiO_2, it competes with CO_2 as electron acceptor, mitigating the CO_2 reduction efficiency. The reaction can be expressed as:

$$O_2 + e^- \rightarrow O_2^- \tag{9.11}$$

where the reduction potential is $E^0 = -0.28$ V vs. NHE at pH = 7 [50]. The electron scavenging effect by O_2 is supported by the experimental results of the work of Li et al. [30] that no CO_2 photoreduction product was observed when the reactor influent gas was changed to a mixture of O_2/CO_2 (1:1 v/v) upstream of the water bubbler. The result implies that CO_2 photoreduction is not favorable in an O_2-rich environment (e.g., for atmospheric CO_2 reduction). Rather, the technology is more appropriate for CO_2 mitigation from combustion exhausts where the O_2 concentration is normally less than a few percent or even close to zero in the case of oxy-fuel combustion.

Another important aspect in the CO_2 photoreduction process is that O_2 should be produced from water to close the redox cycle, as shown in reaction 9.2 of Figure 9.1. To verify this, the concentrations of O_2 and N_2 in the effluent gas were also monitored for the experiment using 0.5%Cu/TiO_2–SiO_2 as the catalyst [30]. Background O_2 was detected in the reactor effluent gas (300 ~ 400ppm) at the beginning of the test, possibly because the reactor was not vacuumed before purging with the CO_2- H_2O mixture and because of the low concentration of impurity gases in the CO_2 cylinder. Hence, a better indicator of O_2 production from the photocatalytic reaction is the volumetric ratio of O_2/N_2 in the effluent gas.

As shown in Figure 9.9, the O_2 / N_2 ratio gradually increased with the irradiation time, reached a maximum value over 2–4 h, and then decreased slowly thereafter. The time dependence of the O_2 / N_2 ratio is well-correlated with those of CO and CH_4 production, confirming the production of O_2 from the dissociation of H_2O. Because O_2 behaves as an electron scavenger, the production of O_2 from CO_2 photoreduction process may reduce the probability of electron transfer to CO_2. On the other hand, O_2 may oxidize the produced CO or CH_4 and promote backward reactions. These may be some of the reasons that the rate of photocatalytic CO_2 reduction is typically very low. Designs of reactor systems that can separate the CO_2 reduction site and O_2 production site similar to a galvanic cell may be helpful. However, this is difficult to realize in a gas-solid reaction system for CO_2 reduction with water vapor over a solid catalytic bed.

9.6.3   In Situ DRIFTS Analysis on Surface Chemistry

Improving CO_2 photoreduction efficiency and designing more efficient catalysts require fundamental understanding of the reaction mechanism and correlation

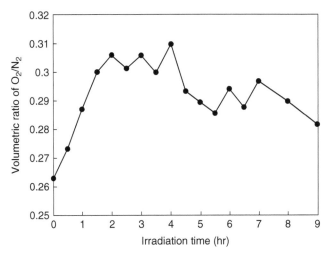

Figure 9.9 Time dependence of the ratio of O_2/N_2 in the effluent gas with 0.5% Cu/TiO_2-SiO_2 as the catalyst. Reproduced with permission from ref. 30.

of the catalytic activity with the material surface property. Diffuse reflectance infrared Fourier transform spectroscopy (DRIFTS) is a power tool to investigate surface-adsorbed species and reaction intermediates on the catalyst surface. Li and co-workers [46,51] used in situ DRIFTS to explore CO_2 activation and photoreduction on TiO_2 catalysts. A Praying Mantis DRIFTS accessory and a reaction chamber (Harrick Scientific, HVC-DRP) were installed in a Nicolet 6700 spectrometer (Thermo Electron) equipped with a liquid nitrogen-cooled HgCdTe (MCT) detector. The dome of the DRIFTS cell has two KBr windows allowing IR transmission and a third quartz window allowing transmission of irradiation from a light source such as a Xe lamp.

In situ DRIFTS spectra were taken for CO_2 interaction on He-pretreated Cu/TiO_2 surface [51]. As shown in Figure 9.10a, exposure of the catalyst to CO_2 in the dark rapidly led to the generation of primary carboxylate (CO_2^-, at 1673 and 1248 cm^{-1}, O-end bonded with Ti^{4+}), bicarbonate (HCO_3^-, at 1412, and 1223 cm^{-1}), bidentate carbonate (b–CO_3^{2-}, at 1555 and 1348 cm^{-1}, bonded with $Ti^{4+} - O_2^-$), and a new bidentate carbonate possibly adsorbed on Cu species (b″–CO_3^{2-}, at 1592 and 1294 cm^{-1}). The formation of $Ti^{4+}-CO_2^-$ suggests that CO_2 can be activated in such a way that excess charge, being trapped at specific Ti^{3+} sites activated by thermal annealing, migrates to adsorbed CO_2 spontaneously through a dissociative electron attachment process.

The intensity of the CO_2^- species decreased gradually, and it almost disappeared after 20 min. In the meantime, a small peak at 2110 cm^{-1} assigned to CO coordinated with Cu^+ ions evolved, suggesting the formation of CO from

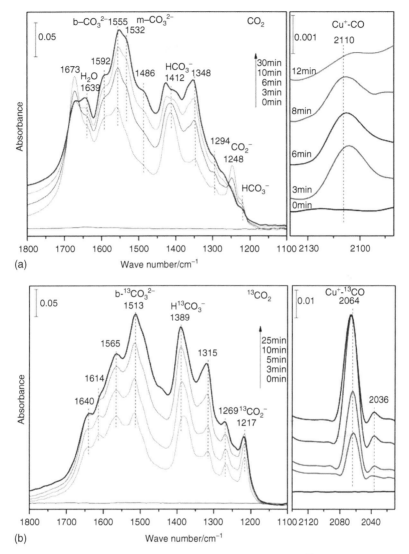

Figure 9.10 In situ DRIFTS spectra of (a) CO_2 and (b) $^{13}CO_2$ interaction with He-pretreated Cu/TiO_2 in the dark. In spectra (c) of CO_2 interaction with He-pretreated Cu/TiO_2, CO_2 was first introduced to the DRIFTS cell for 30 min in the dark at 25°C; then the cell was operated as a closed system with UV-vis light turned on for another 30 min. Reproduced with permission from ref. 51.

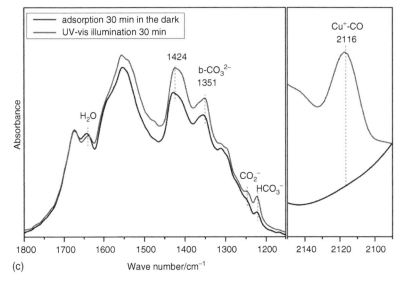

Figure 9.10 (*Continued*)

metastable CO_2^-. To confirm this spontaneous dissociation of CO_2 to CO on the defective Cu/Ti(He) surface, isotopically labeled $^{13}CO_2$ adsorption was performed under the same condition [51]. The result of $^{13}CO_2$ interaction on He-pretreated Cu/TiO_2 is shown in Figure 9.10b, which follows a trend similar to that of $^{12}CO_2$ in Figure 9.10a. According to the harmonic equation [52], the theoretical vibration frequencies of ^{13}CO and $^{13}CO_2^-$ (bonded with Cu^+ and Ti^{4+}, respectively) would shift to 2063 cm^{-1} and 1219 cm^{-1}. The characteristic bands of ^{13}CO and $^{13}CO_2^-$ in Figure 9.10b match the theoretical shifts very well. The result of labeled carbon studies confirms that the formed CO bound to the Cu^+ site is derived from CO_2 and that CO_2 is indeed activated and dissociated on defective Cu/TiO_2 even in the dark at room temperature.

Figure 9.10c shows the difference in IR spectra when the Cu/TiO_2 catalyst was exposed to CO_2 in the dark and subsequently under photo-illumination [51]. After 30 min in dark, there were no observable peaks of CO_2^- species at 1248 cm^{-1} and $Cu^+ - CO$ at 2116 cm^{-1}, although they did appear upon the exposure to CO_2 at the beginning. Subsequent photo-illumination induced the reappearance of CO_2^- and $Cu^+ - CO$ species. Meanwhile, the trace of adsorbed H_2O at 1640 cm^{-1} slightly decreased in intensity, while HCO_3^- at 1424 and 1221 cm^{-1} increased in intensity. The result suggests that a photoinduced electron is transferred to adsorbed CO_2, and that surface-adsorbed H_2O participates in the CO_2 reduction (as an electron donor).

The results obtained in this IR study further confirm that CO_2 is photocatalytically reduced to CO, which is widely reported in experiments measured by GC. This work also suggests that HCO_3^- and CO_2^- species are possible intermediates during the CO_2 photoreduction process, although the literature indicates the difficulty of one-electron reduction of CO_2 on TiO_2.

9.7 CONCLUSIONS AND FUTURE RESEARCH RECOMMENDATIONS

Photocatalytic reduction of CO_2 with water is an important sustainable energy technology that not only reduces CO_2 emissions but generates renewable and carbon-neutral fuels. CO and CH_4 are found to be the major products from CO_2 reduction using TiO_2-based nanocomposite photocatalysts when water vapor serves as the electron donor. The rate of CO_2 conversion by TiO_2 is low but can be improved by several means:

1. Incorporation of metal or metal ion species such as copper to enhance electron trapping and transfer to the catalyst surface
2. Application of high-surface-area support such as mesoporous silica to enhance better dispersion of TiO_2 nanoparticles and increase reactive surface sites
3. Doping of nonmetal ions such as iodine in the lattice of TiO_2 to improve the visible light response and charge carrier separation
4. Pretreatment of the TiO_2 catalyst in a reducing environment like helium to create surface defects to enhance CO_2 adsorption and activation

Combinations of these different strategies may result in synergistic effects and much higher CO_2 conversion efficiency. The Cu/TiO_2–SiO_2 nanocomposite [30] and the iodine-doped TiO_2 (including I/TiO_2 [39] and Cu-I/TiO_2 [43]) are among the best catalysts reported for CO_2 photoreduction with water in terms of overall CO_2 conversion efficiency and visible light activity, respectively. Brookite-phase TiO_2 is the least studied, but defective brookite has shown promising activity in CO_2 photoreduction that is higher than that of anatase TiO_2.

Further improvement in CO_2 reduction efficiency and better understanding of the reaction mechanisms and product selectivity are demanded in the challenging area of CO_2 photoreduction.

Future studies are recommended in the following directions:

1. Investigation of brookite-containing mixed-phase TiO_2 nanocrystals such as anatase-brookite mixtures that may promote interfacial charge transfer between different crystal phases
2. Minimization of the negative effect of oxygen
3. Further understanding of the reaction intermediates and pathways through in situ DRIFTS analysis of the catalyst surface
4. Study of the stability of the catalyst for long-term performance

REFERENCES

1. C.M. White, B.R. Strazisar, E.J. Granite, J.S. Hoffman, H.W. Pennline (2003). *J. Air and Waste Management Assoc.*, 53: 645–715.

2. M. Anpo, H. Yamashita, Y. Ichihashi, S. Ehara (1995). *J. Electroanal. Chem.*, 396: 21–26.

3. I.H. Tseng, W.C. Chang, J.C.S. Wu (2002). *Appl. Catal. B-Environ.* 37: 37–48.

4. J.C.S. Wu, H.M. Lin, C.L. Lai (2005). *Appl. Catal. A-Gen.*, 296: 194–200.

5. T. Inoue, A. Fujishima, S. Konishi, K. Honda (1979). *Nature*, 277: 637–638.

6. V.P. Indrakanti, J.D. Kubicki, H.H. Schobert (2009). *Energy & Env. Science*, 2: 745–758.

7. M. Anpo, H. Yamashita, Y. Ichihashi, Y. Fujii, M. Honda (1997). *J. Phys. Chem. B*, 101: 2632–2636.

8. K. Ikeue, S. Nozaki, M. Ogawa, M. Anpo (2002). *Catal. Today*, 74: 241–248.

9. S.S. Tan, L. Zou, E. Hu (2006). *Catal. Today* 115: 269–273.

10. H. Yamashita, Y. Fujii, Y. Ichihashi, S.G. Zhang, K. Ikeue, D.R. Park, K. Koyano, T. Tatsumi, M. Anpo (1998). *Catal. Today*, 45: 221–227.

11. Y. Kohno, T. Tanaka, T. Funabiki, S. Yoshida (2000). *Phys. Chem. Chem. Phys.*, 2: 2635–2639.

12. C.C. Lo, C.H. Hung, C.S. Yuan, J.F. Wu (2007). *Solar Energy Materials and Solar Cells*, 91: 1765–1774.

13. H. Fujiwara, H. Hosokawa, K. Murakoshi, Y. Wada, S. Yanagida, T. Okada, H. Kobayashi (1997). *J. Phys. Chem. B*, 101: 8270–8278.

14. P.W. Pan, Y.W. Chen (2007). *Catal. Comm.*, 8: 1546–1549.

15. A.J. Morris, G.J. Meyer, E. Fujita (2009). *Acc. Chem. Res.*, 42: 1983–1994.

16. K. Koci, L. Obalova, Z. Lacny (2008). *Chem. Papers*, 62: 1–9.

17. K. Koci, L. Obalova, L. Matejova, D. Placha, Z. Lacny, J. Jirkovsky, O. Solcova (2009). *Appl. Catal. B-Environ.*, 89: 494–502.

18. Z.B. Zhang, C.C. Wang, R. Zakaria, J.Y. Ying (1998). *J. Phys. Chem. B*, 102: 10871–10878.

19. M. Anpo, H. Yamashita, K. Ikeue, Y. Fujii, S.G. Zhang, Y. Ichihashi, D.R. Park, Y. Suzuki, K. Koyano, T. Tatsumi (1998). *Catal. Today*, 44: 327–332.

20. Y. Shioya, K. Ikeue, M. Ogawa, M. Anpo (2003). *Appl. Catal. A-Gen.*, 254: 251–259.

21. A.L. Linsebigler, G.Q. Lu, J.T. Yates (1995). *Chem. Rev.*, 95: 735–758.

22. J.C.S. Wu, C.Y. Yeh (2001). *J. Mat. Res.*, 16: 615–620.

23. O.K. Varghese, M. Paulose, T.J. LaTempa, C. A. Grimes (2009). *Nano Lett.*, 9: 731–737.

24. H.Y. Chuang, D.H. Chen (2009). *Nanotechn.*, 20: 10.

25. C.S. Chou, R.Y. Yang, C.K. Yeh, Y. J. Lin (2009). *Powder Techn.*, 194: 95–105.

26. H. Yamashita, H. Nishiguchi, N. Kamada, M. Anpo, Y. Teraoka, H. Hatano, S. Ehara, K. Kikui, L. Palmisano, A. Sclafani, M. Schiavello, M.A. Fox (1994). *Res. on Chem. Intermediates*, 20: 815–823.

27. Y.H. Xu, D.H. Liang, M.L. Liu, D.Z. Liu (2008). *Mat. Res. Bull.*, 43: 3474–3482.

28. K. Koci, K. Mateju, L. Obalova, S. Krejcikova, Z. Lacny, D. Placha, L. Capek, A. Hospodkova, O. Solcova (2010). *Appl. Catal. B-Environ.*, 96: 239–244.

29. O. Ishitani, C. Inoue, Y. Suzuki, T. Ibusuki (1993). *J. Photochem. and Photobiol. A-Chem.*, 72: 269–271.

30. Y. Li, W.N. Wang, Z.L. Zhan, M.H. Woo, C.Y. Wu, P. Biswas (2010). *Appl. Catal. B-Environ.* 100: 386–392.

31. C.Y. Zhao, A.J. Krall, H.L. Zhao, Q.Y. Zhang, Y. Li (2012). *Int. J. Hydrogen Energy*, 37: 9967–9976.

32. N.L. Wu, M.S. Lee (2004). *Int. J. Hydrogen Energy*, 29: 1601–1605.

33. N. Sasirekha, S.J.S. Basha, K. Shanthi (2006). *Appl. Catal. B-Environ.*, 62: 169–180.

34. R. Dholam, N. Patel (2009). *Int. J. Hydrogen Energy*, 34: 5337–5346.

35. B.F. Xin, P. Wang, D.D. Ding, J. Liu, Z.Y. Ren, H.G. Fu (2008). Appl. *Surface Science*, 254: 2569–2574.

36. C.Y. Zhao, L.J. Liu, Q.Y. Zhang, J. Wang, Y. Li (2012). *Catal. Science & Techn.*, 2: 2558–2568.

37. G.S. Wu, J.L. Wen, J.P. Wang, D.F. Thomas, A.C. Chen (2010). *Mat. Lett.*, 64: 1728–1731.

38. M. Pelaez, A.A. de la Cruz, E. Stathatos, P. Falaras, D.D. Dionysiou (2009). *Catal. Today*, 144: 19–25.

39. Q.Y. Zhang, Y. Li, E.A. Ackerman, M. Gajdardziska-Josifovska, H.L. Li (2011). *Appl. Catal. A-Gen.*, 400: 195–202.

40. S. Tojo, T. Tachikawa, M. Fujitsuka, T. Majima (2008). *J. Phys. Chem. C*, 112: 14948–14954.

41. Z.Q. He, X. Xu, S. Song, L. Xie, J.J. Tu, J.M. Chen, B. Yan (2008). *J. Phys. Chem. C*, 112: 16431–16437.

42. K.X. Song, J.H. Zhou, J.C. Bao, Y.Y. Feng (2008). *J. Am. Ceramic Soc.*, 91: 1369–1371.

43. Q.Y. Zhang, T.T. Gao, J.M. Andino, Y. Li (2012). *Appl. Catal. B: Env.*, 123–124: 257–264.

44. R. Asahi, T. Morikawa, T. Ohwaki, K. Aoki, Y. Taga (2001). *Science*, 293: 269–271.

45. T. Ohno, T. Mitsui, M. Matsumura (2003). *Chem. Lett.*, 32: 364–365.

46. L.J. Liu, H.L. Zhao, J.M. Andino, Y. Li (2012). *ACS Catal.*, 2: 1817–1828.

47. I.H. Tseng, J.C.S. Wu, H.Y. Chou (2004). *J. Catal.*, 221: 432–440.

48. H. Irie, K. Kamiya, T. Shibanuma, S. Miura, D.A. Tryk, T. Yokoyama, K. Hashimoto (2009). *J. Phys. Chem. C*, 113: 10761–10766.

49. J.A. Rodriguez, J. Evans, J. Graciani, J.B. Park, P. Liu, J. Hrbek, J.F. Sanz (2009). *J. Phys. Chem. C*, 113: 7364–7370.

50. A. Fujishima, T.N. Rao, D.A. Tryk (2000). *J. Photochem. and Photobiol. C: Photochem. Reviews*, 1: 1–21.

51. L. Liu, C. Zhao, Y. Li (2012). *J. Phys. Chem. C*, 116: 7904–7912.

52. C.C. Yang, Y.H. Yu, B. van der Linden, J.C.S. Wu, G. Mul, *J. Am. Chem. Soc.* 132 (2010) 8398–8406.

CHAPTER 10

Photocatalytic Reduction of CO_2 to Hydrocarbons Using Carbon-Based AgBr Nanocomposites Under Visible Light

MUDAR ABOU ASI, CHUN HE*, QIONG ZHANG, ZUOCHENG XU, JINGLING YANG, LINFEI ZHU, YANLING HUANG, YA XIONG, and DONG SHU

10.1 INTRODUCTION

There is a general agreement in the scientific community that the rise in atmospheric CO_2, the most abundant greenhouse gas, derives from anthropogenic sources such as the burning of fossil fuels, and that this atmospheric rise in CO_2

*Indicates the corresponding author.

Green Carbon Dioxide: Advances in CO₂ Utilization, First Edition.
Edited by Gabriele Centi and Siglinda Perathoner.
© 2014 John Wiley & Sons, Inc. Published 2014 by John Wiley & Sons, Inc.

results in global climate change [1–3]. Therefore, methods for photochemical conversion CO_2 into fuel-type products could offer an attractive way to decrease the atmospheric concentrations and simultaneously develop alternative energy sources [4]. However, new developments in CO_2 reduction methods are required to meet this challenge. The synthesis of the novel nanostructured catalysts for CO_2 reduction has opened new possibilities for creating and mastering advanced photocatalytic materials for this purpose. One way to accomplish this conversion method is through the light-driven reduction of CO_2 to hydrocarbons with electrons and protons derived from water [5–9].

10.2 MECHANISM OF PHOTOCATALYTIC REDUCTION FOR CO_2

In 1972, Fujishima and Honda discovered the photocatalytic splitting of water on TiO_2 electrodes [10]. This event marked the beginning of a new era in heterogeneous photocatalysis. A great deal of research has focused recently on the photocatalytic reduction of CO_2 using various semiconductor catalysts to form different products [11–13]. The photocatalytic technique makes use of semiconductors to promote reactions in the presence of light radiation [14]. Unlike metals, which have a continuum of electronic states, semiconductors exhibit a void energy region, or band gap, that extends from the top of the filled valance band to the bottom of the vacant conduction band when exposed to light radiation. Figure 10.1a illustrates the excitation of an electron from the valence band to the conduction band initiated by light absorption with energy equal to or greater than the band gap of the semiconductor.

The generation of electron/hole pairs (e^-/h^+) and its reverse process are shown in Eqs. 10.1 and 10.2, respectively.

$$\text{Photocatalyst} + \text{visible light} \rightarrow e^- (CB) + h^+ (VB) \qquad (10.1)$$

$$e^- (CB) + h^+ (VB) \rightarrow \text{Energy} \qquad (10.2)$$

The lifetime of this excited electron–hole pair is only a few nanoseconds [15], but this is adequate for promoting the redox reactions. The initial excitation and electron transfer make the chemical reactions in the photocatalytic process possible. The initial process of electron and hole formation and charge separation is followed by several different possible pathways. Migration of electrons and holes to the semiconductor surface is followed by transfer of photo-induced electrons to adsorbed molecules or to solvent. The electron-transfer process is more efficient if the species are adsorbed on the surface [16]. At the surface, the semiconductor can donate electrons to acceptors (CO_2). In turn, holes can combine with electrons from donor species (H_2O). The rate of charge transfer depends on the band-edge position of the band gap and the redox potential of the adsorbed species, respectively [17]. Electron–hole pair recombination prevents them from transferring to the surface to react with adsorbed molecules. Recombination can occur in the volume of the semiconductor particle (internal electron–hole pair recombination). The optimal characteristics required for an effective photocatalyst include the following [18]:

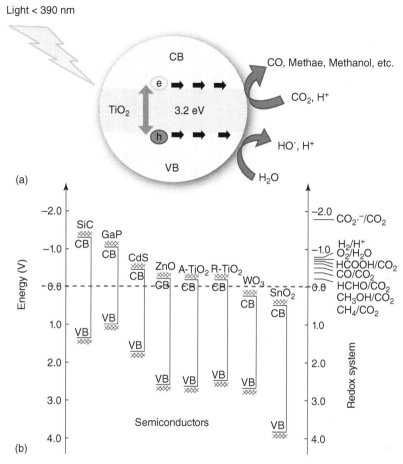

Figure 10.1 (a) Photoreaction mechanism of CO_2 reduction on TiO_2. (b) Conduction band and valence band potentials of semiconductor vs. energy levels of the redox couples.

1. The redox potential of the photogenerated valence-band hole must be sufficiently positive for the hole to act as an acceptor.
2. The redox potential of the photogenerated conductance-band electron must be sufficiently negative for the electron to act as a donor.
3. The photocatalyst must not be prone to photocorrosion or produce toxic by-products.
4. The photocatalyst should be commercially and economically available.

10.3 CARBON DIOXIDE REDUCTION

CO_2 is a relatively inert and stable compound, and thus its reduction is quite challenging. Most conversion and removal methods rely on high energy input (high-temperature and/or-pressure conditions) [19–27]. Conversely, the photocatalytic process occurs under relatively mild conditions (e.g., low energy input), especially when the reaction is driven by solar energy or other easily obtained light sources. The use of solar energy is a particular advantage because of the continuous and readily available power supply.

In addition to reducing CO_2 emissions to the atmosphere, photocatalytic conversion of CO_2 can also produce valuable chemicals that increase the added value of the process in comparison to conventional CO_2 removal methods. For these valuable products, which can be obtained from CO_2 usually at higher thermodynamic energy with respect to CO_2, it is necessary to supply this energy externally, either through reaction with a higher-energy reactant or through the direct transfer of energy, for example, through an electron transfer. Semiconductors that have their conduction and valence bands placed suitably to transfer sacrificial electrons to CO_2 and oxidize inexpensive reductants (e.g., water) can accomplish this process [28–31]. Photocatalytic reduction of CO_2 (an endothermic process) can be thus accomplished using the most abundant sources of energy and hydrogen, namely, sunlight and water, respectively.

CO_2 reduction is thus a challenging and difficult task because of the stability of CO_2, being the most oxidized carbon compound. High reduction potential is required for electrochemical activation of CO_2, that is, -1.9 V vs. NHE for one-electron reduction of CO_2 giving unstable $CO_2^{\cdot-}$, which is difficult to handle [32–34].

However, these difficulties can be solved by introducing multi-electron transfer in the process reduction of CO_2. The reactions in Table 10.1 indicate the potential energy for some of the expected reactions.

One of the earliest reports on photoelectrochemical reduction of CO_2 was published by Halmann in 1978 [35]. An electrochemical cell, composed of a single crystal p-type GaP cathode, a carbon anode, and an aqueous buffer solution as an electrolyte, in which oxygen-free CO_2 was continuously bubbled, was used. When the GaP crystal was illuminated (mercury lamp) and a voltage bias applied, the photocurrent was detected with the formation of formic acid, formaldehyde, and methanol in the electrolyte solution.

TABLE 10.1 CO_2 Conversion Reactions and Their Potential Energies

Reaction	Potential Energy
$CO_2 + e^- \rightarrow CO_2^{\cdot-}$	$E° = -1.90\,V$
$2H^+ + 2e^- \rightarrow H_2$	$E° = -0.61\,V$
$H_2O \rightarrow 1/2O_2 + 2H^+ + 2e^-$	$E° = -0.61\,V$
$CO_2 + 2H^+ + 2e \rightarrow HCOOH$	$E° = -0.61\,V$
$CO_2 + 2H^+ + 2e^- \rightarrow CO + H_2O$	$E° = -0.53\,V$
$CO_2 + 4H^+ + 4e^- \rightarrow HCHO + H_2O$	$E° = -0.48\,V$
$CO_2 + 6H^+ + 6e^- \rightarrow CH_3OH + H_2O$	$E° = -0.38\,V$
$CO_2 + 8H^+ + 8e^- \rightarrow CH_4 + H_2O$	$E° = -0.24\,V$

Many research groups have investigated the use of different compound semiconductors to achieve higher visible light catalytic activities. For example, Canfield and Frese [36] achieved CO_2 reduction to methanol using p-GaAs and p-InP photoelectrodes in a CO_2-saturated solution of Na_2SO_4. Similarly, formation of methanol together with formic acid and formaldehyde using single-crystal p-GaP and p-GaAs photoanodes, was reported by Blajeni et al. [37]. Eggins and co-workers [38] reported visible light photoreduction of CO_2 in the presence of colloidal CdS in an aqueous solution of tetramethylammonium chloride, yielding glyoxylic acid as well as formic and acetic acids and HCHO. Fujiwara and co-workers [39] studied the use of ZnS nanocrystals as visible light-driven photocatalysts, demonstrating that excess metal ions enhanced the photocatalytic response. The use of ZnS crystallites was also reported by Kuwabata et al. [40], obtaining methanol as the main product.

Barton and co-workers reported the selective reduction of CO_2 to methanol using a catalyzed p-GaP-based photoelectrochemical cell [41]. In 1979, Inoue and co-workers [42] investigated the use of semiconductor powders suspended in CO_2-saturated water for CO_2 reduction by a Xe lamp, including TiO_2, ZnO, CdS, SiC, and WO_3. Small amounts of formic acid, formaldehyde, methyl alcohol, and methane were produced. Inoue proposed that the conversion of CO_2 to methane was a multi-step reduction process started by photo-generated electron–hole pairs [42]. The formaldehyde and methyl alcohol yield was highest in the presence of SiC, which was attributed to the relative position of the SiC conduction band with respect to the $HCHO/H_2CO_3$ redox potential [43].

Figure 10.1b shows the band edge positions of the different semiconductor materials with respect to the redox potentials for the different chemical species. The SiC conduction band edge lies at a higher position (more negative) than the $HCHO/H_2CO_3$ redox potential, which is believed to be responsible for the high rates of product formation. When WO_3 with a conduction band at a position lower than the $HCHO/H_2CO_3$ redox potential was used as catalyst, the results indicate no methyl alcohol formation, confirming the influence of band edge positions on CO_2 reduction.

In subsequent work, Halmann et al. [43] reported the use of strontium titanate catalyst powder suspended in an aqueous solution in which CO_2 was bubbled. Formic acid, formaldehyde, and methanol under natural sunlight illumination were observed as the reaction products. Because of its conduction band edge positioned at higher energy compared to the redox potential of CH_3OH/H_2CO_3, strontium titanate could effectively reduce CO_2 dissolved in an aqueous electrolyte.

Cook and co-workers [44] reported the reduction of CO_2 to CH_4, C_2H_4, and C_2H_6 as a function of electrolyte pH using a mixture of p-SiC and Cu particles. Their work showed that the addition of Cu cocatalysts to SiC enhances the electron transfer rate from the SiC conduction band to CO_2, because the reduction occurs at Cu particle sites. The use of Cu as a cocatalyst was also reported by Adachi et al. [45]. Cu-loaded TiO_2 powder was suspended in a CO_2 pressurized solution at ambient temperature, with the formation of methane and ethylene under Xe lamp illumination.

Recent developments in photocatalysis technology further indicate CO_2 photoreduction processes as a potentially promising application. The focus of recent developments is the use of visible light as the energy source for semiconductor electron-hole pair generation, followed by transfer of the photoexcited electrons to CO_2 to form products such as carbon monoxide, methane, methanol, formaldehyde, and formic acid [46–53].

A variety of photocatalysts such as TiO_2, CdS, ZrO_2, ZnO, and MgO have been studied. Among them, wide band-gap catalysts such as TiO_2 are considered to be the most convenient candidates in terms of stability [54–56]. Modification of the catalysts by doping with transition metals on the catalyst surface is required to enhance the reaction rate, to increase visible light utilization, and to control the selectivity of products. These surface metal nanoparticles act as "charge-carrier traps" suppressing recombination of photoexcited electron–hole pairs [46]. However, the instability of the transition metal-doped titania [57,58] indicates the need to explore the challenging alternative of realizing highly efficient photocatalysts with a narrow band gap for the reduction of CO_2 under visible light.

10.4 AgBr NANOCOMPOSITES

It is well known that silver bromide (AgBr) is a photosensitive material with an indirect band gap of 2.64 eV (470 nm) [57–59] that can be applied in photocatalytic reactions by coupling some supporting materials [58–60]. In the photosensitive process, the AgBr absorbs a photon and generates an electron and a positive hole [59–62]. If the photographic process (i.e., interstitial ions combined with electrons to form silver atoms) is inhibited, the generated electron and hole can be used in the photocatalytic process.

To effectively suppress the photographic process, the electron–hole pair should be quickly separated and the generated electron must be transferred to some supporting materials [59,61,63]. Our recent work on the photocatalytic reduction of CO_2 has demonstrated that AgBr coupled with TiO_2 maintains its catalytic activity by injecting electrons to the conduction band of TiO_2 upon excitation with visible light [58]. Therefore, the charge transfer between the nanostructure interfaces is very important to determine the photoreduction efficiency of CO_2 and the stability of the photocatalyst [64,65].

Recently, carbon-based nanocomposites have been found to be effective in improving the charge transfer between the nanostructured interfaces and exhibit a high catalytic activity due to their intrinsic physical and chemical properties [64–68]. Shaban and Khan [69] firstly demonstrated the high photoactivity of carbon-doped TiO_2 in the photoconversion of water to H_2 and O_2. Akhavan and Ghaderi [70] prepared a graphite oxide-titania (GO/TiO_2) composite with a sol–gel technique and found an enhanced rate of pollutant degradation compared to bare anatase TiO_2. Kamat and his research team [71] successfully prepared a multifunctional photocatalyst by anchoring two-dimensional (2D) carbon to oxide semiconductor (TiO_2, ZnO) and metal nanoparticles (Au, Pt).

The main focus of previous studies was on the photocatalytic mechanism and optimization of the photocatalysts for efficient CO_2 treatment. Less attention was given to analyzing the role of nanostructured interfaces. Nano-size effects in semiconductor nanoparticles on the photoreduction of CO_2 have been only recently studied.

Carbon-based materials that have a large specific surface area, a high porous structure, and low cost such as graphite powder (GP), granular activated carbon (GAC), and graphite oxide (GO) could be good candidates for analyzing these effects and their role in improving the charge transfer in the photoreduction of CO_2. This chapter compares the use of AgBr nanocomposite in different kinds of carbon-based supports such as GP, GO, and GAC. The aim is to investigate the potential of AgBr for improving the CO_2 reduction activity under visible light. It is expected that AgBr could be a good visible light photocatalyst candidate for the reduction of CO_2 in the presence of carbon-based materials, but also that materials have good stability.

This study offers a comprehensive critical analysis of the feasibility, stability, and efficiency of carbon-based AgBr nanocomposites (AgBr/GP AgBr/GO, and AgBr/GAC) as effective photocatalysts for the reduction of CO_2 under visible light ($\lambda > 420$ nm).

10.4.1   Preparation of Catalyst

Carbon-based AgBr nanocomposites were prepared by the deposition-precipitation method in the presence of the cation surfactant cetyltrimethylammonium bromide (CTAB) [72]. An appropriate amount of the carbon-based materials was oxidized with a HNO_3 boiling solution ($HNO_3 : H_2O = 1 : 4$, v/v) for 1 h and then washed with distilled water until the filtrate was neutral. The sample was added to 100 ml of CTAB aqueous solution with a concentration of 0.01 M (10 times above the CMC of CTAB 9.8×10^{-4} M), and the suspension was sonicated for 30 min and then stirred magnetically for 30 min. Then 0.0012 mol of $AgNO_3$ in 2.3 ml of ammonia hydroxide (25 wt.% NH_3) was quickly added to the suspension.

In this process, cationic surfactant CTAB was adsorbed onto the surface of carbon-based supporting materials under alkaline conditions to limit the number of nucleation sites for the formation of AgBr aggregates, leading to well-dispersed AgBr on carbon-based materials. In addition, the amount of bromide ion from CTAB is more than sufficient to precipitate Ag^+ from the added $AgNO_3$ in aqueous solution. The resulting mixture was stirred at room temperature for 12 h. The product was filtered, washed with deionized water, dried at 75°C, and subsequently thermally treated in a muffle furnace at 500°C in the presence of N_2 for 3 h.

10.4.2   Characterization of Carbon-Based AgBr Photocatalysts

XRD data of the fresh prepared carbon-based AgBr nanocomposites show that GP, GO, and GAC have a hexagonal structure (a sharp diffraction line at around $2\theta = 26.4°$, corresponding to 002 plane). Crystalline AgBr in the fresh nanocomposites shows reflections at around $2\theta = 26.7°$, $30.9°$, $44.3°$, and $55.0°$, indicating a hexagonal structure for AgBR nanoparticles according to Rodrigues and Martyanov et al. [73]. The synthesis procedure used by us leads to a large precipitation of the silver species, since the amount of bromide species from CTAB is larger than that of silver species.

XRD patterns of carbon-based AgBr nanocomposites after use for 5 h under visible light irradiation (150 W Xe lamp) were almost the same as those of fresh nanocomposites. The diffraction lines related to metallic Ag ($38.2°$, $44.4°$, $64.4°$, and $77.4°$) were not detected in either fresh or used nanocomposites. Earlier studies reported that the irradiation led to the appearance of metallic Ag, for example, 1.2 wt.% Ag was formed after 14-h irradiation in $AgBr/TiO_2$ [74]. However, other reports indicate the photostability of the systems, without formation of Ag in $AgCl/TiO_2$ [75] and AgI/TiO_2 [76] photocatalysts.

Figure 10.2a shows TEM images of the carbon-based AgBr nanocomposites. It can be seen that AgBr nanoparticles with a size of approximately 10 nm are deposited on the surface of carbon-based materials, indicating that AgBr nanoparticles (small black dots) are well dispersed on the carbon-based supports.

To obtain further information on the structure at the interface between the AgBr nanoparticle and the carbon supports, the nanocomposites were charaterized by HRTEM (Fig. 10.2b). It can be seen that the interplanar spacing (0.338 nm) corresponds to the (002) planes of GP, GO, and GAC, while the interplanar spacing on AgBr (0.288 nm) matches the (200) planes of AgBr, further confirming the contact between the interface of carbon-based supporting materials and AgBr nanoparticles.

10.4.3 Photocatalytic Reduction Activity of Carbon-Based AgBr Nanocomposites

Photocatalysis reduction was carried out in a stainless steel vessel with valves for evacuation and gas feeding, in which an O-ring-sealed glass window was placed at the top for admitting light irradiation. A 150 W Xe lamp (Shanghai Aojia Lighting Appliance Co. Ltd.) with UV cutoff filter (providing visible light with $\lambda > 420$ nm) was used for irradiation. In a typical batch, 0.5 g of prepared carbon-based AgBr nanocomposites was suspended in 100 ml of 0.2 M $KHCO_3$ solution in the vessel. Prior to the irradiation, pure CO_2 (99.99%), via a flow controller, was passed through the solution for 30 min to remove the oxygen, and then the flow controller was closed, maintaining a pressure of 7.5 MPa inside the reactor.

During reaction, the powder was continuously agitated by a magnetic stirrer to prevent sedimentation of the catalyst. The photocatalytic reaction was operated for 5 h at room temperature. After illumination, small aliquots of the suspension were withdrawn by syringe, filtered through Millipore membranes, and then analyzed. Gas samples were taken with a gas-tight syringe through a septum. Reaction products in liquid phase were analyzed with a gas chromatograph (Agilent HP6890N) equipped with a flame ionization detector (FID) and a HP-5 capillary column (30 m × 320 μm × 0.50 mm). The products in gas phase were analyzed

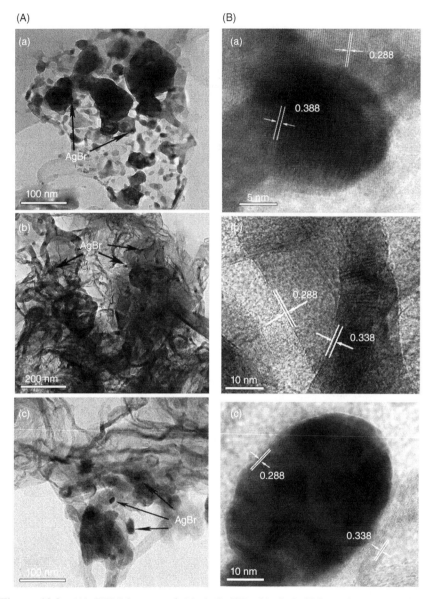

Figure 10.2 (A) TEM images of (a) AgBr/GP, (b) AgBr/GO, and (c) AgBr/GAC. (B) HRTEM images of (a) AgBr/GP, (b) AgBr/GO, and (c) AgBr/GAC.

by GC/MS Hewlett-Packard (HP) 6890 gas chromatography with a HP 5973 mass detector. A 60 m length × 0.32 mm I.D 1.8 m film thickness HP-VOC column (Agilent Scientific, USA) was used. The oven was programmed as follows: the initial temperature of the column was held at 40°C for 2 min, followed by a ramp of 5°C min^{-1} to 120°C (1-min hold), injection model splitless, for a 1-µl sample.

The photocatalytic activity of prepared samples was evaluated for 5 h in the reduction of CO_2 in the presence of water under visible light irradiation. Gas chromatographic analysis of the reaction products shows the formation of methane and CO in gas phase and methanol and ethanol in liquid phase. Other products such as formic acid, formaldehyde, and hydrogen are not detectable with the GC analysis used by us. The detected products (methanol, ethanol, methane, and CO) were not present in blank tests (dark operations or without the presence of the photocatalysts), indicating that both the presence of visible light irradiation and the photocatalyst are necessary for the photocatalytic reduction of CO_2 in the presence of water.

Figure 10.3 shows the dependence of CO_2 reduction yield on the different carbon-based materials. It can be seen that the product yield under visible light (5 h) increased in the order

$$AgBr/GP > AgBr/GO > AgBr/GAC$$

Experimental results demonstrate that AgBr/GP is an effective catalyst compared with AgBr/GO and AgBr/GAC and produces a maximum yield of methane, methanol, ethanol, and CO of 155.09, 93.83, 16.02, and 38.77 $\mu mol\ g^{-1}$, respectively.

The interesting behavior of AgBr supported on zeolite [73] and titanium dioxide [72,74] has been reported, and it is thus useful to investigate their behavior in comparison with AgBr nanocomposites on carbon materials. Therefore, we investigate their efficiency. The results of the behavior of $AgBr/TiO_2$ and AgBr/zeolite are also shown in Figure 10.3. Even though the performances of $AgBr/TiO_2$ and

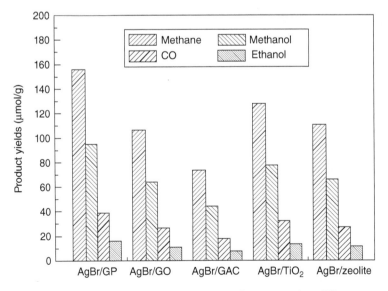

Figure 10.3   Yield of photocatalytic products for AgBr supported on different materials.

AgBr/zeolite samples are confirmed to be interesting, the behavior of AgBr/GP in the photoreduction of CO_2 is better in terms of all products. There is thus a higher activity, while selectivity is not largely influenced.

Since the photoreduction process of CO_2 involves H· radicals and carbon dioxide anion radicals formed by electron transfer from the conduction band [65,77–79], the pH value plays a crucial role in the photoreduction of CO_2. Figure 10.4 indicates that the product yields with carbon-based AgBr nanocomposites are dependent on the pH value, increasing up to a pH of about 8 and then decreasing upon the further increase in the pH. The better performances in the photoreduction of CO_2 under neutral and weak alkaline pH can be explained as follows:

1. The OH^- ions in aqueous solution could act as strong hole scavengers and reduce the recombination of hole-electron pairs [80], beneficial to facilitate the reduction of CO_2. In contrast, H^+ in aqueous solution could be involved in the electron competition with CO_2 reduction (i.e., $H^+ + e^- \rightarrow$ H·), leading to lower yields of hydrocarbons [78].

2. More CO_2 can be dissolved in basic solution to form HCO_3^- ions (i.e., $CO_2 + OH^- \rightarrow HCO_3^-$) than in pure water or acidic solution [81], resulting in a higher yields of hydrocarbons in the photoreduction process.

3. The neutral and weak alkaline pH values are beneficial to the adsorption of CO_2 on the catalyst shown by the zeta potential result of carbon-based AgBr nanocomposites (Fig. 10.5).

Zeta potential measures (Fig. 10.5) show that at pH < 5.5 the carbon-based AgBr nanocomposites results positively charged (with minor dependence on the type of carbon support), while above this pH threshold the nanocomposites results negatively charged.

The higher photocatalytic activity achieved in the neutral and weak alkaline pH range is thus the result of the combined effect of the higher concentration of OH^- ions but lower electrostatic repulsive force.

10.4.4   Stability of Carbon-Based AgBr Nanocomposites and Electron Transfer Mechanism

For the practical application of these photocatalysts not only the efficiency but also their stability is crucial, since AgBr is a photosensitive material. Thus both the carbon-based AgBr nanocomposites and AgBr alone were tested in consecutive experiments to analyze their stability in the photocatalytic reduction of CO_2 in aqueous solution. The results reported in Figure 10.6A indicate that carbon-based AgBr nanocomposites, differently from AgBr alone, maintained photocatalyst stability during five repeated uses.

For example, AgBr/GP retained its yield of methane, methanol, ethanol, and CO in five repeated uses at $150.98 \pm 5.09, 89.84 \pm 4.80, 15.56 \pm 0.39$ and 36.30 ± 2.40 µmol/g, respectively. The total yields of the photocatalytic products on AgBr/GP after the five repeated uses remained about 92% of the first run.

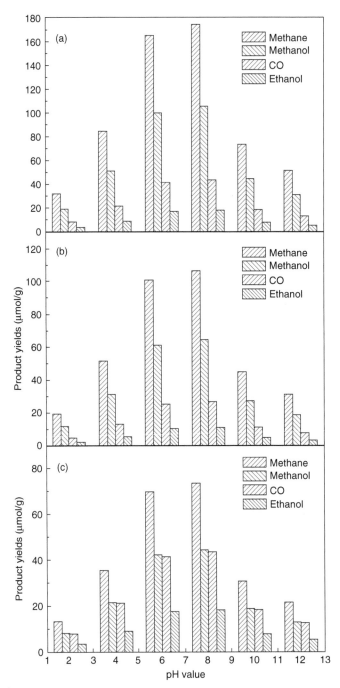

Figure 10.4   Effect of pH on the product yield in the photocatalytic reduction of CO_2 (5 h, visible light irradiation) using (a) AgBr/GP, (b) AgBr/GO, and (c) AgBr/GAC.

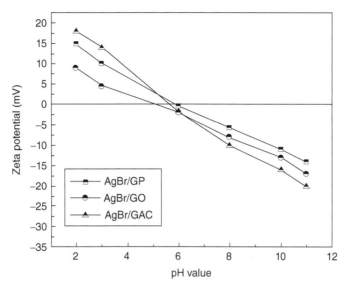

Figure 10.5 Zeta potential for suspension of carbon-based AgBr nanocomposites in 0.02 M KCl solution.

The results indicated that AgBr/GP is an effective and chemically stable catalyst. However, AgBr demonstrated a significant decline of product yield after three repeated uses, from 36.73 to 0 μmol g^{-1} for methane, from 32.16 to 3.71 μmol g^{-1} for methanol, from 6.49 to 1.46 μmol g^{-1} for ethanol, and from 9.18 to 0 μmol g^{-1} for CO (Fig. 10.6A). The total yields of the photocatalytic products on AgBr after the three repeated uses decreased to about 6% of the first run.

The used AgBr and AgBr/GP after repeated uses were examined by XRD. Figure 10.6B shows the XRD patterns of the fresh and used AgBr. It can be seen that crystalline Ag is present in the used AgBr (diffraction lines at $2\theta = 38.2°$; see Fig. 10.6Bb) compared to fresh AgBr (Fig. 10.6Ba), indicating the formation of metallic Ag through the photographic process under visible light irradiation. This result indicated that the decreased activity of pure AgBr is ascribed to the decomposition of the catalyst. In used AgBr/GP XRD results do not show changes, indicating that AgBr nanoparticles well-dispersed on GP can maintain their stability and photocatalytic activity. In this photocatalytic process, the transfer of photoexcited electrons from the conduction band of well-dispersed AgBr to carbon-based materials is beneficial for the stability of AgBr. Therefore, in practice, the stability of carbon-based AgBr nanocomposites would be a significant advantage for CO_2 photocatalytic reduction under visible light.

A possible mechanism is proposed to explain the reasons of the high photocatalytic activity and stability of the as-prepared carbon-based AgBr nanocomposite photocatalyst (Fig. 10.7). The photogenerated electrons in AgBr (produced upon excitation by visible light) are transferred to the surface of the carbon support, and then the transferred electrons (possibly localized at carbon defect sites) react with the CO_2 and water adsorbed on the surface of the carbon.

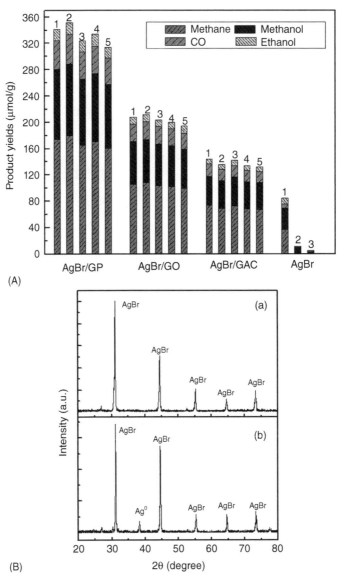

Figure 10.6 (A) Yield of photocatalytic products in carbon-based AgBr nanocomposites as a function of the number of batch runs. (B) XRD pattern of (a) fresh AgBr and (b) used AgBr after visible light irradiation.

This can explain why carbon-based AgBr nanocomposites exhibit a higher photocatalytic activity (in CO$_2$ reduction) than pure AgBr as well as a dependence on the type of carbon material. In addition, the process of electron transfer is faster than the electron–hole recombination between valence band and conduction band of AgBr, thus plenty of CB-electrons (AgBr) can be stored on the surface of

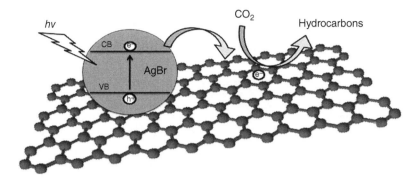

Figure 10.7   Schematic of photoexcitation process under visible light.

carbon-based supporting materials to be reacted later with CO_2. Thus the timely electron capture on the surface of carbon-based supporting materials plays an important role in the stability of carbon-based AgBr nanocomposites.

10.5 CONCLUSIONS

In this chapter, the catalytic activity of carbon-based AgBr nanocomposites for CO_2 reduction to hydrocarbons under visible light was investigated. The carbon-based AgBr nanocomposites were prepared by the deposition-precipitation method in the presence of CTAB. These photocatalytic activities were evaluated by the reduction yield in the presence of CO_2 and water under visible light ($\lambda > 420$ nm). The experimental results showed that AgBr/GP has a relatively higher activity under visible light irradiation. Moreover, it was found that carbon-based AgBr nanocomposites were stable under consecutive tests under visible light irradiation because of the transfer of photoexcited electrons from the conduction band of well-dispersed AgBr to GP.

This investigation demonstrated that carbon-based AgBr nanocomposites are interesting materials in the CO_2 photoreduction to hydrocarbons under visible light irradiation. Their behavior provides new clues to a fundamental theory for an efficient CO_2 conversion and fixation but also for storage of solar energy and resolution of environmental problems as well.

ACKNOWLEDGMENTS

The authors wish to thank the National Natural Science Foundation of China (No. 20877025, 21273085), the National Natural Science Foundation of Guangdong Province (No. S2013010012927), the Fundamental Research Funds for the Central Universities (No. 13lgjc10), Project from Guangzhou City (2010Z2-C1009) and Ministry of Science and Technology of China (No. 10C26214414753), the Research Fund Program of Guangdong Provincial

Key Laboratory of Environmental Pollution Control and Remediation Technology (No. 2011K0003), and the Scientific Research Foundation for the Returned Overseas Chinese Scholars from State Education Ministry for financially supporting this work.

REFERENCES

1. Wolff, E.W. (2011). *Phil. Trans. Math. Phys. Eng. Sci.*, 369: 2133–2147.

2. Al-Fatesh Ahmed. S.A.; Fakeeha Anis, H. (2011). *Res. J. Chem. Eev.*, 15: 259–268.

3. Flower David, J.M.; Sanjayan Jay, G. (2007). *Int. J. L. C. A.*, 12: 282–288.

4. Nguyen, T.V.; Wu, J.C.S. (2008). *Sol. Energ. Mat. Sol. C.*, 92: 864–872.

5. Centi, G.; Perathoner, S. (2011). *Coord. Chem. Rev.*, 255: 1480–1498.

6. Qin, S.Y.; Xin, F.; Liu, Y.D. (2011). *J. Colloid Interface Sci.*, 356: 257–261.

7. Nguyen, T.V.; Wu, J.C.S. (2008). *Appl. Cat. A: General*, 335: 112–120.

8. Wu, J.C.S; Lin, H.M.; Lai, C.L. (2005). Appl. Cat. *A: General*, 296: 194–200.

9. Wu, J.C.S.; Lin, H.M. (2005). *Int. J. Photoenergy*, 7: 115–119.

10. Fujishima, A.; Honda, K. (1972). *Nature*, 238: 37–38.

11. Liu, Q.; Zhou, Y.; Kou, J.H.; Chen, X.Y.; Tian, Z.P.; Gao, J.; Yan, S.C,; Zou, Z.G. (2010). *J. Am. Chem. Soc.*, 132: 14385–14387.

12. Roy, S.C.; Varghese, O.K.; Paulose, M.; Grimes, C.A. (2010). *ACS. Nano*, 4: 1259–1278.

13. Liu, R.X.; Yu, Y.C.; Yoshida, K.; Li, G.; Jiang, H.X.; Zhang, M.H.; Zhao, F.Y.; Fujita, S.I.; Arai, M. (2010). *J. Catal.*, 269: 191–200.

14. Bhatkhande, D.S.; Pangarkar, V.G.; Beenackers, A. (2002). *Review. J. Chem. Technol. Biotechnol.*, 77: 102–116.

15. Bussi, J.; Ohanian, M.; Vazquez, M.; Dalchiele, D.A (2002). *J. Environ. Eng.*, 128: 733–739.

16. Linsebigler, A.L.; Lu, G.; Yates, J.T. (1995). *Chem. Rev*, 95: 735–739.

17. Chang, S.M.; Doong, R.A. (2004). *J. Phys. Chem. B.*, 108: 18098–18103.

18. Rufino, M.; Navarro, Y.; Consuelo, l.G.; del Valle, F.; Jos, A.; Villoria, d.l.M.; Jos, L.G.F. (2009). *ChemSusChem.*, 2: 471–485.

19. Materic, V.; Smedley, S.I. (2011). *Ind. Eng. Chem. Res.*, 50: 5927–5932.

20. Monazam, E.R.; Shadle, L.J.; Siriwardane, R. (2011). *AIChE J.*, 57: 3153–3159.

21. Garcia-Labiano F.; Rufas, A.; Luis, D.D.F. (2011). *Fuel*, 90: 3100–3108.

22. Jin, W.Q.; Zhang, C.; Chang, X.F. (2008). *Environ. Sci. Tech.*, 42: 3064–3068.

23. Garcia-Serna J.; Garcia-Merino E.; Cocero, M. J. (2007). *J. Supercritical Fluids*, 43: 228–235.

24. Hershkowitz, F.; Deckman, H.W.; Frederick, J.W. (2009). *GHGT-9 Book Series: Energy Procedia.*, 1: 683–688.

25. Descamps, C.; Bouallou, C.; Kanniche, M. (2008). *Energy*, 33: 874–881.

26. Montes-Hernandez, G.; Renard, F.; Geoffroy, N. (2007). *J Crystal Growth*, 308: 228–236.

27. Kohno, Y.; Hayashi, H.; Takenaka, S.; Tanaka, T.; Funabiki, T.; Yoshida, S (1999). *J. Photochem. Photobiol. A.*, 126: 117–123.

28. Praus, P.; Kozak, O.; Koci, K. (2011). *J. Colloid Interface Sci.*, 360: 574–579.

29. Li, H.L.; Lei, Y.G.; Huang, Y. (2011). *J. Nat. Gas Chem.*, 20: 145–150.

30. Lo, C.C.; Hung, C.H.; Yuan, C.S. (2007). *Sol. Energ. Mat. Sol. C.*, 91: 1765–1774.

31. Rappe, A.K.; Ford, K.B. (2007). *Abstr. Pap. Am. Chem. Soc.*, 234: 114.

32. Cybula, A.; Klein, M.; Zielinska-Jurek, A. (2012). *Physicochem. Probl. Miner. Process.*, 48: 159–167.

33. Jin, Z.X.; Ismail, M.N.; Jr, D.M.C.; Warzywoda, J.; Jr, A.S. (2011). *J. Photochem. Photobiol. A: Chem.*, 221: 77–83.

34. Fatimah I.; Wang, S.B.; Wulandari, D. (2011). *Appl. Clay Sci.*, 53: 553–560.

35. Halmann, M. (1978). *Nature*, 275: 115–116.

36. Canfield, D.; Frese, K.W (1983). *J. Electrochem. Soc.*, 130: 1772–1773.

37. Blajeni, B.A.; Halmann, M.; Manassen, J. (1983). *Sol. Energy. Mater.*, 8: 425–440.

38. Eggins, B.R.; Irvine, J.T.R.; Murphy, E.P.; Grimshaw, J. (1988). *J. Chem. Soc. Chem. Commun.*, 16: 1123–1124.

39. Fujiwara, H.; Hosokawa, H.; Murakoshi, K.; Wada, Y.; Yanagida, S. (1998). *Langmuir.*, 14: 5154–5159.

40. Kuwabata, S.; Nishida, K.; Tsuda, R.; Inoue, H.; Yoneyama, H. (1994). *J. Electrochem. Soc.*, 141: 1498–1503.

41. Barton, E.E.; Rampulla, D.M.; Bocarsly, A.B. (2008). *J. Am. Chem. Soc*, 130: 6342–6344.

42. Inoue, T.; Fujishima, A.; Konishi, S.; Honda, K. (1979). *Nature*, 277: 637–638.

43. Halmann, M.; Ulman, M.; Blajeni, B.A. (1983). *Energy*, 31: 429–431.

44. Cook, R.L.; Macduff, R.C.; Sammells A.F. (1988). *J. Electrochem. Soc.*, 135: 3069–3070.

45. Adachi, K.; Ohta, K.; Mijuma, T (1994). *Sol. Energy*, 53: 187–190.

46. Li, Y.; Wang, W.N.; Zhan, Z.l.; Woo, M.H. (2010). *Appl. Catal. B: Environ.*, 100: 386–392.

47. Ahmed, N.; Shibata, Y.; Taniguchi, T. (2011). *J. Catal.*, 279: 123–135.

48. Zang, Y.; Farnood, R. (2008). *Appl. Catal. B: Environ.*, 79: 334–340.

49. Xue, L.M.; Zhang, F.H.; Fan, H.J (2011). *Env. Biotech. Mat. Eng.*, *PTS 1–3 Book Series*: *Adv. Mat. Res.*, 183–185: 1842–1846.

50. Liu, Y.Y.; Huang, B.B.; Dai, Y. (2009). *Catal. Commun.*, 11: 210–213.

51. Yoong, L.S.; Chong, F.K.; Dutta, B.K. (2009). *Energy*, 34: 1652–1661.

52. Lisachenko Andrei, A.; Mikhailov Ruslan V.; Basov Lev, L. (2007). *J. Phy. Chem. C.*, 111: 14440–14447.

53. Pan, P.W.; Chen, Y.W. (2007). *Catal. Commun.*, 8: 1546–1549.

54. Koci, K.; Obalova, L.; Lacny, Z. (2008). *Chemical Papers.*, 62: 1–9.

55. Kitano, M.; Matsuoka, M.; Ueshima, M.; Anpo, M. (2007). *Appl. Catal. A-Gen.*, 325: 1–14.

56. Usubharatana, P.; Mcmartin, D.; Veawab, A.; Tontiwachwuthikul, P. (2006). *Ind. Eng. Chem. Res.*, 45: 2558–2568.

57. Hailstone, R. (1999). *Nature*, 402: 856–857.

58. AbouAsi, M.; Su, M.H.; Xia, D.; He, C.; Lin, L.; Deng, H.Q.; Xiong, Y.; Li, X.Z. (2011). *Catal. Today.*, 175: 256–263.

59. He, C.; Sasaki, T.; Zhou, Y.; Shimizu, Y.; Masuda, M.; Koshizaki, N. (2007). *Adv. Funct. Mater.*, 17: 3554–3561.

60. Zang, Y.; Farnood. R. (2008). *Appl. Catal. B. Environ.*, 79: 334–340.

61. Zhang, L.S., Wong, K.H., Yip, H.Y.; Hu, C.; Yu, J.C.; Chan, C.Y.; Wong, P.K. (2010). *Environ. Sci. Technol.*, 44: 1392–1398.

62. Rodrigues, S.; Uma, S.; Martyanov, I.N.; Klabunde, K.J. (2005). *J. Catal.*, 233: 405–410.

63. Zang, Y.J.; Farnood, R.; Currie, J. (2009). *Chem. Eng. Sci.*, 64: 2881–2886.

64. Leary, R.; Westwood, A. (2011). *Carbon.*, 49: 741–772.

65. Xia, X.H.; Jia, Z.H.; Yu, Y.; Liang, Y.; Wang, Z.; Ma, L.L. (2007). *Carbon.*, 45: 717–721.

66. Eder, D. (2010). *Chem. Rev.*, 110: 1348–1385.

67. Gao, B.; Peng, C.; Chen, G.Z.; Li, G.P. (2008). *Appl. Catal. B. Environ.*, 85: 17–23.

68. Farrow, B.; Kamat, P.V. (2009). *J. Am. Chem. Soc.*, 131: 11124–11131.

69. Shaban, Y.A.; Khan, S.U.M. (2007). *Chem. Phys.*, 339: 73–85.

70. Akhavan, O.; Ghaderi, E. (2009). *J. Phys. Chem. C.*, 113: 20214–20220.

71. Prashant, V.; Kamat, J. (2010). *Phys. Chem. Lett.*, 1: 520–527

72. Elahifard, M.R.; Rahimnejad, S.; Haghighi, S.; Gholami, M.R. (2007). *J. Am. Chem. Soc.*, 129: 9552–9553.

73. Rodrigues, S.; Uma, S.; Martyanov, I.N.; Klabunde, K.J. (2005). *J. Catal.*, 233: 405–410.

74. Hu, C.; Hu, X.X.; Wang, L.S.; Qu, J.H.; Wang, A.M. (2006). *Environ. Sci. Technol.*, 40: 7903–7907.

75. Huo, P.W.; Yan, Y.S.; Li, S.T.; Li, H.M.; Huang, W.H. (2010). *Desalination.*, 256: 196–200.

76. Vieira, F.; Cisneros, I.; Rosa, N.G.; Trindade, G.M.; Mohallem, N.D.S. (2006). *Carbon*, 44: 2590–2592.

77. Zhao, Z.H.; Fan, J.; Liu, S.H.; Wang, Z.Z. (2009). *Chem. Eng. J.*, 151: 134–140.

78. Usubharatana, P.; McMartin, D.; Veawab, A.; Tontiwachwuthikul, P. (2006). *Ind. Eng. Chem. Res.*, 45: 2558–2568.

79. Bhatkhande, D.S.; Pangarkar, V.G.; Beenackers A. (2002). *Review. J. Chem. Technol. Biotechnol.*, 77: 102–116.

80. Indrakanti, V.P.; Kubicki, J.D.; Schobert, H.H. (2009). *Energy. Environ. Sci.*, 2: 745–758.

■■■■■■■ CHAPTER 11

Use of Carbon Dioxide in Enhanced Oil Recovery and Carbon Capture and Sequestration

SUGURU UEMURA*, SHOHJI TSUSHIMA, and SHUICHIRO HIRAI

11.1 INTRODUCTION

Energy and environmental problems are important issues facing the world today [1,2]. Some of the solutions include renewable energy generation and energy conservation technologies, but both still need considerable time before being applied

*Indicates the corresponding author.

Green Carbon Dioxide: Advances in CO₂ Utilization, First Edition.
Edited by Gabriele Centi and Siglinda Perathoner.
© 2014 John Wiley & Sons, Inc. Published 2014 by John Wiley & Sons, Inc.

commercially because of cost and energy supply issues. Consequently, fossil fuels are still expected to dominate energy supplies in the short and medium term owing to their considerable quantitative contribution to primary energy. However, many large oil fields are running out, and the number of newly discovered oil deposits is decreasing. This, in turn, leads to oil price rises and the development of new oil fields in frontier regions such as remote areas, polar regions, and deep sea areas. As a result, the risk associated with oil production is increasing. Several incidents, such as the Gulf of Mexico oil spill, have already occurred and can cause severe environmental destruction.

The use of fossil fuels has led to increasing atmospheric concentrations of carbon dioxide (CO_2), and the greenhouse effect induced by this additional CO_2 has led to climate change [3]. In addition to energy supply needs, the reduction of CO_2 emissions also presents a global challenge to mitigate the reduction in energy resources and global warming.

To solve both problems, providing a stable energy supply and reduction in CO_2 emissions, techniques such as enhanced oil recovery (EOR) and carbon capture and sequestration (CCS) are becoming increasingly important [4–10]. EOR increases oil production by using CO_2, thus achieving both a stable energy supply and CO_2 reduction simultaneously. In contrast, CCS reduces CO_2 emissions even for non-oil producers. CO_2 is a key substance in both EOR and CCS, and consequently it is important to understand the thermofluid behavior of CO_2-oil and CO_2-water systems in porous media. In this chapter, the background, fundamental mechanisms, and challenges associated with EOR and CCS are introduced.

11.2 ENHANCED OIL RECOVERY

11.2.1 Oil Production Stages

There are several stages in the oil production process, as shown in Figure 11.1a, including primary recovery, secondary recovery, and enhanced oil (or tertiary) recovery. Generally, oil production follows this process, but sometimes EOR is applied while the secondary recovery stage is skipped.

A schematic image of each oil production is shown in Figure 11.1b and c and Figure 11.2. The primary recovery stage is oil production, which depends on natural driving energy. The crude oil is trapped in the porous media of the oil reservoir. Gas or water causes natural pressure in the reservoir, which forces oil to the production well. During oil production, the oil reservoir begins to lose pressure, and a pump is used to continue the oil production. Finally, the oil reservoir is depleted because of the loss of driving energy.

Production efficiency strongly depends on the characteristics of each oil field, such as pressure, viscosity, and porous structure. The average production efficiency of an oil reservoir is approximately 20%. Some oil reservoirs achieve higher efficiencies (approximately 50%) by maintaining the high pressure derived from gas or water. In heavy oil reservoirs, the efficiency decreases to around 10%. To remove the remaining oil, secondary recovery and EOR are performed.

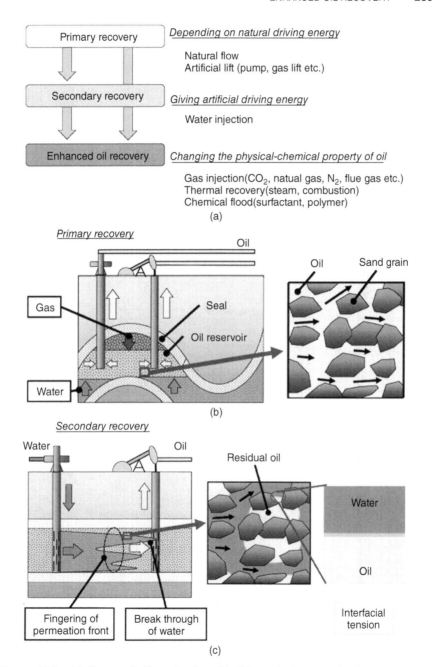

Figure 11.1 (a) Stages of oil production. (b) Schematic images of primary recovery. (c) Schematic images of secondary recovery.

EOR(Enhanced oil recovery)

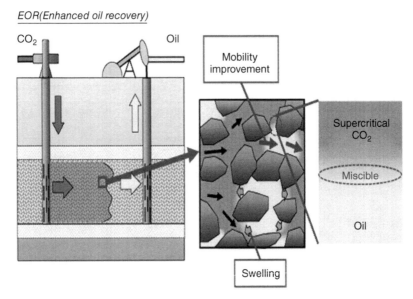

Figure 11.2 Schematic images of EOR.

Secondary recovery is a technique that uses an artificial driving energy. Currently, most oil production uses water or brine injection to maintain the pressure in an oil reservoir, as those materials can be sourced easily and cheaply. The average oil recovery efficiency is approximately 35% and is strongly related to the flow pattern in the oil field. The 65% residual oil occurs because water injection creates an immiscible two-phase flow in the oil reservoir. The flow pattern in the reservoir is dominated by its porous structure, capillary pressure, flow speed, and water/oil viscosity ratio [11]. The process of water permeation creates a fingering structure, and when the water reaches the production well, the water can no longer force oil into the production well [12]. A higher capillary pressure is also induced in the small pore-throat structures, so the water only permeates the large pore-throat structures and oil is trapped in the small pore-throat structures by interfacial tension.

EOR allows for high oil production efficiency by changing the physical-chemical properties of the oil. The process uses gas (CO_2, hydrocarbon gas, or acidic gas), steam, polymer, or surfactant to enhance oil production. Recently, gas injection and thermal recovery have become the mainstream techniques in EOR [5]. Of the different gases, CO_2 injection has attracted attention owing to its higher oil recovery. CO_2 dissolves well in the oil, creates a swelling effect and viscosity reduction, and is miscible with the oil. In the SACROC oil field in the United States, oil recovery of approximately 50% has been achieved [13].

11.2.2 Physicochemical Mechanism of CO_2 EOR

CO_2 is highly soluble in oil, that is, hydrocarbons, and it causes a range of changes in the characteristics of the oil. When CO_2 dissolves into oil, the CO_2-oil mixture

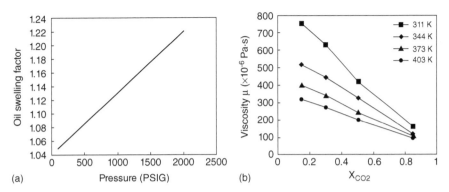

Figure 11.3 (a) Oil swelling factor vs. pressure [14]. (b) Viscosity reduction caused by CO_2 at 21 MPa. Adapted from ref. 15.

displays a swelling effect (Fig. 11.3a), with the swelling factor proportional to the CO_2 pressure [14]. The swelling factor is defined as the ratio of the volume of CO_2-saturated oil to the volume of the initial pure oil. The driving energy in the oil reservoir is spontaneously induced by the pressure increase.

The dissolution of CO_2 into the oil also causes a reduction in oil viscosity (Fig. 11.3b) [15]. The mobility of oil is improved, and the fingering phenomenon observed during secondary recovery is suppressed. The lower viscosity of the CO_2-oil mixture is achieved at higher CO_2 mole fractions, but this means that a lot of CO_2 is required to reduce the oil viscosity. Thermal recovery is a better technique to reduce viscosity, because viscosity is drastically changed by temperature. Thermal recovery is an effective technique for heavy oil, but a large amount of energy is required to heat the oil.

CO_2 and oil become mutually miscible fluids at specific temperature and pressure. The dissolution of the CO_2 causes the disappearance of the interface, leading to enhanced oil production through recovery of the trapped oil.

11.2.3   Phase Equilibrium of CO_2 and Oil Binary Mixture

Miscible displacement is an important mechanism in CO_2 EOR, but miscibility requires specific temperature and pressure conditions. Figure 11.4a shows the vapor–liquid equilibrium of a CO_2-oil mixture [16]. The mole fractions of the gas and liquid phases depend on pressure and temperature. In the liquid phase, a nearly linear relationship, which is described by Henry's law, is observed at low pressures, while the gas phase is the CO_2-rich component. At higher pressures, the mole fractions of the two phases become more similar, and eventually both phases have the same mole fraction at a specific temperature and pressure. This temperature and pressure is the critical point, and beyond this point the CO_2-oil mixture forms a supercritical phase in which the CO_2 and oil become mutually miscible, the interfacial tension is zero, and there is no interface between the gas and liquid phases.

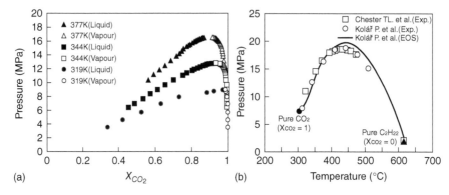

Figure 11.4 (a) Vapor–liquid equilibrium of CO_2–n-decane ($C_{10}H_{22}$) mixture [16]. (b) Critical point of CO_2–n-decane mixture. Reprinted with permission from refs. 17,18.

The critical point is dependent on the temperature, pressure, and CO_2 mole fraction. The critical point of a CO_2-oil mixture can be determined with the method shown in Figure 11.4b [17,18]. In practice, several equation of state (EOS) models are generally used to predict the relationship between pressure, volume, and temperature (PVT) at high pressures, for example, the Soave–Redlich–Kwong EOS, or the Peng–Robinson EOS. The results from these EOS models are compared with the experimental data for the hydrocarbon mixtures, and the EOS is then modified as required [19–22].

11.2.4 Minimum Miscibility Pressure

The miscible point of a CO_2-pure hydrocarbon mixture can be estimated experimentally or from several EOS. However, the determination of the miscible point of a CO_2-crude oil mixture is more difficult because the oil consists of many different components. To decide whether an EOS model can be applied, the minimum miscibility pressure (MMP) is an important index [23,24].

The MMP is the lowest pressure at which a gas can become miscible with a given reservoir oil at the reservoir temperature, and it is estimated by a slim tube test. The slim tube is a 5- to 120-ft-long coiled tube made of stainless steel or Hastelloy C tubing, with a typical outer diameter of 0.25 in. The tube is filled with 100- to 200-mesh sand or glass beads of a specific mesh size. At the beginning of each test, the tube is saturated with the crude oil sample at a given temperature. CO_2 injection is then performed at several test pressures, and the oil recovery efficiency is determined as a function of the injected volume. The efficiency is proportional to the injection pressure in the low pressure range (Fig. 11.5). When the peak efficiency occurs at a given pressure, the MMP is estimated from the point of intersection of the two trend lines.

The MMP has an important role in determining whether CO_2 EOR will be used. If the MMP is lower than the pressure of the oil reservoir, CO_2 EOR can be easily

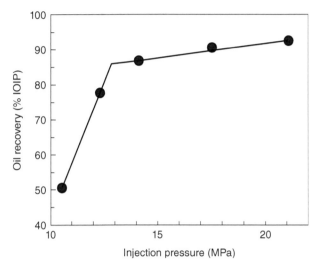

Figure 11.5 Estimation of MMP in crude oil by the slim tube test. Reprinted with permission from ref. 23.

applied and CO_2 miscible displacement in the reservoir is expected to realize high oil recoveries. If the MMP is higher than the reservoir pressure, immiscible CO_2 EOR can be undertaken, but the recovery efficiency is lower than for miscible CO_2 EOR. Although high-pressure injection is also an option, it may cause destruction of the oil reservoir.

11.2.5   Implementation of EOR

EOR has attracted attention owing to the high oil recovery efficiency, and thus many EOR projects are being conducted. The CO_2 EOR projects are mainly being carried out in the United States and Canada with CO_2 from synfuel plants or large CO_2 gas fields. One of the best-known projects is the Weyburn EOR project in Canada, which uses both EOR and CCS [25]. In this project, 5000 t-CO_2/day is injected, and overall 200 Mt of CO_2 will be stored with transportation for 330 km by pipeline. The life of the oil field will be extended by 25 years, and an additional 1.3 billion barrels of oil will be recovered.

It is important to note that even if the CO_2-oil mixture shows good miscibility, EOR will not be applied to all oil reservoirs because it is strongly affected by economics and crude oil prices. When CO_2 EOR is used in an oil reservoir, large amounts of pure CO_2 need to be obtained, which involves substantial costs such as CO_2 separation, CO_2 transport, and CO_2 injection installations. Unless the CO_2 can be purchased cheaply, or high oil recoveries can be realized, EOR is not economically feasible. Currently, the use of natural CO_2 sources is the cheapest method, and consequently CO_2 EOR projects are mainly occurring in the United States because there are many CO_2 gas fields. If EOR is to be linked with industrial CO_2 sources, significant reductions in the cost of CO_2 capture are necessary. Also, further clarification of the fundamental mechanisms of the CO_2-oil system dynamics in porous media is also important to allow significant improvement in oil recovery efficiency.

11.3   CARBON CAPTURE AND SEQUESTRATION

11.3.1   Background and Basis of CCS

EOR can allow both oil production and CO_2 reduction, but oil fields do not exist in all countries. To reduce CO_2 emissions, CCS is an important technology. With this technology, CO_2 from large stationary sources (e.g., fossil fuel-based power plants and cement manufacturing plants) is captured and stored underground (Fig. 11.6). Geological CO_2 sequestration can quantitatively contribute to a reduction in CO_2 emissions and is considered to be a technically feasible method for CO_2 mitigation [26–29] because the basic EOR technology can be adapted for CO_2 injection.

Suitable candidate sites for CO_2 sequestration include geological formations such as deep coal seams, depleted hydrocarbon reservoirs, and deep saline aquifers. Of these options, deep saline aquifers offer enormous storage potential. Field tests of geological sequestration of CO_2 have been performed in many parts of the world in recent years [30–32], and valuable experience has been gained from these projects. The SACS project in Sleipner, Norway, is a famous CCS project that is underway. At this time, 10 Mt of CO_2 has been stored in the aquifer [33,34].

Suitable sites require a highly impermeable layer (known as a caprock structure) to prevent CO_2 leakage from the storage reservoirs. In the Sleipner project, repeat seismic surveys have imaged migration of the injected CO_2 within the reservoir

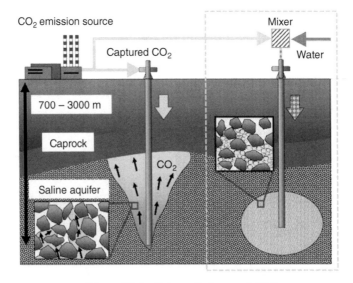

Figure 11.6   Schematic image of CCS.

[33]. This movement can occur because the density of CO_2 is lower than the density of water, so CO_2 migrates upward in the aquifer because of buoyancy. Geological CO_2 sequestration involves several physical and chemical trapping mechanisms, including structural, residual, dissolution, and mineral trapping. Those mechanisms involve immiscible two-phase flow in porous media, the diffusion of CO_2 in water (or brine), and chemical reactions between the CO_2 and minerals. There are several issues that still need to be resolved, such as estimation of storage potential, prediction of CO_2 migration, and storage cost. The risk of CO_2 leakage from storage reservoirs also needs to be evaluated for public acceptance.

11.3.2 CCS with Micronized CO_2

The Kanto region is one of Japan's largest industrial areas and has a large saline aquifer with a monoclinal structure and a storage potential of 27.5 Gt-CO_2 [35]. The CO_2 would be present in the liquid state after injection because the geological temperature gradient around the Kanto region is mild compared with elsewhere in Japan [36]. To achieve public acceptance of CCS and to minimize the CO_2 transport cost, a stable and caprock-independent geological storage technique is required.

A new stable geological sequestration technique using liquid CO_2 in a water emulsion has been proposed, as shown in Figure 11.6 [37,38]. The CO_2 emulsion, which has a droplet diameter the same size as the throat diameter of the pore structure in the aquifer, can be trapped in a stable manner owing to the capillary effect (interfacial tension). This technique improves the CO_2 storage stability, allows the injection of a CO_2 emulsion, and does not rely on a caprock structure. An enhanced CO_2 storage potential is expected to be an additional benefit from this change in the permeation process.

11.3.3 Experimental CO_2 Micronization

The formation and stability of a CO_2 emulsion in water can be examined experimentally with the apparatus shown in Figure 11.7a. The emulsion formation apparatus consists of a static mixer, a circulation pump, and an observation section. The liquid CO_2 was micronized using a static mixer placed in a closed circulation channel. The CO_2 emulsion was observed through observation windows made of sapphire glass. The size of the micronized liquid CO_2 was measured using a dynamic light scattering technique.

The circulation channel was filled with water and added surfactant. The surfactant was added to assist the micronization of liquid CO_2 because water and liquid CO_2 are immiscible. The circuit pressure and temperature were controlled at 7.5 MPa and approximately 22°C, respectively.

After the system was set up, liquid CO_2 was injected into the pressurized circulation channel with the injection pump. After the injection of the liquid CO_2, the water overflowed from the back pressure regulating valve. The liquid CO_2 injection volume could be measured by determining the volume of the displaced water, as this was the same volume as the injected liquid CO_2. The ratio of water to liquid CO_2 was 3:2. The water and liquid CO_2 mixture was circulated for 3 min at

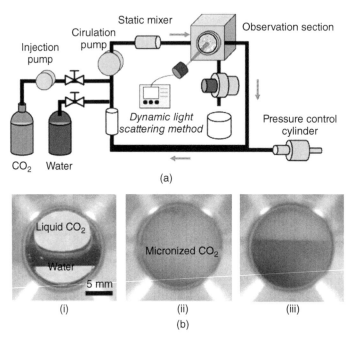

Figure 11.7 (a) Experimental apparatus. (b) Water and liquid CO_2 mixtures: (i) without surfactant; (ii) micronized liquid CO_2 with surfactant; (iii) creaming of micronized liquid CO_2.

a flow rate of 500 ml/min. The diameter of the micronized liquid CO_2 was measured after the mixing stopped, as the measuring procedure used Brownian motion. The diameter distribution and average diameter of the micronized liquid CO_2 were measured.

11.3.4 Experimental Results

The CO_2 emulsion formation process was observed through the sapphire glass windows. Figure 11.7b shows the mixing of liquid CO_2 and water without surfactant. The liquid CO_2 was micronized during the mixing process, but the lifetime of the microscale CO_2 droplets was only about 2 min. Separation of the water and liquid CO_2 phases was then observed because of instability.

The liquid CO_2 emulsion was formed in the presence of a surfactant. The micronized liquid CO_2 was immediately stable and had a long lifetime. Creaming was caused by gravity, and a cloudy state existed for 2 h, where small CO_2 droplets were dispersed in the lower phase. The diameter of the CO_2 emulsion droplets was estimated by measuring the rising speed of the creaming front. The droplet diameter was calculated from Stokes's terminal velocity, at 4.5 μm (the rising speed of the creaming front was 2.7 mm/h).

11.3.5   Droplet Diameter Distribution in the CO_2 Emulsion

The time variation in the average droplet diameter of the CO_2 emulsion, obtained by the dynamic light scattering method, is shown in Figure 11.8a. The measuring point was the center of the sapphire glass window. Immediately after mixing stops, the initial droplet diameter is dependent on the surfactant concentration $C_s (= V_{surfactant}/V_{CO_2})$. When $C_s = 3.5\%$, the initial average droplet diameter was approximately 0.2 μm and the droplet diameter increased logarithmically with time.

To trap the liquid CO_2 in the sandstone aquifer, the CO_2 droplet needs to be micronized to the same size as the sandstone gaps. Figure 11.8b shows the gap diameter distributions of two sandstones, Berea and Tako. There are dispersion peaks for both sandstones at 1 and 10 μm. The CO_2 emulsion droplet diameter distributions at $C_s = 3.5\%$ are shown in Figure 11.8c, showing that the liquid CO_2 is micronized with droplets in the 0.1–10 μm range. The distribution shifts slightly toward larger-diameter droplets over time, but the distribution range is still consistent with the scale of the gap diameters, indicating that liquid CO_2 can be micronized to the same scale as the sandstone gap diameter.

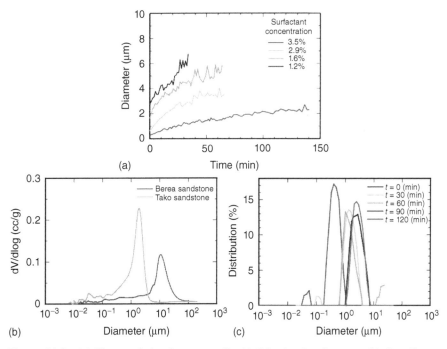

Figure 11.8   (a) Time variation in average liquid CO_2 droplet diameter. (b) Gap diameter distributions of Berea and Tako sandstones. (c) Time variation of liquid CO_2 diameter distribution.

11.4   FUTURE TASKS

For geological sequestration to occur with a CO_2 emulsion, the CO_2 must permeate the sandstone and be trapped in the gaps. Coalescence of the CO_2 emulsion is suggested because the average droplet diameter of the emulsion increases with time in the pore as shown in Figure 11.8a–c. Thus CO_2 droplets that increase to an appropriate size will be trapped in the gaps. As shown in Figure 11.8a, the initial size of the emulsion droplet and its growth rate can be controlled by the surfactant concentration. The micronized liquid CO_2 is expected to maintain its size in the sandstone aquifer as the porous structure limits aggregation and coalescence. To confirm the stability of the average droplet diameter, further research with a longer observation period is required. The distribution of the CO_2 emulsion in sandstone also needs to be investigated with X-ray computed tomography [39] or MRI [40]. Finally, both storage stability and comparisons of CO_2 storage efficiency between bulk CO_2 and the CO_2 emulsion are important issues.

11.5   SUMMARY

CO_2 is becoming a key factor in both energy and environmental problems. In this field, huge amounts of CO_2 are used for EOR and CCS, but there are still several issues that need to be resolved, including recovery or storage efficiency, the cost of CO_2 capture, transport, and injection, and the CO_2 leakage risk. More research is required on fundamental mechanisms of the dynamics of EOR and CCS to allow significant improvements in the efficiency and safety of these techniques.

REFERENCES

1. World Energy Outlook 2010 (Executive summary). International Energy Agency; 2010.
2. World Energy Outlook 2011 (Executive summary). International Energy Agency; 2011.
3. IPCC fourth assessment report: Climate change 2007. IPCC; 2007.
4. Herzog HJ (2011). Scaling up carbon dioxide capture and storage: From megatons to gigatons. *Energy Economics*, 33: 597–604.
5. Adasani AA, Bai B (2011). Analysis of EOR projects and updated screening criteria. *J. Petrol. Science and Eng.*, 79: 10–24.
6. Alvarado V, Manrique E (2010). Enhanced oil recovery: An update review. *Energies*, 3: 1529–1575.
7. Owenn NA, Inderwildi OR, King DA (2010). The status of conventional world oil reserves—Hype or cause for concern? *Energy Policy*, 38: 4743–4749.
8. Bachu, S. (2000). Sequestration of CO_2 in geological media: criteria and approach for site selection in response to climate change, *Energy Conversion & Management*, 41: 953–970.
9. Gale, J. (2004). Geological storage of CO_2: What do we know, where are the gaps and what more needs to be done? *Energy*, 29: 1329–1338.

10. Praetorius B, Shumacher K (2009). Greenhouse gas mitigation in a carbon constrained world: The role of carbon capture and storage. *Energy Policy*, 37: 5081–5093.

11. Lenormand R, Touboul E, Zarcone C (1988). Numerical models and experiments on immiscible displacement in porous media. *J. Fluid Mechanics*, 189: 165–187.

12. Peters EJ, Flock DL (1981). The onset of instability during two-phase immiscible displacement in porous media. *Soc. Petrol. Eng.*, 21: 249–258.

13. Dicharry RM, Perryman TL, Ronquille JD (1973). Evaluation and design of a CO_2 miscible flood project-SACROC Unit, Kelly-Snyder Field. *J. Petrol.Techn.*, 25: 1309–1318.

14. U.S. Department of Energy, Advanced Resources International: Basin Oriented Strategies for CO_2 Enhanced Oil Recovery : PERMEAN BASIN, 4–4.

15. Cullick AS, Mathis ML (1984). Densities and viscosities of mixtures of carbon dioxide and n-decane from 310 to 403 K and 7 to 30 MPa. *J. Chem. Eng. Data*, 29: 393–396.

16. Jiménez-Gallegos R, Galicia-Luna LA, Elizalde-Solis O (2006). Experimental vapor–liquid equilibria for the carbon dioxide + octane and carbon dioxide + decane systems. *J. Chem. & Eng. Data*, 51: 1624–1628.

17. Kolář P, Kojima K (1996). Prediction of critical points in multicomponent systems using the PSRK group contribution equation of state. *Fluid Phase Equilibria*, 118: 175–200.

18. Chester, T.L. and Haynes, B.S. (1997). Estimation of pressure-temperature critical loci of CO_2 binary mixtures with methyl-tert-butyl ether, ethyl acetate, methyl-ethyl ketone, dioxane and decane, *J. Supercritical Fluids*, 11: 15–20.

19. Nagarajan, N. and Robinson, R. L. Jr. (1986). Equilibrium phase compositions, phase densities, and interfacial tensions for CO_2 + hydrocarbon systems. 2. CO_2 + n-decane, *J. Chem. and Eng. Data*, 31: 168–171

20. Peng DY, Robinson DB (1976).A new two-constant equation of state. *Ind. & Eng. Chem. Fund.*, 15: 59–64.

21. Adrian T, Wendland M, Hasse H, Maurer G (1998). High-pressure multiphase behaviour of ternary systems carbon dioxide–water–polar solvent: review and modeling with the Peng–Robinson equation of state. *J. Supercritical Fluids*, 12: 185–221.

22. Alfradique, M. F. and Castier, M. (2007). Critical points of hydrocarbon mixtures with the Peng–Robinson, SAFT, and PC-SAFT equations of state, *Fluid Phase Equilibria*, 257: 78–101.

23. Dong M, Huang S, Dyer SB, Mourits FM (2001). A comparison of CO_2 minimum miscibility pressure determinations for Weyburn crude oil. *J. Petrol. Science and Eng.*, 3: 13–22.

24. Elsharkway, A. M. (1996). Measuring CO_2 minimum miscibility pressures: Slim-tube or rising-bubble method?, *Energy & Fuels*, 10: 443–449.

25. Wilson M, Monea M, editor. IEA GHG Weyburn CO_2 Monitoring & Storage Project Summary Report 2000–2004. 7th International Conference on Greenhouse Gas Control Technologies; 2004 Sep 5–9; Vancouver, Canada: IEAGHG; 2004.

26. Koide, H., Takahashi, M., Tsukamoto, H., Shindo, Y. (1995). Self-trapping mechanisms of carbon dioxide in the aquifer disposal, *Energy Conversion & Management*, 36: 505–508.

27. Bachu, S., Adams, J.J. (2003), Sequestration of CO_2 in geological media in response to climate change: capacity of deep saline aquifers to sequester CO_2 in solution, *Energy Conversion & Management*, 44: 3151–3175.

28. Bachu, S. (2007). CO_2 storage capacity estimation: Methodology and gaps. *Int. J. Greenhouse Gas Control*, 1: 430–443.

29. Kongsjorden, H., Kårstad, O., Torp, T.A. (1998), Saline aquifer storage of carbon dioxide in the Sleipner project, *Waste Management*, 17: 303–308.

30. Michael, K., Golab, A., Shulakova, V., Ennis-King, J., Allinson, G., Sharma, S., Aiken, T. (2010), Geological storage of CO_2 in saline aquifers—A review of the experience from existing storage operations, *Int. J. Greenhouse Gas Control*, 4: 659–667.

31. Spetzler, J., Xue, Z., Saito, H., Nishizawa, O. (2008). Case story: time-lapse seismic crosswell monitoring of CO_2 injected in an onshore sandstone aquifer. *Geophys. J. Int.*,172: 214–225.

32. Bradshaw, J., Bachu, S., Bonijoly, D., Burruss, R., Holloway, S., Christensen, N.P., Mathiasseng, O.M. (2007). CO_2 storage capacity estimation: Issues and development of standards *Int. J. Greenhouse Gas Control*, 1: 62–68.

33. Trop, T.A., Gale, J. (2004). Demonstrating storage of CO_2 in geological reservoirs: The Sleipner and SACS projects, *Energy*, 29: 1361–1369.

34. Bickle, M., Chadwick, A., Huppert, H.E., Hallworth, M., Lyle, S. (2007). Modelling carbon dioxide accumulation at Sleipner: Implications for underground carbon storage, *Earth and Planetary Science Letters*, 225: 164–176.

35. Takahashi, T., Ohshumi, T., Nakayama, K., Koide, K., Miida, H. (2009). Estimation of CO_2 aquifer storage potential in Japan, *Energy Procedia*, 1: 2631–2638.

36. Uyeda, S., Horai, K. (1963). Studies of the thermal state of the earth. The eighth paper: terrestrial heat flow measurements in Kanto and Chubu Districts, *Bull. Earthquake Res. Inst.*, 41: 83–107.

37. Koide, H., Xue, Z. (2009). Carbon microbubbles sequestration: A novel technology for stable underground emplacement of greenhouse gases into wide variety of saline aquifers, fractured rocks and tight reservoirs, *Energy Procedia*, 1: 3655–3662.

38. Uemura, S., Tsushima, S., Hirai, S. (2009). Super-atomization of liquid CO_2 for stable geological storage, *Energy Procedia*, 1: 3087–3090.

39. Wellington, S.L., Vinegar, H.J. (1987). X-ray computerized tomography, *J. Petrology Techn.*, 39: 885–898.

40. Suekane S, Ishii T, Tsushima S, Hirai S. Migration of CO_2 in porous media filled with water (2006). *J. Thermal Science and Techn.*, 1: 1–11.

■ INDEX

Green Carbon Dioxide: Advances in CO_2 Utilization, First Edition.
Edited by Gabriele Centi and Siglinda Perathoner.
© 2014 John Wiley & Sons, Inc. Published 2014 by John Wiley & Sons, Inc.